Epilepsy in Children and Adolescents

Edited by

Albert P. Aldenkamp
Department of Neuropsychology
"Meer en Bosch" Epilepsy Centre
Heemstede, The Netherlands

Wil O. Renier
Department of Child Neurology
University of Nijmegen
Nijmegen, The Netherlands

Fritz E. Dreifuss
Department of Neurology
University of Virginia
Charlottesville, Virginia

Theo P.B.M. Suurmeijer
Department of Medical Sociology
Northern Centre for Healthcare Research
University of Groningen
Groningen, The Netherlands

CRC Press
Boca Raton   New York   London   Tokyo

Library of Congress Cataloging-in-Publication Data

Epilepsy in children and adolescents / edited by Albert P. Aldenkamp
  ...[et al.].
     p.   cm.
  Includes bibliographical references and index.
  ISBN 0-8493-7659-9(alk. paper)
  1. Epilepsy in children.   2. Epilepsy in adolescence.
  I. Aldenkamp, A. P.,   1951-
  RJ496.E6E6515   1995
  618.92′853—dc20   95-90097
                                                                               CIP

   This book contains information obtained from authentic and highly regarded sources. Reprinted material is quoted with permission, and sources are indicated. A wide variety of references are listed. Reasonable efforts have been made to publish reliable data and information, but the author and the publisher cannot assume responsibility for the validity of all materials or for the consequences of their use.
   Neither this book nor any part may be reproduced or transmitted in any form or by any means, electronic or mechanical, including photocopying, microfilming, and recording, or by any information storage or retrieval system, without prior permission in writing from the publisher.
   All rights reserved. Authorization to photocopy items for internal or personal use, or the personal or internal use of specific clients, may be granted by CRC Press, Inc., provided that $.50 per page photocopied is paid directly to Copyright Clearance Center, 27 Congress Street, Salem, MA 01970 USA. The fee code for users of the Transactional Reporting Service is ISBN 0-8493-7659-9/95/$0.00+$.50. The fee is subject to change without notice. For organizations that have been granted a photocopy license by the CCC, a separate system of payment has been arranged.
   CRC Press, Inc.'s consent does not extend to copying for general distribution, for promotion, for creating new works, or for resale. Specific permission must be obtained in writing from CRC Press for such copying.
   Direct all inquiries to CRC Press, Inc., 2000 Corporate Blvd., N.W., Boca Raton, Florida 33431.

© 1995 by CRC Press, Inc.

No claim to original U.S. Government works
International Standard Book Number 0-8493-7659-9
Library of Congress Card Number 95-90097
Printed in the United States of America 1 2 3 4 5 6 7 8 9 0
Printed on acid-free paper

Editorial

Epidemiological studies show that epilepsy is the most common neurological problem in childhood and adolescence. The life-time prevalence of epilepsy (the percentage of people that suffer from, or have suffered from epilepsy during some period in their life) is 3.2% (Hauser et al., 1983). The yearly incidence rate varies roughly around 100 per 100.000 per year and point prevalence rates (the percentage of people that actually suffer from the disease) are estimated to be approximately 0.75% of the population. This illustrates that epilepsy is a condition with a high prevalence, that is, for example, 10 times higher than multiple sclerosis. The lifetime incidence is much higher in childhood and epilepsy must therefore be considered a childhood disease.

The difference between the 0.75% point prevalence and the 3.2% life-time prevalence illustrates, that, although epilepsy is a chronic disease, a large number of patients 'outgrow' their condition. In fact, two subgroups exist. In the largest group, seizure remission is achieved soon after seizure onset. Seizure remission is achieved in 42-47% within one year after seizure onset (Goodridge & Shorvon, 1983). In the second group (probably 10% of all patients with epilepsy), epilepsy is therapy-resistant and becomes a chronic condition (with an average duration of 20 to 30 years). In this latter group, the risk for social and emotional disabilities is increased and many of the chapters in our book give special attention to this subgroup.

The primary aim of our book is to provide up-to-date information for all involved in the cure and care of children and adolescents with epilepsy. Each chapter can be read by itself. Therefore some overlap between the chapters could not be avoided. The first part of the book (chapters 1 to 6) describes the clinical manifestation of 'epilepsy' in children and adolescents and focuses on diagnosis and classification. The second and third parts (chapters 7-12) give an extended overview of the current possibilities for drug treatment and surgical interventions. Both parts attempt to give a comprehensive approach in which the prevention of adverse effects of the treatment is given due attention. Part four and five (chapters 13-18) consider the impact of epilepsy on daily life function and the treatment of epilepsy as a chronic condition. The latter condition requires at least a partial shift of attention from cure to care.

Hauser WA, Annegers JF, Anderson VE. Epidemiology and genetics of epilepsy. In: Ward AA, Penry JK, Purpura D, et al., editors. Epilepsy. New York: Raven, 1983: 274.

Goodridge DMG, Shorvon SD. Epileptic seizures in a population of 6000: II Treatment and prognosis. Brit. Med. J., 1983; 287: 645-647.

Acknowledgments

We are grateful for the substantial work of many people, in particular Tineke van El-Visser and Dagmar Tan-Kildentoft, both secretaries to the Neuropsychological Department of 'Meer en Bosch' Epilepsy Centre for their enthusiasm and their constant and tolerant struggle with manuscripts and time-schedules and Ans Draijer for her computer assistance.

The editors:

Albert P. Aldenkamp
Fritz E. Dreifuss
Wil O. Renier
Theo P.B.M. Suurmeijer

The Editors

Albert P. Aldenkamp, Ph.D. is Head of the Neuropsychological Department of the 'Meer & Bosch' Epilepsy Centre in the Netherlands.

Dr. Aldenkamp trained in Child Neuropsychology and Educational Psychology at the University of Groningen. His Ph.D. thesis in the University of Amsterdam was on the Treatment of Cognitive impairments of Epilepsy in Children. He published over 200 papers, chapters and abstracts in the field of Epileptology and Paediatric Neuropsychology. One of these papers is the Dutch Handbook on Paediatric Neuropsychology. He was Full Professor of Child Psychology at the University of Leiden and is currently the Secretary General of the Dutch League Against Epilepsy, the Secretary General of the 2nd European Congress of Epileptology, member of the Scientific Committee of the 21st International Epilepsy Congress in Sydney and referee for several international journals.

His recent research work is on the cognitive side-effects of antiepileptic drug treatment, the relations between EEG abnormalities and cognitive dysfunction, memory impairment and specific forms of learning impairment in children with epilepsy. Other areas of research were the use of neuroimaging techniques in relation to cognitive impairment and the use of computerized neuropsychological assessment. Dr. Aldenkamp was involved in the development of FePsy, a neuropsychological assessment system that is now used in more than 15 countries worldwide, in more than 200 centres.

For his work he received several recognitions, such as the Ambassador of Epileptology Award.

Fritz E. Dreifuss, M.B., F.R.C.P., F.R.A.C.P. is the Thomas E. Worell Jr. Eminent Scholars Research Professor of Neurology and Epileptology at the University of Virginia.

A graduate of Otago University, New Zealand, Dr. Dreifuss trained in Neurology at the National Hospital, Queen Square, London and joined the University of Virginia in 1959. He is Director of the Comprehensive Epilepsy Program and published over 300 papers, chapters and abstracts in his field and other aspects of Paediatric Neurology as well as authoring and editing several books.

He has held the presidency of the Epilepsy Foundation of America, the American Epilepsy Society, the International League Against Epilepsy and served on the Commission for Control of Epilepsy and Its Consequences. He continues on many committees in all three organizations. Active research has included Classification of Seizures and the Epilepsies, the identification and management of childhood epilepsy syndromes, and he has a long-standing interest in the pharmacological treatment of epilepsy in children which has contributed to his understanding of ethosuximide, clonazepam, nitrazepam, cinromide, valproate, vigabatrin, oxcarbazepine, and felbamate. Special recognitions include membership in Alpha-Omega-Alpha Honor Medical Society, the Lennox Award, the Milken Family Foundation Award, and the EFA 25th Anniversary Award.

Willy Omer Renier, M.D., Ph.D., is staff member of the Institute of Neurology, the Department of Child Neurology and Lecturer in Epilepsy at the University Hospital of Nijmegen, the Netherlands.

He received his M.D. degree cum laude and started a department of Neurology, Psychiatry and Clinical Neurophysiology in the General Hospital of Kortrijk Belgium in 1973. He participated in the foundations of other health care facilities for neurological and mental illness. He was affiliated to the Court of Justice for Youth Criminality. His Ph.D. thesis was on X-linked Mental Retardation.

Dr. Renier published more than 200 scientific papers and chapters in handbooks. Apart from studies on extrapyramidal disorders in children and neurometabolic and neurodegenerative diseases in infancy and childhood, a special topic was the refractory epilepsies. Studies have been published on progressive myoclonus epilepsy (PME), mitochondrial encephalomyopathies with PME, Lafora epilepsy, immunological aspects of malignant childhood epilepsies, and status epilepticus. Learning and behavioral disorders in children with epilepsy, selfhelp groups in epilepsy care, neurogenetics, clinimetrics and pharmacoepidemiology in epilepsy are recent research topics.

He has been member of scientific and organising committees of several national and international congresses on epilepsy and Paediatric Neurology. He is the President of the Dutch League Against Epilepsy. He is member of the Commission on Paediatrics of the International League Against Epilepsy.

Theo P.B.M. Suurmeijer, Ph.D. is associate professor of Medical Sociology at the University of Groningen, the Netherlands.

His Ph.D. thesis was on the educational problems in children with epilepsy. He has a long-standing interest in the field of 'quality of life' and 'quality of care' in elderly, chronically ill (especially epilepsy), and their families and networks.

He is author of many national and international publications. He has presented many invited presentations during national and international congresses, meetings and seminars. He is member of several professional boards and organizations. He is the chairman of the Dutch Case Register (E.R.G.).

Contributors

Albert P. Aldenkamp, Department of Neuropsychology, 'Meer en Bosch' Epilepsy Centre, Achterweg 5, NL-2103 SW, Heemstede, The Netherlands

Gus A. Baker, University Department of Neurosciences, Walton Centre for Neurology and Neurosurgery, Rice Lane, Liverpool L9 1AE, United Kingdom

Martin J. Brodie, Epilepsy Research Unit, Department of Medicine and Therapeutics, Western Infirmary, Glasgow, G11 6NT, Scotland

Stephen W. Brown, The David Lewis Centre, Cheshire and Manchester Royal Infirmary. Alderley Edge, Cheshire SK9 7UD, United Kingdom

Fritz E. Dreifuss, Department of Neurology, School of Medicine, University of Virginia, Charlottesville, VA 22908, USA

Michael Dublin, Western Galilee Regional Hospital, P.O. Box 21, Nahariya 22100, Israel

Bernard Echenne, Service de Neuropédiatrie, Centre Gui de Chauliac, 34295 Montpellier Cedex 5, France

Jean A. Hannah, Epilepsy Research Unit, Department of Medicine and Therapeutics, Western Infirmary, Glasgow G11 6NT, Scotland

Ann Jacoby, Centre for Health Services Research, University of Newcastle upon Tyne, 21 Claremont Place NE2 4AA, United Kingdom

Judith Manelis, Western Galilee Regional Hospital, P.O. Box 21, Nahariya 22100, Israel

Antonio Martins da Silva, Servico de Neurofisiologia, Hospital Geral de Santo Antonio, 4050 Porto, Purtugal

Harry Meinardi, Department of Neurology, University of Nijmegen, Radboud Ziekenhuis, P.O. Box 9101, NL-6500 HB Nijmegen, The Netherlands

Jan Overweg, Northern Outpatient Departments, 'Meer en Bosch' Epilepsy Centre, Achterweg 5, 2103 SW, Heemstede, The Netherlands

Wouterina C.G. Overweg-Plandsoen, Academic Medical Center, Department of Neurology, P.O. Box 2700, NL-1100 DE Amsterdam, The Netherlands

Wil O. Renier, University of Nijmegen, Radboud Ziekenhuis P.O. Box 9101, NL-6500 HB Nijmegen, The Netherlands

Kimberlee J. Sass, Yale University School of Medicine, Section of Neurosurgery, 333 Cedar Street, New Haven, Con. 06510, USA

Herbert Silfvenius, Department of Neurosurgery, University Hospital, S 901 85, Umeå, Sweden

Matti Sillanpää, Department of Child Neurology, University of Turku Hospital TYKS, 20520 Turku, Finland

Graeme J. Sills, Epilepsy Research Unit, Department of Medicine and Therapeutics, Western Infirmary, Glasgow G11 6NT, Scotland

Theo P.B.M. Suurmeijer, Northern Centre for Healthcare Research (NCH), University of Groningen, A. Deusinglaan 1, NL-9713 AV Groningen, The Netherlands

Pamela J. Thompson, Chalfont Centre for Epilepsy, Chalfont St. Peter, Gerrards Cross, Bucks SL9 4RJ, United Kingdom

Michael Westerfeld, Yale University School of Medicine, Section of Neurosurgery, 333 Cedar Street, New Haven, Con. 06510, USA

Contents

1 **Classification of epilepsies in childhood and adolescence** . 1
   Fritz E. Dreifuss
   Introduction 1
   Nonepileptic seizures 1
   Classification of seizures 2
   Classification of epilepsies or epileptic syndromes 5
   Definitions 5
      Definitions of terms 5
      Benign childhood epilepsy with centro-temporal spikes .. 8
      Childhood epilepsy with occipital paroxysms 8
      Benign neonatal familial convulsions 9
      Benign neonatal convulsions 9
      Benign myoclonic epilepsy in infancy 9
      Severe myoclonic epilepsy in infancy (Dravet) 9
      Childhood absence epilepsy (pyknolepsy) 10
      Juvenile absence epilepsy 10
      Juvenile myoclonic epilepsy (impulsive petit mal) 10
      Epilepsy with GTCS on awakening 10
      West syndrome (infantile spasms, Blitz-Nick-Salaam
         Krämpfe) 11
      Lennox-Gastaut syndrome 11
      Epilepsy with myoclonic-astatic seizures 12
      Epilepsy with myoclonic absences 12
      Early myoclonic encephalopathy 12
      Neonatal seizures 12
      Epilepsy with continuous spike-waves during slow sleep . 13
      Acquired epileptic aphasia (Landau-Kleffner syndrome) .. 13
      Malformations 13
      Proven or suspected inborn errors of metabolism 14
   Conclusions 14
   References 15

2 **The epidemiology of epilepsy in children and adolescents** 17
   Jan Overweg
   Wouterina C.G. Overweg-Plandsoen
   Introduction 17
   Guidelines for epidemiological studies 18
      Proposed definitions 18
      Measurements indices 19
   Prevalence and incidence of the major types of epileptic
      syndromes 20
      Epileptic syndromes in neonates 22
      Epileptic syndromes in infancy and childhood 22
      Epileptic syndromes in childhood 24
      Epileptic syndromes in childhood and adolescence 26
   In conclusion 29
   References 29

3	**Seizures in the newborn**	33
	Wil O. Renier	
	Introduction	33
	Investigations	34
	Treatment	35
	Conclusion	39
	References	39
4	**The malignant epilepsies of childhood and adolescence**	43
	Wil O. Renier	
	Introduction	43
	Early infantile epileptic encephalopathies with suppression-bursts	44
	Infantile spasms	46
	Myoclonic epilepsy of infancy	49
	Lennox-Gastaut syndrome and related myoclonic epilepsies of childhood	50
	The Landau-Kleffner syndrome and the epilepsia partialis continua	54
	The progressive myoclonus epilepsies	55
	Conclusion	56
	References	56
5	**The benign epilepsies of childhood and adolescence**	59
	Judith Manelis	
	Michael Dublin	
	Benign partial epilepsies	59
	Benign partial epilepsy with centro-temporal spikes (Rolandic Epilepsy)	60
	Benign epilepsy of childhood with occipital paroxysms	64
	Benign partial epilepsy with affective symptoms (Benign Psychomotor Epilepsy)	66
	Benign partial epilepsy with extreme somato-sensory evoked potentials	68
	Benign partial seizures of adolescence	69
	Reading Epilepsy	70
	Idiopathic generalized epilepsies	72
	Childhood Absence Epilepsy	72
	Juvenile Absence Epilepsy	76
	Juvenile Myoclonic Epilepsy	77
	Epilepsy with Grand Mal on Awakening	79
	References	81
6	**Neurophysiological aspects of epilepsy in children and adolescents**	83
	António Martins da Silva	
	Introduction	83
	EEG patterns in age-related generalized epilepsies	83
	The EEG in idiopathic generalized epilepsies	83

       The EEG in cryptogenic and symptomatic generalized
          epilepsies 84
      EEG patterns in localization-related epilepsies 86
          The EEG in idiopathic localization-related epilepsies 86
          The EEG in symptomatic localization-related epilepsies .. 87
      The EEG in state-related epilepsies 89
          EEG in awakening epilepsies 89
          The EEG in focal seizures, related to sleep 90
      Conclusion 91
      References 96

7 **Established and new antiepileptic drugs: an overview** ... 101
   Jean A. Hannah
   Graeme J. Sills
   Martin J. Brodie
      Introduction 101
      Established antiepileptic drugs 102
          Carbamazepine 104
          Clobazam 107
          Clonazepam 107
          Ethosuximide 108
          Phenobarbital 109
          Phenytoin 110
          Primidone 112
          Valproic acid 113
      New antiepileptic drugs 115
          Felbamate 115
          Gabapentin 116
          Lamotrigine 117
          Oxcarbazepine 118
          Vigabatrin 119
      Conclusion 122
      References 122

8 **Why and when to start treatment, when to stop treatment** 131
   Bernard Echenne
      Introduction 131
      Basic principles of treatment 131
      When to start treatment 134
      When to stop treatment 135
      References 138

9 **Clinical aspects of antiepileptic treatment** 141
   Stephen W. Brown
      Why treat epilepsy? 141
          Aim of treatment 141
          Consumer attitudes 142
          First medical contact 142
      Basic clinical aspects of antiepileptic medication 143
          Indications 143
          General principles of prescribing 149
          Examples 149

  Dose timing & formulations 151
  Monitoring treatment 151
 Polytherapy - 'rational' and irrational 153
  How irrational polytherapy develops 154
  Is there rational polytherapy? 154
 Other aspects of drug management 155
  Rectal diazepam & paraldehyde 155
 Other non-surgical treatments available 156
  Ketogenic diet 156
  Vagal nerve stimulation 156
  Psychological treatments 157
 References 157

10 Cognitive side-effects of antiepileptic drugs 161
Albert P. Aldenkamp

 Introduction 161
 Identification of relevant studies 163
 Description of studies that do not allow valid inferences ... 164
  Polytherapy studies 164
  Posttest-only studies 164
  Studies that provide insufficient information 165
 Normal-volunteer studies 166
 Monotherapy studies in patients with epilepsy 169
 Results of the investigated studies; inferences about the
  cognitive side-effects of antiepileptic drugs 174
  Phenobarbital 174
  Phenytoin 174
  Carbamazepine 177
  Valproate 177
 Conclusions 177
 References 178

11 Current state of affairs; epilepsy surgery in children and adolescents 183
Herbert Silfvenius

 Introduction 183
 Candidates for cortical excision 184
 Candidates for palliative surgery 185
 Need for PES 186
 Etiologies 186
 Age at surgery 187
 Clinical diagnosis 188
 Neuroradiology and imaging 189
 Electrophysiological evaluation 190
 Neuropsychology/neuropsychiatry 190
 Psychosocial evaluation 191
 Contraindications for PES 191
 Preparations for surgery 192
 Anaesthesia and other measures 192
 Local cortical excisions 193
  Temporal lobe excision 193
  Extratemporal excisions 194

  Frontal lobe excisions 194
  Excisions in the motor cortex 194
  Excisions in the sensory cortex 195
  Excisions in the parietal & occipital lobes 195
  Excisions in the language area 195
 Radical cortical excisions and hemispherectomies 195
 Other surgical techniques 196
  Multiple subpial transsection 196
  Chronic invasive electrical stimulation 197
  Stereotaxy functional surgery 197
  Non-invasive epilepsy surgery 197
  Palliative PES, corpus callosum sections 197
  Re-operations 198
 Evaluating seizure outcome after surgery 198
  Outcome from temporal lobe excisions 198
  Extra-temporal excisions 199
  Hemispherectomies 199
  Multiple subpial transsection 199
  Surgical results for Lennox-Gastaut syndrome and Infantile spasms 199
  Epilepsia partialis continua & status epilepticus, surgical results 200
  Results from re-operations 200
  Corpus callosum sections 200
 Broader perspectives on PES outcome 200
 Surgical complications 201
 Histopathology 202
 Postoperative follow up 202
 Personal experience of PES 202
 General comments 203
 References 203

12 **Neuropsychological aspects of epilepsy surgery** 211
Michael Westerveld
Kimberlee J. Sass
 Epilepsy surgery in children 211
 Early surgical intervention: controversies and rationale 212
  Controversies 212
  Rationale for Surgical Intervention 214
 Surgical procedures 216
  Temporal Lobectomy 216
  Neuropsychology of Hemispherectomy 217
  Neuropsychology of Disconnection Procedures 218
 Role of child neuropsychology in an epilepsy surgery program 219
  Patient Selection 219
  Outcome Assessment 219
  Developmental Research 220
 References 220

13 The impact of epilepsy on cognitive development and learning behaviour 225
Albert P. Aldenkamp
Introduction 225
Type of learning disability 226
Research on cognitive factors 227
   The impact of seizure activity on cognitive function and learning 227
   Localization of epileptogenic foci - effects on cognition and learning 229
   Cognitive side-effects of antiepileptic treatment 230
Issues of assessment and treatment 231
Socioeconomic status 233
References 234

14 The impact of epilepsy on behaviour and emotional development 239
Pamela J. Thompson
Introduction 239
Seizure related factors 239
   Aetiology 239
   Seizure frequency 240
   Seizure type 240
Treatment 241
   Antiepileptic drugs 241
   Surgery 241
Environmental 242
   The family 242
   The school 244
   Society 244
Developmental stage 245
Positive approaches 246
   Aetiology 246
   Treatment 246
   The family 246
   The school 247
   Society 247
References 248

15 The impact of epilepsy on social integration and 'quality of life': family, peers, and education 251
Theo P.B.M. Suurmeijer
Introduction 251
Family and illness 252
Multi-dimensionality of chronic illness, multidimensionality of problems 253
Family functioning and the child with epilepsy 254
Child and epilepsy: peers and education 259
   Social participation 259
   Education 261
Final remarks 264
References 266

16 **Counselling and rehabilitation; the clinician point of view** 271
   Matti Sillanpää
   Introduction 271
   The start of treatment 272
     Drug treatment 272
     Follow-up visits 272
   Social rehabilitation and counselling 273
     Methods of counselling 273
     Social competence 273
     Information to the family members 274
     Education and work 274
   References 276

17 **Assessment of quality of life in children and adolescents with epilepsy** 279
   Gus A. Baker
   Ann Jacoby
   Introduction 279
     What is meant by the term 'quality of life'? 280
     Why is it important to measure quality of life? .. 280
   Methodological issues in measuring quality of life ... 280
     Quality of life assessments subjective or objective? 281
     Generic versus disease-specific measures? 281
     Psychometric properties of QOL measures 282
     Standardisation of measures 282
     What domains should be measured? 282
     Issues specific to measuring quality of life in children and adolescents 283
   Application of QOL measures in studies of children and adolescents with epilepsy 283
     Descriptive studies 284
     Experimental studies involving children and adolescents .. 285
   Relevance of QOL research to clinical practice 286
   Conclusions 287
   References 287

18 **Assessment of quality of epilepsy care from child to adult; the clinimetric point of view** 291
   Harry Meinardi
   Introduction 291
   Seizure frequency as effect parameter 292
   Clinimetric approach 292
   Seizure severity 293
   Neurotoxicity and systemic toxicity 293
   Measure of medication strength 294
   Quality of life 294
   References 296

1 Classification of epilepsies in childhood and adolescence

FRITZ E. DREIFUSS

Department of Neurology, School of Medicine, University of Virginia
Charlottesville, USA

INTRODUCTION

Epileptic seizures have been described from the earliest recordings of medical events. In ancient times, seizures were a source of anxiety and horror to the on-looker which begat the mythology and misunderstandings which dog the sufferer from these phenomena to the present time.[1] Yet seizures are but a symptom of any number of disorders of the nervous system. The nervous system has a very limited repertoire of response to inimical stimuli. It either ceases to function, in which case one has a paralysis or an increase in the number of electrical discharges on the basis of unwanted and untoward depolarizations occurs, leading to seizures, the typology of which depends on the area involved at the onset and of the areas mobilized in the spread of the discharge. Such symptoms of neurological dysfunction do not necessarily represent the condition known as epilepsy which is a term used for recurrent seizures which arise apparently spontaneously, thereby excluding such conditions as recurrent hypoglycaemia or hypocalcemia or eclampsia, conditions which will result in seizures in anybody. The difference between epileptic seizures and nonepileptic seizures is that the epileptic brain is peculiarly predisposed to recurrent seizures which are thus precipitated under circumstances which would not be epileptogenic in the average individual not so predisposed. They are probably not unprovoked, but the provoking factor is not always evident. It may be a hormonal trigger, an emotional one, a photic or auditory event, a deprivation from normal amounts of sleep or exposure to, or withdrawal from, certain drugs or alcohol. Recurrent episodes of fever may also precipitate seizures in those who have the predisposition. The process of precipitation of epileptic seizures is known as ictogenesis whereas the production of the underlying predisposition, be it innate or acquired, is known as epileptogenesis. The predisposition is known as the epilepsy and the precipitated event, in those so predisposed, is the epileptic seizure.

NONEPILEPTIC SEIZURES

Obviously it is important to distinguish between epileptic and nonepileptic seizures because the latter can be eliminated by avoiding or eliminating the causative condition. A large number of childhood disorders may be confused with epileptic seizures as illustrated in Table 1. The distinction between epileptic and nonepileptic seizures is crucial in the thoughtful management of patients because a course of antiepileptic medication is a protracted one in which the patient is exposed to substances whose use may be fraught with unwanted side effects.

Table 1. Disorders which may be confused with epileptic seizures

Reduced cardiac output
- Vasovagal syncope
- Cardiac syncope
- Reflex syncope

Breath holding
- Apneic spells
- The thwarted child syndrome
- Reflex

Hyperventilation

Periodic Attacks
- Hydrocephalic attacks
- Vertigo
- Paroxysmal choreoathetosis (maybe Kinesiogenic)
- Shuddering in infants
- Headache, abdominal pain, vomiting
   (periodic syndrome in children, usually migraine)

Sleep Syndromes
- Narcolepsy
- Sleep terrors
- Hypnogogic hallucinations
- Hypnogogic myoclonus
- Sleep paralysis
- Somnambulism

Pseudoseizure
- Hysteria
- Malingering
- Von Muenchausen Syndrome
- Von Muenchausen Syndrome by proxy

Rage attacks

CLASSIFICATION OF SEIZURES

There are many reasons why seizures have been classified. One is the localizational importance of discerning the origin and spread as exemplified by the outward expression of the individual seizure. For descriptive purposes it is also important to adopt a uniform terminology for communicating with colleagues, using a descriptive short-hand so that the recipient of the communication sees in his mind's eye that which the observer conveys to him. This becomes particularly important when data are pooled between various centers, engaged in research. Moreover, uniformity of terminology is becoming especially important for pooling data which may lead to finding the placement of an underlying aberration in the human genome. Early attempts at classification were purely descriptive, though as early as 375 A.D., Galen spoke of epileptic seizures which were on the basis of idiopathic processes and those which were secondary to disease elsewhere and which he termed symptomatic. Childhood seizures were first described by Reynolds in 1861 as 'eclampsia' by which was meant that the seizures were the result of febrile illnesses.[2] Poupart described absence in 1705 and in 1772 Tissot[3] classified epileptic seizures and this classification was further elaborated by Bernard Sachs[4] including the concept that epilepsy is composed of an ongoing predisposing condition or diathesis and that the individual epileptic seizures are triggered by a concatenation of factors considered as precipitating or triggering factors. This consideration seems to have been forgotten over the subsequent 200 years, leading to the proposition that epilepsy is a liability to unprovoked seizures which may be a procrustean attempt to exclude febrile convulsions in the epilepsy rubric. Sachs divided the childhood epilepsies into eclamptic and epileptic and the latter were further divided into partial and generalized as well as lesional and idiopathic. He considered idiopathic seizures on a heritable basis to have a bad prognosis and symptomatic seizures to have an even worse prognosis. In his book on infantile cerebral paralysis, Sigmund Freud[5] related epilepsy and cerebral palsy. Both Sachs and Tissot realized that the prognosis of epileptic seizures ultimately depended on the underlying cause of the epilepsy, causing the seizures and was not inherent in the convulsions themselves which is a different conclusion than had been arrived at by Gowers[6] who felt that the repetitiveness of a convulsive disorder carried within it the seeds of a progressive disease. Subsequent classification efforts were aimed at localization of seizures, which related the physiological stimulation experiments of Fritsch and Hitzig and of Ferrier and formed the basis of Hughlings Jackson's localizational descriptions. These in turn formed the basis of early attempts at epilepsy surgery, culminating in the present surge of this procedure for the amelioration of seizure disorders. Electroencephalography was developed during this time and subsequently became elaborated with split-screen techniques of seizures and electroencephalographic phenomenology for objective documentation of individual seizure types which in turn lead to the present seizure classification[7] (Table 2).

Table 2. Classification of Seizures.

I. Partial (focal, local) seizures

Partial seizures are those in which, in general, the first clinical and electroencephalographic changes indicate activation of a system of neurons limited to part of one cerebral hemisphere. A partial seizure is classified primarily on the basis of whether or not consciousness is impaired during the attack. When consciousness is not impaired, the seizure is classified as a simple partial seizure. When consciousness is impaired, the seizure is classified as a complex partial seizures. Impairment of consciousness may be the first clinical sign, or simple partial seizures may evolve into complex partial seizures. In patients with impaired consciousness, aberrations of behaviour (automatisms) may occur. A partial seizure may not terminate, but instead progress to a generalized motor seizure. Impaired consciousness is defined as the inability to respond normally to exogenous stimuli by virtue of altered awareness and/or responsiveness.

There is considerable evidence that simple partial seizures usually have unilateral hemispheric involvement and only rarely have bilateral hemisphere involvement; complex partial seizures, however, frequently have bilateral hemispheric involvement. Partial seizures can be classified into the following three fundamental groups:

A. Simple partial seizures
   (consciousness not impaired)
   1. With motor symptoms
   2. With somatosensory of special sensory symptoms
   3. With autonomic symptoms
   4. With psychic symptoms
B. Complex partial seizures
   (with impairment of consciousness)
   1. Beginning as simple partial seizures and progressing to impairment of consciousness
      a. With no other features
      b. With features as in A. 1-4
      c. With automatisms
   2. With impairment of consciousness at onset
      a. With no other features
      b. With features as in A. 1-4
      c. With automatisms
C. Partial seizures secondarily generalized

II. Generalized seizures (convulsive or nonconvulsive)

Generalized seizures are those in which the first clinical changes indicate initial involvement of both hemispheres. Consciousness may be impaired and this impairment may be the initial manifestation. Motor manifestations are bilateral. The ictal electroencephalographic patterns initially are bilateral, and presumably reflect neuronal discharge which is widespread in both hemispheres.

A. 1. Absence seizures
   2. Atypical absence seizures
B. Myoclonic seizures
C. Clonic seizures
D. Tonic seizures
E. Tonic - Clonic seizures
F. Atonic seizures

III. Unclassified epileptic seizures

Includes all seizures that cannot be classified because of inadequate or incomplete data and some that defy classification in hitherto described categories. This includes some neonatal seizures, e.g., rhythmic eye movements, chewing, and swimming movements.

Adapted from Epilepsia 22:489-501, 1981

CLASSIFICATION OF EPILEPSIES OR EPILEPTIC SYNDROMES

The Commission on Classification of the International League Against Epilepsy having proposed the presently extant Classification of Epileptic Seizures, regarding these seizures as symptoms of a potentially identifiable underlying disorder, a so-called epilepsy or epileptic syndrome, took into consideration factors including anatomical substrates, etiology, and age factors, family history, natural history of the seizure disorder and response to medication and developed a classification of individual epileptic syndromes which has contributed to and benefited from an understanding of childhood epilepsies.[8]

Modelling of the epilepsies using various animal models, as well as sophisticated neurophysiological, neurochemical, and pharmacological techniques, tissue slice preparations with the application of putative excitatory and inhibitory neurotransmitters and both extracellular and intracellular recordings, as well as individual neuronal culture preparations have greatly advanced the study of the epilepsies at the molecular or membrane level. In addition, the realization that epilepsy is a system disease led to studies of models of secondary epileptogenesis and kindling as explaining semiology of individual seizures and their underlying disease states. The study of molecular biology has allowed inclusion in the human genome of some of the epileptic syndromes which are presently defined as 'idiopathic' and which have a heritable basis. It has also become apparent that whereas the seizure type is the product of the area of the nervous system involved, the causative factors have implications reaching into areas of genetic higher cortical function and intelligence all of which contribute to the concept of epileptic syndrome which defines the natural history of the condition under consideration (Table 3).

DEFINITIONS

Definitions of terms

Seizures are categorized as either partial or generalized. Partial (focal or localization-related) seizures arise in specific loci in the cortex, which carry with them identifiable signatures, either subjective or objective, which may range from disorders of sensation or thought to convulsive movements of a part of the body which may become generalized. Simple partial seizures are those in which consciousness is preserved and probably arise predominantly from six-layered isocortex, remaining localized sufficiently long to allow specific symptoms to be discerned.

Table 3. International Classification of Epilepsies and Epileptic Syndromes

1. **Localization-related (focal, local, partial) epilepsies and syndromes**

   1.1 Idiopathic (with age-related onset)
   At present, the following syndromes are established but more may be identified in the future.
   - benign childhood epilepsy with centro-temporal spike
   - childhood epilepsy with occipital paroxysms
   - primary reading epilepsy

   1.2 Symptomatic
   This category comprises syndromes of individual variability which is mainly based on anatomical localization, clinical features, seizure types and etiological factors (if known).

   1.2.1 Epilepsy is characterized by simple partial seizures with the characteristics of seizures:
   - arising from frontal lobes
   - arising from parietal lobes
   - arising from temporal lobes
   - arising from occipital lobes
   - arising from multiple lobes
   - locus of onset unknown

   1.2.2 Characterized by complex partial seizures, that is attacks with alteration of consciousness often with automatisms.
   Characterized by seizures:
   - arising from frontal lobes
   - arising from parietal lobes
   - arising from temporal lobes
   - arising from occipital lobes
   - arising from multiple lobes
   - locus of onset unknown

   1.2.3 Characterized by secondarily generalized seizures with seizures:
   - arising from frontal lobes
   - arising from parietal lobes
   - arising from temporal lobes
   - arising from occipital lobes
   - arising from multiple lobes
   - locus of onset unknown

   1.3 Unknown as to whether the syndrome is idiopathic or symptomatic

-table 3 continues next page

2. **Generalized epilepsies and syndromes**

2.1 Idiopathic (with age-related onset - listed in order of age)
- benign neonatal familial convulsions
- benign neonatal convulsions
- benign myoclonic epilepsy in infancy
- childhood absence epilepsy (pyknolepsy)
- juvenile absence epilepsy
- juvenile myoclonic epilepsy (impulsive petit mal)
- epilepsy with grand mal (GTCS) seizures on awakening

Other generalized idiopathic epilepsies, if they do not belong to one of the above syndromes can still be classified as generalized idiopathic epilepsies.

2.2 Cryptogenic or symptomatic (in order of age)
- West syndrome (infantile spasms, Blitz-Nick-Salaam Krämpfe)
- Lennox-Gastaut syndrome
- Epilepsy with myoclonic-astatic seizures
- Epilepsy with myoclonic absences

2.3 Symptomatic
2.3.1 Non-specific etiology
- Early myoclonic encephalopathy

2.3.2 Specific syndromes
- Epileptic seizures may complicate many disease states.

Under this heading are included those diseases in which seizures are a presenting or predominant feature.

3. **Epilepsies and syndromes undetermined whether focal or generalized**

3.1 With both generalized and focal seizures
- Neonatal seizures
- Severe myoclonic epilepsy in infancy
- Epilepsy with continuous spike-waves during slow wave sleep
- Acquired epileptic aphasia (Landau-Kleffner-syndrome)

3.2 Without unequivocal generalized or focal features

All cases with generalized tonic-clonic seizures where clinical and EEG findings do not permit classification as clearly generalized or localization-related such as in many cases of sleep-grand mal.

4. **Special Syndromes**

4.1 Situation-related seizures (Gelegenheitsanfälle)
- Febrile convulsions
- Isolated seizures or isolated status epilepticus
- Seizures occurring only when there is an acute metabolic or toxic event due to, for example, alcohol, drugs, eclampsia, non-ketotic hyperglycemia, uremia, etc.

Adapted from Epilepsia 30:389-399, 1989

At other times they spread rapidly and become elaborate in their manifestations and may even generalize. Complex partial seizures are those in which consciousness is impaired and they may be the first indication of a seizure or they may follow upon simple partial seizures through involvement of limbic cortex leading to impairment of consciousness at the onset. Motor activity during impaired consciousness may be manifested in the form of automatisms. Either temporal or frontal lobe structures may be involved in complex partial seizures. These may at times be difficult to distinguish from absence seizures especially if they in the frontal cortex where there is less likely to be a postictal confusion, than if they arise in temporal cortex. Generalized seizures involve large volumes of brain from the outset and are usually bilateral in their initial manifestations, including electroencephalographic findings. These may range from absence seizures, characterized only by impairment of consciousness, to generalized tonic-clonic seizures in which widespread convulsive activity takes place. Myoclonic, tonic and clonic seizures may also occur in generalized attacks. Epileptic spasms, usually occurring in infancy, represent another variety of seizures which are usually generalized but may also present focally. As mentioned earlier, individual epileptic seizures are the symptoms which present to the physician, but they may be indicative of an epileptic disorder or syndrome characterized by a cluster of signs and symptoms, customarily occurring together some of which are very specific, for example benign rolandic epilepsy, juvenile myoclonic epilepsy and the Lennox-Gastaut syndrome or the West syndrome. Some of these individual syndromes will be further amplified in a definitional sense.

Benign childhood epilepsy with centro-temporal spikes

This is a syndrome of brief, simple, partial, hemifacial motor seizures, frequently having associated somatosensory symptoms, which have a tendency to evolve into GTCS.[9-12] Both seizure types are often related to sleep. Onset is between 3 and 13 years of age (peak, 9-10), and recovery before ages 15-16. Genetic predisposition is frequent, and there is male predominance. The EEG has blunt high-voltage centro-temporal spikes, often followed by slow waves that are activated by sleep and tend to spread or shift from side to side.

Childhood epilepsy with occipital paroxysms

This syndrome is, in general respects, similar to the previous one.[13] The seizures start with visual symptoms (amaurosis, phosphenes, illusions, or hallucinations), and are often followed by a hemiclonic seizure or automatisms. In a quarter of the cases, the seizures are immediately followed by migrainous headache. The EEG has paroxysms of high-amplitude spike-waves or sharp waves recurring rhythmically on the occipital and posterior temporal areas of one or both hemispheres, but only when the eyes are closed. During seizures, the occipital discharge may spread to the central or temporal region. At present no definite statement on prognosis is possible.

Benign neonatal familial convulsions

These are rare, dominantly inherited disorders, manifesting mostly on the second and third days of life, with clonic or apneic seizures and no specific EEG criteria. History and investigations reveal no etiological factors. About 14% of these patients later develop epilepsy. This syndrome appears to be associated with a gene resident on chromosome 20Q.[14]

Benign neonatal convulsions

These are very frequently repeated clonic or apneic seizures occurring around the fifth day of life, without known etiology or concomitant metabolic disturbance. Interictal EEG often shows alternating sharp theta waves. There is no recurrence of seizures, and the psychomotor development is not affected.

Benign myoclonic epilepsy in infancy

This form is characterized by brief bursts of generalized myoclonus that occur during the first or second year of life in otherwise normal children, who often have a family history convulsions or epilepsy.[15] EEG recording show generalized spike-waves occurring in brief bursts during the early stages of sleep. These attacks are easily controlled by appropriate treatment. They are not accompanied by any other types of seizures, although GTCS may occur during adolescence. The epilepsy may be accompanied by a relative delay of intellectual development and minor personality disorders.

Severe myoclonic epilepsy in infancy (Dravet)

This condition which carries a poor prognosis usually begins during the first year of life, frequently with febrile convulsions, followed by a retardation in developmental milestones and associated myoclonic disorder.[16] There is frequently a family history of febrile seizures.

Childhood absence epilepsy (pyknolepsy)

This syndrome occurs in children of school age (peak manifestation, ages 6-7) with a strong genetic predisposition in otherwise normal children.[17-19] It appears more frequently in girls than in boys. It is characterized by very frequent (several to many per day) absences. The EEG reveals bilateral, synchronous symmetrical spike-waves, usually 3/s, on a normal background activity. During adolescence, GTCS often develop. Otherwise, absences may remit or, more rarely, persist as the only seizure type.

Juvenile absence epilepsy

The absences of this syndrome are the same as in pyknolepsy, but absences with retropulsive movements are less common.[20] Age of manifestation is around puberty. Seizure frequency is lower than in pyknolepsy, with absences occurring less frequently than every day, mostly sporadically. Association with GTCS is frequent, and they precede the absence manifestations more often than in childhood absence epilepsy, often occurring on awakening. Not infrequently, the patients also have myoclonic seizures. Sex distribution is equal. The spike-waves are often faster than 3/s. Response to therapy is excellent.

Juvenile myoclonic epilepsy (impulsive petit mal)

This syndrome appears around puberty and is characterized by seizures with bilateral, single or repetitive, arrhythmic, irregular myoclonic jerks, predominantly in the arms.[21-23] Some patients may suddenly fall from a jerk. No disturbance of consciousness is noticeable. The disorder may be inherited and sex distribution is equal. Often, there are GTCS and, less often, infrequent absences. The seizures usually occur shortly after awakening, and are often precipitated by sleep deprivation. Interictal and ictal EEG have rapid, generalized, often irregular spike-waves and polyspike-waves; there is no close phase correlation between EEG spikes and jerks. Frequently, the patients are photosensitive. Response to appropriate drugs is good.

Epilepsy with GTCS on awakening

This is a syndrome with onset mostly in the second decade of life. The grand mal seizures are presumably mainly GTCS and occur exclusively or predominantly (over 90% of the time) shortly after awakening, regardless of the time of day, or in a second seizure peak in the evening period of relaxation. If there are other seizures, these are mostly absences or myoclonic, as in juvenile myoclonic epilepsy. Seizures may be precipitated by sleep deprivation and other external factors. Genetic predisposition is relatively frequent. The EEG shows one of the patterns of idiopathic generalized epilepsy. There is a significant correlation with photo-sensitivity.

West syndrome (infantile spasms, Blitz-Nick-Salaam Krämpfe)

Usually, West syndrome consists of a characteristic triad: infantile spasms, arrest of psychomotor development, and hypsarrhythmia, although one element may be missing.[24-26] Spasms may be flexor, extensor, lightning, or nods but most commonly are mixed. Onset peaks between 4 and 7 months and is always before 1st year. Boys are more commonly affected, and the prognosis is generally poor. West syndrome may be separated into two groups. The symptomatic group is characterized by the previous existence of brain damage signs (psychomotor retardation, neurological signs, radiological signs, or other types of seizures) or by a known etiology. The smaller, idiopathic group is characterized by the absence of previous signs of brain damage and of known etiology. The prognosis is partly based on early therapy with adrenocorticotropic hormone (ACTH) or oral steroids. However, it principally depends on the symptomatic or idiopathic cases which have had favorable prognoses without psychic impairment and later epilepsy when treated early. In recent years, it has become evident that infantile spasms may be symptomatic not only of tuberous sclerosis but also of other developmental disorders some of which may be potentially amenable to surgical therapy of abnormal brain tissue and therefore more sophisticated diagnostic techniques such as PET scan may be required to identify these.[27]

Lennox-Gastaut syndrome

This syndrome manifests itself in children from 1 to 8 years of age, but appears mainly in preschool-age children.[28-30] The most common seizure types are tonic-axial, atonic, and absence seizures, but other types such as myoclonic, GTCS, or partial are frequently associated with this syndrome. Seizure frequency is high, and status epilepticus frequent (stuporous states with myoclonias, tonic, and atonic seizures). The EEG usually has abnormal background activity, slow spike-waves of less than 3/s and, often, multifocal abnormalities. During sleep, bursts of fast rhythms (around 10/s) appear. In general, there is mental retardation. Seizures are difficult to control, and the development is mostly unfavorable. In 60% of cases, the syndrome occurs in children suffering from a previous encephalopathy, but it is primary in other cases.

Epilepsy with myoclonic-astatic seizures

Manifestation begins between 7 months and 6 years, mostly from 2 to 5 years, with (except if beginning in the first year) twice as many boys affected.[31] There is frequently hereditary predisposition and usually a normal developmental background. The seizures are myoclonic, astatic, myoclonic-astatic, absences with clonic and tonic components, and tonic-clonic. Status frequently occurs. Tonic seizures develop late in the course of unfavorable cases. The EEG, initially often normal except for 4-7/s rhythms, may have irregular fast spike-wave or poly-spike wave. Course and outcome are variable.

Epilepsy with myoclonic absences

This syndrome is clinically characterized by absences accompanied by severe bilateral rhythmical clonic jerks, often associated with a tonic contraction.[32] On the EEG they are always accompanied by bilateral, synchronous, and symmetrical discharge of rhythmical spike-wave at 3/s, similar to childhood absence. These seizures occur many times a day. Awareness for the jerks may be maintained. Associated seizures are rare. Age of onset is about 7 years and there is a male preponderance. Prognosis is less favorable than in pyknolepsy due to resistance to therapy of the seizures, mental deterioration, and possible evolution to other types of epilepsy such as Lennox-Gastaut syndrome.

Early myoclonic encephalopathy

The principal features of this syndrome are onset before 3 months of age, initially fragmentary myoclonus, then erratic partial seizures, massive myoclonias, or tonic spasms.[33] The EEG is characterized by suppression-burst activity, which may evolve into hypsarrhythmia. The course is severe, psychomotor development is arrested, and death may occur in the first year. Familial cases are frequent and suggest the influence of one or several congenital metabolic errors, but there is no constant genetic pattern. The status of early infantile epileptic encephalopathy with suppression bursts, described by Ohtahara in relation to early myoclonic encephalopathy, is at present unclear, especially in view of its ictal features and its frequent evolution into a syndrome indistinguishable from West syndrome.

Neonatal seizures

Neonatal seizures differ from those of older children and adults. The most frequent neonatal seizures are described as subtle because the clinical manifestations are frequently overlooked. These include tonic, horizontal deviation of the eyes with or without jerking, eyelid blinking or fluttering, sucking, smacking, or other buccal-lingual oral movements, swimming or pedalling movements, and occasionally, apneic spells. Other neonatal seizures occur as tonic extension of the limbs, mimicking decerebrate or decorticate posturing. These are particularly seen in premature infants. Multifocal clonic seizures, characterized by clonic movements of a limb, which may migrate to other body parts or other limbs, or focal clonic seizures, which are much more localized, may occur. In the latter, the infant is usually not unconscious. Rarely, myoclonic seizures may occur, and the EEG pattern is frequently that of suppression-burst activity. The tonic seizures have a poor prognosis because they frequently accompany intraventricular hemorrhage. The myoclonic seizures also carry a poor prognosis because they are frequently a part of the early myoclonic encephalopathy syndrome.

Epilepsy with continuous spike-waves during slow sleep

This syndrome results from the association of various seizure types, partial or generalized, occurring during sleep, and atypical absences when awake.[34] Tonic seizures do not occur. The characteristic EEG pattern consists of continuous diffuse spike-waves during slow wave sleep, which is seen after the onset of seizures. Duration varies from months to years. The prognosis is guarded because of the appearance of neuropsychologic disorders, despite the usually benign evolution of seizures.

Acquired epileptic aphasia (Landau-Kleffner syndrome)

The Landau-Kleffner syndrome is a childhood disorder associating an acquired aphasia, multifocal spikes, and spike and wave discharges.[35] Epileptic seizures and behavioral and psychomotor disturbances occur in two-thirds of the patients. There is verbal auditory agnosia and rapid reduction of spontaneous speech. The seizures are generally generalized convulsive or partial motor. They are rare, and remit before the age of 15 years, as do the EEG abnormalities.

Malformations

Conditions such as Aicardi syndrome, lissencephaly-pachygyria, and the individual phacomatoses would fall into this category.

Proven or suspected inborn errors of metabolism

In the neonate, these would include nonketotic hyperglycenemia and d-glyceric acidemia with early myoclonic encephalopathy. In the infant, this would include phenylketonuria, Tay-Sachs, and Sandhoff disease and early infantile ceroid lipofuscinosis as well as pyridoxine dependency. In later childhood, the late infantile ceroid lipofuscinosis, juvenile Gaucher's disease, the juvenile form of ceroid lipofuscinosis and the Lafora variety of progressive myoclonic epilepsy may present. In adolescence, progressive myoclonic epilepsy at the Lundborg type and Ramsay-Hunt syndrome and cherry red spot myoclonus syndrome (sialidosis with isolated deficiency in neuraminidase) are forms of progressive myoclonic epilepsy. The mitochondrial encephalopathies with abnormalities in lactate-pyruvate metabolism frequently present as progressive myoclonic epilepsies including the Ramsay-Hunt syndrome.[36]

CONCLUSIONS

It is quite clear that where childhood epilepsies are concerned, the classifications of epileptic seizures and of the epilepsies is extraordinarily useful. The former to a large degree determines the optimal therapeutic management regimen. From the latter, one can derive prognostic information from which to devise a long range management plan. The syndrome might dictate that no pharmacological interactions is necessarily indicated as in the case of febrile convulsions and frequently in benign focal seizures of childhood. Additionally, it may indicate how long a course of treatment has to be employed according to the predicted natural history of the condition under consideration. One has to bear in mind that no classification is either perfect or definitive. One patient at different times may exhibit different syndromes. Thus, a syndrome may evolve into Lennox-Gastaut syndrome and what may begin as febrile convulsions may end as a progressive limbic epilepsy with mesial temporal sclerosis. Constant reevaluation of the management plan must therefore take place shaping the treatment to the condition at hand as this changes over time. The classification will add to knowledge concerning the pathophysiology of the epilepsies and the increasing knowledge will lead to changes in the classification.

REFERENCES

1. Temkin O. The Falling Sickness. A History of Epilepsy from the Greeks to the Beginnings of Modern Neurology, ed. 2 Baltimore, John Hopkins Press, 1971.
2. Reynolds JR. Epilepsy: its symptoms, treatment and relation to other chronic convulsive diseases. London, Churchill, 1861.
3. Tissot SA. Traite de l'epilepsie faisant le tome troisieme du traite des nerfs et de leurs maladies. Paris, P.F. Didot, 1772.
4. Sachs B. A treatise on the nervous system of children for physicians and students. New York, William Wood & Co, 1985.
5. Freud S. Infantile cerebral paralysis translated by L.A. Russin. Coral Gables, University of Miami Press, 1968.
6. Gowers WR. Epilepsy and other chronic convulsive disorders. London, Churchill, 1881.
7. Commission on Classification and Terminology of the International League Against Epilepsy. Proposal for revised clinical and electroencephalographic classification of epileptic seizures. Epilepsia, 1981; 22: 489-501.
8. Commission on Classification and Terminology of the International League Against Epilepsy. Proposal for revised Classification of epilepsies and epileptic syndromes. Epilepsia, 1989; 30: 389-399.
9. Nayrac P, Beaussart M. Les pointe-ondes prerolandique: Expression EEG tres particuliere. Rev. Neurol., 1958; 99: 201-206.
10. Beaussart M. Benign epilepsy of children with rolandic (centro-temporal) paroxysmal foci. Epilepsia, 1972; 13: 795-811.
11. Beaumanoir A, Ballist T, Varfis G, et al. Benign epilepsy of childhood with rolandic spikes. Epilepsia, 1974; 15: 301-315.
12. Loiseau P, Beaussart, M.: The seizures of benign childhood epilepsy with rolandic paroxysmal discharges. Epilepsia, 1973, 14: 381-389.
13. Gastaut H. A new type of epilepsy: benign partial epilepsy of childhood with occipital spike-waves. In: Advances in Epileptology, XIIIth Epilepsy International Symposium. Raven Press, 1982; 18-25.
14. Leppert M, Anderson VE, Quattlebaum T, et al. Benign familial neonatal convulsions linked to genetic markers on chromosome 20. Nature, 1988; 337: 647-648.
15. Dravet C, Bureau M, Roger J. Benign Myoclonic Epilepsy in Infants. In: Roger J, Dravet C, Bureau M, et al. editors. Epileptic Syndromes in Infancy, Childhood and Adolescence (2nd edition). London. John Libbey, 1989; 67-74.
16. Dravet C, Bureau M, Roger J. Severe myoclonic epilepsy in infants. In: Roger J, Bureau, M., Dravet, C, et al. editors. Epileptic Syndromes in Infancy, Childhood and Adolescence (2nd edition), London: John Libbey, 1989.
17. Drury I, Dreifuss FE. Pyknoleptic petit mal. Acta Neurol. Scand., 1985; 72: 353-362.
18. Loiseau P. Childhood Absence Epilepsy. In: Roger J, Dravet C, Bureau M, et al, editors. Epileptic Syndromes in Infancy, Childhood and Adolescence. London: John Libbey Eurotext, 1985; 106-120.
19. Penry JK, Porter RJ, Dreifuss FE. Simultaneous recording of absence seisures with videotape and electro-encephalography: A study of 374 seizures in 48 patients. Brain, 1975; 98: 427-440.

20. Wolf P. Juvenile Absence Epilepsy. In: Roger J, Dravet C, Bureau M, et al, editors. Epileptic Syndromes in Infancy, Childhood and Adolescence. London: John Libbey, Eurotext, 1985; 242-246.
21. Janz D, Christian W. Impulsive-Petit mal. Dtsch. Z. Nervenheilkd, 1957; 176: 346-386.
22. Asconape J, Penry JK. Some clinical and EEG aspects of benign juvenile myoclonic epilepsy. Epilepsia, 1984; 25: 108-114.
23. Dreifuss FE. Juvenile myoclonic epilepsy: characteristics of a primary generalized epilepsy. Epilepsia, 1989; 30 Suppl. 4: S1-S7.
24. West WJ. On a peculiar form of infantile convulsions. Lancet, 1841; 1: 724-725.
25. Jeavons PM, Bower BD. Infantile spasms: A review of the literature and a study of 112 cases in Clinics in Developmental Medicine, No.15, London: Spastics Society and Heinemann, 1964.
26. Kellaway P, Hrachovy RA, Frost JD, et al. Precise characteristics and quantification of infantile spasms. Ann. Neurol., 1979; 6: 214-218.
27. Dulac O, Chugani HT, Della Bernardina B, editors. Infantile Spasms and West Syndrome, London: Saunders WB, 1994.
28. Lennox WG, Davis JP. Clinical correlates of the fast and the slow spike and wave electroencephalogram. Trans. Amer. Neurol. Assoc, 1949; 74: 194-197.
29. Lennox WG. The slow-spike-wave EEG and its clinical correlates. In: Lennox WG. Epilepsy and Related Disorders Vol. 1, Boston, Toronto: Little, Brown and Co., 1966; 156-170.
30. Gastaut H, Roger J, Soularyrol R, et al. Childhood epileptic encephalopathy with diffuse slow spike-waves (otherwise known as 'Petit mal variant') or Lennox Syndrome. Epilepsia, 1966; 7: 139-179.
31. Doose H, Gerken H, Leonhardt R, et al. Centrencephalic myoclonic-astatic petit mal. Neuropadiatric, 1970; 2: 59-78.
32. Tassinari CA, Bureau M. Epilepsy with myoclonic absences, In: Roger J, Dravet C, Bureau M, et al. editors. Epileptic Syndromes in Infancy, Childhood and Adolescence. London: John Libbey Eurotext, 1985; 121-129.
33. Ohtahara S, Ishida T, Oka E, et al. On the age-dependent epileptic syndromes: the early infantile encephalopathy with suppression-burst. Brain and Development, 1976; 8: 270-288.
34. Tassinari CA, Bureau M, Dravet C, et al. Epilepsy with continuous spike and waves during slow sleep. In: Roger J, Dravet C, Bureau M, et al, editors. Epileptic Syndromes in Infancy, Childhood and Adolescence, London: John Libbey Eurotext, 1985.
35. Landau WM, Kleffner FR. Syndrome of acquired aphasia with convulsive disorder in children. Neurology, 1957; 7: 523-550.
36. Berkovic SF, Andermann F, Carpenter S, et al. Progressive myoclonus epilepsies: Specific cases and diagnosis. New Eng. J. Med, 1986; 315: 296-305.

2 The epidemiology of epilepsy in children and adolescents

JAN OVERWEG
"Meer en Bosch", Epilepsy Center, Heemstede, The Netherlands

WOUTERINA C.G. OVERWEG-PLANDSOEN
Academic Medical Center, Department of Neurology, Amsterdam, The Netherlands

The yearly incidence of epilepsy in children and adolescents ranges from 50 to 100 per 100.000 with the highest incidence in the first year of life. Reliable epidemiological data on the epilepsies and epilepsy syndromes in children and adolescents can only be obtained if proper and sharp definitions are used by every investigator. However, most epidemiological studies focus on epilepsy as such; only a minority have provided information of specific epilepsy syndromes. As the classification of the epilepsies and epileptic syndromes has been published only recently, epidemiological data, based on this classification, will become available in the next decennium. At present the only reliable figures on incidence and prevalence are known for febrile seizures, West syndrome, childhood absence epilepsy, juvenile myoclonic epilepsy and benign partial seizures of adolescence. The prevalences and incidences of the other syndromes are largely unknown. According to the recent classification, a short description and, if available, epidemiological data, are given for the epilepsies and epileptic syndromes in childhood and adolescence.

INTRODUCTION

All children with epilepsy are suffering from seizures but not all children with seizures have epilepsy. Seizure classification in epidemiological studies is difficult. Sander and Shorvon[1] pointed out that in many studies the majority of patients are reported to have generalized seizures, rates as high as 88% being published.[2-9] In each of these studies, generalized seizures accounted for the majority of cases. However, in studies in which classification has been carried out by teams of specialists using EEG, a much higher proportion of partial seizures has been found.[10,11] Often classification is carried out by non-specialist trained physicians and even by non-medically qualified practitioners. The classifications are usually based on clinical data, often without the information from the EEG. Inclusion of febrile convulsions and single seizures in some studies makes comparison between studies difficult and often the criteria, used for categorizing the patients, are not given. As a result of this inaccurate classification, partial seizures are particularly under-reported. The classification of the International League Against Epilepsy[12] (ILAE) requires both clinical and EEG data. Nonetheless, EEG-data are not available in the majority of the epidemiological studies. The use of a standardized classification and of sharp definitions will improve the quality of comparative data, and will give more accurate estimates of prevalence and incidence of seizure types and epilepsy syndromes.

In the present situation there is little agreement about the incidence and prevalence of the epilepsies. Reported rates for all epilepsies, and for all ages combined, ranges from 2.2/100.000 to 100/100.000/year[13] for incidence and from 230/100.000[14] to 2800/100.000[15] for prevalence. The various investigators employ different definitions of epilepsy, different methods of identifying patients and different inclusion criteria. The classification of the epileptic seizures[12] and of the epilepsies and epileptic syndromes,[16] proposed by the International League Against Epilepsy is often not used or not used in a proper way. The classification of the epileptic seizures is relevant for a meaningful communication between professionals. Moreover, an exact classification may help to plan a more differentiated and effective pharmacological treatment of epilepsy and establish variations in prognostic outcome. The classification of the epileptic seizures and of the epilepsies and epileptic syndromes has been described in chapter one of this volume. The definitions, formulated by the Commission of Epidemiology and Prognosis of the International League against Epilepsy[17] will be used in this Chapter.

GUIDELINES FOR EPIDEMIOLOGICAL STUDIES

The Commission on Epidemiology and Prognosis of the ILAE has given a set of guidelines for future epidemiological research.[17] Since the proposed guidelines represent, to a large extent, a consensus between epidemiologists and epileptologists, it is worthwhile to adopt these guidelines for epidemiological studies on epilepsy. Before discussing the epidemiology of epileptic seizures and of the epilepsies in childhood, it is appropriate to give definitions of epileptic seizures and the epilepsies, as well as definitions of variables affecting epilepsy. As long as a definition is present, even if it is not consistent across studies, it is possible to identify how the different definitions affect the result.[13]

Proposed definitions

Epileptic seizure: a clinical manifestation, presumed to be due to an abnormal and excessive discharge of a set of neurons of the brain. The clinical manifestation consists of sudden and transitory abnormal phenomena which may include alterations of consciousness, motor, sensory, autonomic or psychic events, perceived by the patient or an observer. *Epilepsy*: a condition characterized by recurrent (2 or more) epileptic seizures, unprovoked by any immediate identified cause. Multiple seizures occurring within a 24 hour period are considered a single event. An episode of status epilepticus is considered one single event. Individuals who have had only febrile seizures or only neonatal seizures, are excluded from this category. *Status epilepticus*: a single epileptic seizure of more than 30 minutes duration, or a series of epileptic seizures during which function is not regained between ictal events over a period of more than 30 minutes. *Active epilepsy*: a prevalent case of 'active' epilepsy is defined as a person with epilepsy who has had at least one epileptic seizure within the previous five years, regardless of antiepileptic drug (AED) treatment. A case under treatment is someone with the correct diagnosis of epilepsy receiving (or having received) AED's (global) on prevalence day. *Epilepsy in remission on treatment*: a prevalent case of epilepsy without seizures for five years or more and taking AED's at the time of ascertainment. *Epilepsy in remission off treatment*: a prevalent case of epilepsy without seizures for five years or more and not taking AED's at the time of ascertainment. *Single or isolated seizure*: only one (or more) epileptic seizure(s) occurring within a 24 hour period without any subsequent seizures. *Febrile seizure*: an epileptic seizure, occurring in childhood after one month of age, associated with a febrile illness not caused by an infection of the central nervous system, without prior neonatal seizures or a prior unprovoked seizure and not meeting criteria for other acute symptomatic seizures. *Neonatal seizure*: an epileptic seizure occurring in the first four weeks of life. *Febrile seizure with neonatal seizure*: a child who has experienced one or more neonatal seizures and also experiences one or more febrile seizures. *Non-epileptic events*: clinical manifestations presumed to be unrelated to an abnormal and excessive discharge of a set of neurones of the brain, including; (1) paroxysmal disturbances in brain function: vertigo or dizziness, syncope, sleep and movement disorders, transient global amnesia, migraine, enuresis and (2) pseudo-seizures: nonepileptic sudden behavioral episodes presumed to be of psychogenic origin. These may coexist with true epileptic seizures.

Measurements indices

Several measures have been utilized to describe the frequency of epilepsy. However, often these indices are inappropriately used or used without correct definition. The following indices for prevalence and incidence are recommended.[17] *Point prevalence*: the proportion of patients with epilepsy in a given population at a specified time (usually a specific day - the prevalence day). Inclusion criteria should be specified (i.e. active epilepsy, epilepsy in remission on treatment, epilepsy in remission off treatment). *Period prevalence*: the proportion of patients with epilepsy in a given population during a defined time interval (e.g. one year). Inclusion criteria should be specified. *Lifetime prevalence*: the proportion of patients with a history of epilepsy regardless of treatment or recent seizure activity. This includes patients with active epilepsy or epilepsy in remission, representing all individuals identified with a history of epilepsy at any time. Prevalences which represent the ratio of identified cases to the total population are usually expressed as cases per 1000 persons. *Incidence (or incident number)*: the number of new cases of epilepsy occurring during a given time interval, usually one year, within a specified population. *Incidence rate*: the ratio of new cases to population at risk, usually expressed as cases per 100.000 persons per year. Criteria for defining an incident case must be clearly stated, including specification of whether it is based on data of diagnosis or date of onset. *Incidence density*: the ratio of new cases to a dynamic cohort at risk, usually expressed as cases per 100.000 persons per year. *Cumulative incidence*: the individual's risk of developing epilepsy by a certain time when a specified age is reached.

The most comprehensive investigations that use the classification, as proposed above, were performed in Rochester, Minnesota, by Hauser & Kurland[10] and by Hauser & Hesdorffer.[13] The Rochester investigations showed an annual epilepsy incidence of 48.7/100.000 inhabitants. The highest incidence was found in the first decade of life and particularly in children less than one year of age. There seems to be a general agreement that all detected rates are still underestimates. Studies of incidence and prevalence of epilepsy generally fail to differentiate between different seizure types or seizure syndromes. Three European studies[18-20] and one American group, represented by Hauser,[10,13,21] do report the distribution of seizure types. However, as mentioned above it is difficult to compare the published epidemiological studies of epileptic seizures and of the types of epilepsy because of variations in data selection and inclusion criteria. It is clear that the data, cited for prevalence and yearly incidence for populations studied from a single area[10,20,21] are more reliable than data from selected populations. Populations studied by Juul-Jensen (greater Aarhus) and Hauser (Rochester) are unselected, except in so far of course that a primary care physician had recognized the epilepsy and referred the patient to the hospital.

PREVALENCE AND INCIDENCE OF THE MAJOR TYPES OF EPILEPTIC SYNDROMES

Although epilepsy represents one of the most frequently occurring neurologic problems in children, the incidence and prevalence of the epileptic syndromes, presented in Table 1 are largely unknown. The yearly incidence of epilepsy (recurrent unprovoked seizures) in children and adolescents ranges from 50 to 100/100.000[13] with the highest incidence in the first year of life. Epilepsy manifested by generalized onset seizures accounts for most of the newly diagnosed cases in the first 5 years of life.[12] Most epidemiological studies focus on epilepsy as such, only a minority have provided information about specific epilepsy syndromes. In the epidemiologic category of provoked seizures, the most frequently occurring condition is a febrile illness resulting in febrile convulsions. In the most recent classification of the epileptic syndromes,[16] febrile convulsions fall within the category 'situation-related epilepsies'. The reported cumulative incidence of febrile convulsions varies from 2 to 4% in Europe and the USA[21] increasing to 8% in Japan.[22] The peak incidence of a first febrile convulsion (between 1 and 2%) occurs in the second year of life.[10,22]

Table 1. Epileptic seizures and syndromes in childhood and adolescence

Epileptic syndromes in neonates
    Benign idiopathic familial neonatal convulsions
    Benign idiopathic neonatal convulsions
    Early (or Neonatal) myoclonic encephalopathy (Aicardi-syndrome)
    Early-infantile epileptic encephalopathy with suppression bursts (Ohtahara syndrome)

Epileptic syndromes in infancy and childhood
    Febrile convulsions
    Infantile spasms (West syndrome)
    Benign myoclonic epilepsy in infants
    Severe myoclonic epilepsy in infants
    Myoclonic epilepsy (myoclonic status) in non-progressive encephalopathies
    Epilepsy and inborn errors of metabolism
    Myoclonic astatic epilepsy of early childhood (Doose syndrome)
    The Lennox-Gastaut syndrome

Epileptic syndromes in childhood
    Childhood absence epilepsy
    Epilepsy with myoclonic absences
    Epilepsy with generalized tonic-clonic seizures in childhood
    Idiopathic partial epilepsies in children
    Benign partial epilepsy with centro-temporal spikes
    Benign epilepsy of childhood with occipital paroxysms
    Benign partial epilepsy with affective symptoms
    Benign partial epilepsy with extreme somato-sensory evoked potentials
    Acquired epileptic aphasia (Landau-Kleffner syndrome)
    Epilepsy with continuous spikes and waves during slow sleep or Epilepsy with electrical status epilepticus during slow sleep (ESES)

Epileptic syndromes in childhood and adolescence
    Reading epilepsy
    Photosensitive epilepsy
    Juvenile absence epilepsy
    Juvenile myoclonus epilepsy
    Epilepsy with tonic clonic seizures (grand mal) on awakening
    Benign partial seizures of adolescence
    Non idiopathic partial epilepsies and epileptic syndromes in childhood
    Epilepsia partialis continua (Kojewnikow's syndrome)
    Progressive myoclonus epilepsies in childhood and adolescence

About one third of the children with a first febrile convulsion will suffer from additional convulsions at the time of a subsequent febrile illness[23] and 3 to 6% can be expected to develop epilepsy.[24] There are a few studies available with information on the incidence of specific epilepsy syndromes.[21,25-30] The syndromes, given in Table 1 will be described shortly and, if possible, incidence and prevalence will be given.

Epileptic syndromes in neonates

Benign familial neonatal convulsions occur on the second and third days of life. There is a family history of epilepsy and the outcome is favourable, but secondary epilepsy may develop in 11%. The prevalence can be estimated at about 7% of that of neonatal convulsions.[31] Incidences are not yet given in literature. *Benign idiopathic neonatal convulsions* ('fifth day fits') occur around the fifth day of life. There is no known aetiology but a favourable outcome. Secondary epilepsy occurs in 0.5%. The estimated prevalence varies from 4 to 38% of neonatal convulsions.[31] *Neonatal (or early) myoclonic encephalopathy (Aicardi syndrome)*: this syndrome (NNME) is characterized by the occurrence of erratic and fragmentary myoclonus of early onset, usually in association with other seizure types such as partial motor seizures, massive myoclonus and tonic spasms, the latter usually not appearing before 4 to 5 months of age. The EEG shows a suppression burst pattern persisting after 2 weeks of age.[32] In the majority of reported cases the onset was in the neonatal period and in a few other cases before the age of two months.[32] The erratic myoclonus and the late appearance of tonic spasms separate this syndrome from early infantile epileptic encephalopathy. The incidence and prevalence are not known. *Early-infantile epileptic encephalopathy with suppression-bursts (Ohtahara syndrome)* are atypical infantile spasms of early onset characterized by tonic spasms and by a suppression burst pattern in the EEG.[33] This syndrome appears to be caused by a variety of early and extensive brain insults, especially malformations.[34] The prognosis is serious with early death or marked psychomotor retardation and seizure intractability. The incidence and prevalence are unknown.

Epileptic syndromes in infancy and childhood

Febrile convulsions are events in infancy or childhood that occur between 3 months and 5 years of age, associated with fever but without evidence of intracranial infection or defined cause.[35] Febrile convulsions may be categorized into two groups. The majority are single, brief, bilateral tonic-clonic or clonic seizures occurring in infants or children of normal development. A minority of febrile convulsions are of longer duration (15-20 minutes) or occur repeatedly within 24 hours, may show partial or unila-teral features and may be followed by transient neurological sequelae or even by a permanent neurological handicap. These are described as complex or complicated febrile convulsions. Age is a crucial factor in determining the severity of a febrile convulsion. The reported incidence of simple and complicated febrile convulsions has varied, depending on whether the studies were hospital or community based, and also on whether or not the author had included febrile convulsions in association with infections of the nervous system. Between 2 and 4% of all children in Europe and the United States of America will experience at least one convulsion associated with a febrile illness before the age of 5 years (cumulative incidence).[36] These findings contrast with that reported in Japan (8% or more).[22] The peak incidence of a first convulsion with fever occurs in the second year of life.[10,22] At this age between 1 and 2% of children in western nations will experience a convulsion with fever. The trends in age specific incidences are similar but higher in Japan.[22] About 30% of children with a febrile convulsion will have additional convulsions at the time of a subsequent febrile illness[23] and 3-6% can be expected to develop epilepsy.[24] The prevalence is unknown. *Infantile spasms (West syndrome) and related syndromes:* in the definition of the ILAE of 1989[16] this syndrome consists of a characteristic triad which combines: -infantile spasms, -arrest of psychomotor development, -hypsarrhythmia, although one element may be missing. The majority of the syndromes are symptomatic. A small group of patients[38] described as 'benign epileptic infantile spasms' belongs to the 'idiopathic' type of this syndrome. Incomplete and atypical forms of this syndrome occur; most authors would accept forms without hypsarrhythmia, provided paroxysmal abnormalities of the EEG are present and also forms without mental retardation. Infantile spasms are relatively rare. The incidence has been estimated as 24-42 out of 100.000 births.[39,40] The prevalence is unknown. *Periodic spasms*: the clinical and EEG features of this variant of the West syndrome are series of spasms often asymmetrical or even unilateral, each marked in the EEG by a slow complex with superimposed low amplitude fast rhythms without resumption of interictal activity between individual spasms. Seizures occur mainly on awakening or at the time of transition from REM to slow wave sleep. Incidence and prevalence are unknown. *Benign myoclonic epilepsy in infancy*: in a group of 142 children with various types of myoclonic epilepsy 10 (7%) were suffering from benign myoclonic epilepsy of infancy.[41] Other data on incidence and prevalence are not available, but the syndrome is rare. *Severe myoclonic epilepsy (SME) in infancy*: 42 out of 142 children (30%) with various types of myoclonic epilepsy were suffering from this seizure disorder.[41] Dalla Bernardina et al.[42] found that SME accounted for 7 percent of the epilepsies with an onset before the age of 3. Hurst[43] calculated an incidence of at least one case in 40.000 children before the age of 7 in the general population. *Myoclonic epilepsy (myoclonic status) in non-progressive encephalopathies*: this syndrome is characterized by the repetition of long-lasting myoclonic status. These status, often displayed only polygraphically, induce an important worsening of neuropsychological impairment. There is suggestive similarity with the EEG-records characterizing some cases of Angelman syndrome. The incidence and prevalence are not known. *Epileptic seizures in inborn errors of metabolism*: repeated epileptic seizures are a common manifestation of many disorders of the central nervous system resulting from proven or suspected inborn errors of metabolism. In many of these diseases epileptic seizures are an occasional manifestation of the disorder, however, in some disorders they present an essential symptom. Different seizure types can occur, no characteristic pattern has been described, and the seizures (of which myoclonus represents the most common seizure type) do not represent unitary epilepsy syndromes and therefore it is not reasonable to look for prevalences and incidence. *Myoclonic astatic epilepsy of early childhood (Doose syndrome)* belongs to the epilepsies with generalized seizures. It is polygenetically determined with little monogenetic variability. Clinical characteristics are: mostly a normal development and no neurological deficits before onset of a variety of seizure types such as myoclonic seizures, astatic (atonic) or myoclonic astatic seizures, short absences and mostly tonic-clonic seizures. The prognosis varies from benign, with spontaneous remission, to malignant, with mental deterioration. There is an overlap with other syndromes and the most common differential diagnosis is Lennox-Gastaut syndrome. These findings suggest that the myoclonic astatic epilepsy of early childhood cannot be classified as an entity in its own right. Consequently incidence and prevalence are unknown. *The Lennox-Gastaut syndrome*: in this syndrome multiple seizure forms are present, including, in most cases, tonic and/or atonic seizures and atypical absences, and by a diffuse interictal slow spike-wave activity on EEG. The presence of psychomotor retardation is frequent but not a constant feature. The incidence is poorly known. *Absence epilepsy in early childhood* will affect children under five years of age. Tonic-clonic seizures at onset have a less favourable course of the disease. In the ILAE classification of the epilepsies, this early form has not been distinguished from childhood absence epilepsy.

Epileptic syndromes in childhood

Childhood absence epilepsy (CAE): CAE or pyknolepsy is a relatively rare form of idiopathic generalized epilepsy occurring in previously normal children with a strong genetic disposition. Absence seizures of any kind, apart from myoclonic absences, are the initial seizure type with onset between 3 and 12 years of age and a peak at 6-8 years of age. These seizures occur frequently throughout the day, spontaneously or precipitated by environmental factors and are characterized by a short duration, an abrupt onset and termination, impairment of consciousness with or without other signs and a high seizure frequency. This seizure type tends to remit spontaneously and in 80% of cases the seizures are controlled by valproate or ethosuximide. Forty percent of the children develop tonic-clonic seizures during adolescence (but sometimes later in life). Very few patients have only absence seizures in adult life. The annual incidence of CAE has been estimated at 6.3/100.000 to 8.0/100.000.[29,44,45] This syndrome represents 8% of epilepsy in school-age children.[25] There are major discrepancies concerning the prevalence varying from 2.3% - 37.7%.[46,47] A great deal of misunderstanding about the prevalence is based on the fact that in previous decades all epilepsies with absence seizures were called 'petit mal'. Absence seizures have to be considered as a type of epileptic seizures and not as a syndrome. Conversely, CAE is a real syndrome. *Epilepsy with myoclonic absences (EMA)*: this syndrome is clinically characterized by absences, accompanied by severe bilateral rhythmic clonic jerks, often associated with a tonic contraction. On the EEG these clinical features are always accompanied by bilateral synchronous and symmetric discharges of rhythmic 3/sec spike waves, similar to childhood absence epilepsy. Seizures occur many times a day. Awareness of the jerks may be maintained. Age of onset is around 7 years and there is a male preponderance. Prognosis is less favourable than in pyknolepsy, due to resistance to therapy of the seizures, mental deterioration and possible evolution to other types of epilepsy such as Lennox-Gastaut syndrome. The diagnosis of EMA may be difficult to determine, solely on an anamnestic description. It may be mistaken for partial seizures when myoclonias are asymmetric or when deviation of the trunk is prominent. The tonic component is also misleading. Myoclonias of benign myoclonic epilepsy or juvenile myoclonic epilepsy do not occur in rhythmic clusters and are accompanied by fast polyspike discharges without a strict relationship between myoclonia and EEG discharges. If the clinical pattern of seizures is overlooked, this syndrome may be diagnosed as the typical absence seizure syndrome. Polygraphic recording of the EEG plus surface electromyography of various muscles with special attention for the deltoid muscles bilaterally may be the only way to reach the correct diagnosis. EMA should be considered in all cases of absences with concomitant 3/sec spike and wave discharges that are resistant to conventional valproate monotherapy.[48] The most efficient treatment is a combination of valproate and ethosuximide.[49] The incidence and prevalence are unclear. *Epilepsy with generalized tonic-clonic seizures in childhood* is characterized by tonic-clonic seizures from the onset, associated with absence attacks. Mental deterioration is rare, but more frequent than the cases with seizure onset in adolescence and adulthood. Abnormalities in the interictal EEG are restricted to generalized, bilateral and synchronous discharges of (poly)spike-waves with a normal background rhythm. Incidence and prevalence are unknown. *Idiopathic partial epilepsies in children (IPEC)* are age-dependent epilepsies without anatomical lesions. The children recover without adverse effects before adolescence. Various forms of IPEC are known: (1) Benign childhood epilepsy with Rolandic or centro-temporal spikes, (2) Epilepsy in childhood with occipital paroxysms and some authors have suggested the existence of the following other forms: (3) Benign 'psychomotor' epilepsy with affective symptomatology, (4) Benign partial epilepsy with extreme somatosensory evoked potentials, (5)

Benign frontal epilepsy. Incidence and prevalence of these seizure disorders are largely unknown. *Acquired epileptic aphasia (Landau-Kleffner syndrome, LKS)*: the LKS is a childhood disorder in which an acquired aphasia, multifocal spike and spike-wave discharges in the EEG are associated. This syndrome was identified by Landau and Kleffner in 1957.[50] Some more than 200 cases of this syndrome have been reported since with an increasing number in the last ten years.[48] Epileptic seizures and behavioral and psychomotor disturbances occur in two-thirds of the patients. There is verbal auditory agnosia and rapid reduction of spontaneous speech. The seizures, usually generalized tonic-clonic or partial motor, are rare and remit before the age of 15 years, as do the EEG abnormalities. Typical EEG findings are repetitive spikes (and waves) of great amplitude organized in foci, variable in time and space and preferentially located in the temporal and parieto-occipital regions. Sleep activates diffusing of the paroxysmal discharges. The percentage of cases with typical continuous spikes and waves during slow-wave sleep is hard to establish. The prevalence and incidence of the LKS are unclear. *Epilepsy with continuous spike waves during slow-wave sleep (ESES)*: this syndrome is the result from the association of various seizure types, partial or generalized, occurring during sleep and atypical absences when awake. The characteristic EEG pattern consists of continuous diffuse spike waves during slow-wave sleep, which is noted after onset of seizures. Duration varies from months to years. Despite the usually benign evolution of seizures, prognosis may be less benign because of the appearance of neuropsychologic disorders. The prevalence and incidence of this syndrome are unclear.

Epileptic syndromes in childhood and adolescence

Reading epilepsy is a benign localization-related idiopathic epilepsy syndrome preferentially related to the temporo-parietal region of the language dominant hemisphere but also to other regions, functionally involved in reading. The onset is in adolescence and consistent features are a speech motor (or sensorimotor) aura (and in a minority of patients a visual or ocular aura) provoked by prolonged reading. It is typically enhanced by reading aloud and by texts which are difficult to read (understanding of the text is not required). When the patient continues with reading, the simple partial seizure will evolve into a tonic-clonic seizure. Precipitating mechanisms may include talking, writing or reading music. Spontaneous seizures occur exceptionally. There are no interictal epileptiform EEG abnormalities. The typical discharge in the ictal EEG is a single or a very short volley of steep or sharp waves, spikes or spike and wave complexes, not identical to rolandic spikes. This syndrome may be treated by partial avoidance of stimuli or by treatment with valproate. The prognosis is good. Incidence and prevalence are unknown. *Photosensitive epilepsy* can occur in association with both localization-related and generalized epilepsies and does not represent a unitary syndrome. The term photosensitive should be confined to those subjects who consistently exhibit a generalized discharge of spikes and slow waves in response to intermittent photic stimulation. Such a photoparoxysmal response is strongly associated with epilepsy, particularly of idiopathic generalized type. It is often familial, twice as common in females as in males and presents typically around puberty. Self-induced seizures occur in about one third of subjects and are mostly therapy resistant. In patients with pure photosensitive epilepsy, avoidance of precipitating stimuli may be the therapy required and when drug treatment is desired, valproate is the drug of first choice. About 5% of all patients with epilepsy are photosensitive. *Juvenile absence epilepsy* is an idiopathic generalized epilepsy syndrome with age-related onset. The absence seizures do not differ from those of childhood absence epilepsy but retropulsive movements are less common. The onset is around puberty and the seizure frequency is much lower than in childhood absence epilepsy. Seizures do not occur every day, usually they appear sporadically. If there are also generalized tonic-clonic seizures, these mostly occur on awakening. Patients may have also myoclonic jerks. The background EEG activity is usually normal. The characteristic feature in the ictal and interictal EEG is generalized symmetric spike and wave discharge with frontal accentuation, easily precipitated by sleep deprivation and by hyperventilation. The spike and wave frequency is a little faster than 3 cycles/second. Therapy response to valproate is good. Incidence and prevalence are unclear. *Epilepsy with tonic-clonic seizures (grand mal) on awakening* is a generalized idiopathic syndrome with tonic-clonic seizures, manifesting exclusively or predominantly shortly after awakening (regardless of the time of the day) or in a second peak in the evening during relaxation. Seizure onset is mostly in the second decade. If there are other seizures, these are mostly absence seizures or myoclonic seizures, as in juvenile myoclonic epilepsy. Incidence and prevalence are unknown. *Juvenile myoclonic epilepsy* is a generalized idiopathic syndrome with pre- to postpuberty seizure onset. The seizures are characterized by bilateral, single or repetitive, arrhythmic, irregular myoclonic jerks, predominantly in the arms. Often there are also tonic-clonic seizures and less often absence attacks. These seizures occur predominantly shortly after awakening and are often precipitated by sleep withdrawal. The ictal EEG shows rapid generalized, often irregular (poly)spike waves that also are found interictally. There is no close phase correlation between EEG-spikes and the jerks. Frequently these patients are photosensitive. The response to valproate is good. In selected populations the prevalence is reported to be 4 percent of all patients with epilepsy.[51,52] *Benign partial seizures of adolescence* are characterized by seizures with an onset in the second decade with a peak at 13 to 14 years of age. These simple or complex partial seizures mainly with motor and/or sensory symptoms frequently generalize to a tonic-clonic seizure. Seizures may occur single or in series of two to five within 24 hours. There is a predominance in boys and no family history of epilepsy. The EEG is often normal or with mild abnormalities. When seizure onset is between 10 and 14 years it presents 15 percent of all patients with epilepsy. When the onset is between 15 and 19 years it presents 11 percent. This gives a yearly incidence of 4/100.000, respectively 2/100.000 for these two age groups.[53,54] *Non idiopathic localization related epilepsies and epileptic syndromes in childhood* differ widely in clinical- and EEG-semiology, etiology and prognosis. It is impossible to consider these seizure disorders as a syndrome. Of all childhood epilepsies this form of localization related epilepsy cause most problems in every day practice, because it represents over 40% of the epilepsies in this age-group.[18,55] *Epilepsia partialis continua (Kojewnikow's syndrome)* has to be divided in two electroencephalographic subgroups. The first corresponding to the classical syndrome with a lesion in the rolandic area usually with unknown etiology but with a neurological deficit and partial motor seizures followed after a variable time interval by well-localized myoclonic jerks. The EEG's show focal abnormalities merely in the central area. This syndrome is not progressive. The other type, occurring in previously normal children, is also characterized by partial motor seizures which, however, become rapidly associated with myoclonic jerks. The EEG's show abnormalities in the background activity and focal as well as diffuse paroxysmal abnormalities. This type is progressive with the appearance of a neurological deficit, various other seizure types, mental retardation and inflammatory pathological findings in brain tissue suggesting a chronic 'encephalitis'. Incidence and prevalence are unknown. *Progressive myoclonus epilepsies in childhood and adolescence* are syndromes including massive and segmental myoclonus, myoclonic or tonic-clonic seizures as well as partial seizures, and neurological deterioration consisting of cerebellar impairment and dysfunctions of the higher cerebral functions. These epilepsies are caused by several metabolic deficits as already has been mentioned above. These progressive myoclonus epilepsies account for some 1 percent of all epilepsy cases in childhood.

For a detailed description of the epileptic syndromes and the seizure disorders mentioned above the reader is referred to the 'Guide Bleu' for epileptic syndromes by Roger et al.[56,57] A rough estimate of the percentages of new childhood onset epilepsy disorders is given in Table 2.

Table 2. Rough estimate of the percentages of new childhood onset epilepsy disorders [21]

Yearly percentage of all cases of childhood epilepsy	
Infantile spasms (West syndrome)	2%
Lennox-Gastaut syndrome	1 - 2%
Childhood absence epilepsy (pyknolepsy)	10 - 15%
Juvenile myoclonic epilepsy	5%
Idiopathic localization related epilepsy (benign Rolandic epilepsy)	10%
All other syndromes	66 - 72%

IN CONCLUSION

Epidemiological research for the epilepsies and epileptic syndromes is possible only by using proper definitions. Only then exact data for incidence and prevalence of the epileptic syndromes can be obtained. As has been made clear in this chapter it has been difficult, until now, to obtain these data. Probably this can be explained by the relatively recent (1989) classification for the epileptic seizures and the epileptic syndromes.[16] Five years actually is a short period of time for this classification to 1) become the routine standard for all clinicians in the world, and 2) perform epidemiological research. It is to be expected that in the near future more knowledge will be gained on this subject.

REFERENCES

1. Sander JWAS, Shorvon SD. Incidence and prevalence studies in epilepsy and their methodological problems: a review. J. Neurol. Neurosurg. Psych., 1987; 50: 829-839.
2. Gomez JG, Arciniegas E, Torres J. Prevalence of epilepsy in Bogota, Colombia. Neurology, 1978; 28: 90-94.
3. Granieri E, Rosati G, Tola R, et al. A descriptive study of epilepsy in the district of Copparo, Italy. 1964-1978. Epilepsia, 1983; 24: 502-514.
4. Cruz ME, Barberis P, Schoenberg BS. Epidemiology of epilepsy. In: Poeck K, Freund HJ, Ganshirt H, editors. Neurology. Berlin: Springer Verlag, 1986: 229-239.
5. Haerer AF, Anderson DW, Schoenberg BS. Prevalence and clinical features of epilepsy in a biracial United States population. Epilepsia, 1986; 27: 66-75.
6. Li SC, Schoenberg BS, Wang CC, et al. Epidemiology of epilepsy in urban population of the People's Republic of China. Epilepsia, 1985; 26: 391-394.
7. Stridharan R, Radhakrishnan K, Ashok PP, et al. Epidemiological and clinical study of epilepsy in Benghazi, Libya. Epilepsia, 1986; 27: 60-65.
8. Marino R Jr, Cukiert A, Pinho E. Epidemiologic aspects of epilepsy in Sao Paulo, Brazil. A prevalence rate study. In: Wolf P, Dam M, Janz D, et al., editors. Advances in Epileptology, 1987; 16: 759-764.
9. Koul R, Razdan S, Motta A. Prevalence and pattern of epilepsy

(Lath/Mirgi/Laran) in Rural Kashmir, India. Epilepsia, 1988; 29(2): 116-122.
10. Hauser WA, Kurland LT. The epidemiology of epilepsy in Rochester, Minnesota; 1935-1967. Epilepsia, 1975; 16: 1-66.
11. Zielinski JJ. Epidemiology and medicosocial problems of epilepsy in Warsaw. Final report on research program no. 19-P-58325-F-01. Warszawa: Psychoneurological Institute, 1974.
12. Commission on Classification and Terminology of the ILAE. Proposal for revised clinical and electroencephalographic classification of epileptic seizures. Epilepsia, 1981; 22: 469-501.
13. Hauser WA, Hesdorffer DC. Epilepsy: Frequency, Causes and Consequences. New York, Demos, 1990.
14. Krohn W. A study of epilepsy in northern Norway, its frequency and character. Acta Psychiatr. Neurol. Scand., 1961 (150 Suppl.): 215-225.
15. Chiófale N, Kirschbaum A, Fuentes A, et al. Prevalence of epilepsy in children in Melipilla, Chile. Epilepsia, 1979; 20: 261-266.
16. Commission on Classification and Terminology of the International League against Epilepsy. Proposal for revised classification of Epilepsies and Epileptic Syndromes. Epilepsia, 1989; 30: 389-399.
17. Commission on Epidemiology and Prognosis of the International League against Epilepsy. Guidelines for epidemiologic studies on epilepsy. Epilepsia, 1993; 34(4): 592-596.
18. Gastaut H, Gastaut JL, Goncalves e Silva GE, et al. Relative frequency of different types of epilepsy: a study employing the classification of the International League against Epilepsy. Epilepsia, 1975; 16: 457-461.
19. Alving J. Classification of the epilepsies. An investigation of 402 children. Acta Neurol. Scand., 1979; 60: 157-163.
20. Juul-Jensen P, Foldspang A. Natural history of epileptic seizures. Epilepsia, 1983; 24: 297-312.
21. Hauser WA. The prevalence and incidence of convulsive disorders in children. Epilepsia, 1994; 35 (2 Suppl.): S1-S6.
22. Tsuboi T. Seizures of childhood: a population and clinic based study. Acta Neurol. Scand., 1986; 110 (Suppl.): 22-29.
23. Annegers JF, Blakey SA, Hauser WA, et al. Recurrence of febrile convulsions in a population-based cohort of children. Epilepsy Res., 1990; 5: 209-216.
24. Annegers JF, Hauser WA, Shirts SB, et al. Factors prognostic of unprovoked seizures after febrile convulsions. N. Engl. J. Med., 1987; 316: 493-498.
25. Cavazzuti GB. Epidemiology of different types of epilepsy in school age children of Modena, Italy. Epilepsia, 1980; 21: 57-62.
26. Joensen P. Prevalence, incidence and classification of epilepsy in the Faroes. Acta Neurol. Scand., 1986; 76: 150-155.
27. Tsuboi T. Prevalence and incidence of epilepsy in Tokyo. Epilepsia, 1988; 29: 103-110.
28. Cowan LD, Bodensteiner JB, Leviton A, et al. Prevalence of the epilepsies in children and adolescents. Epilepsia, 1989; 30: 94-106.
29. Loiseau J, Loiseau P, Guyot M, et al. Survey of seizure disorders in the French Southwest. I. Incidence of epileptic syndromes. Epilepsia, 1990; 31: 391-396.
30. Hauser WA, Annegers JF, Gomez M. Incidence of West syndrome in Rochester, Minnesota; 1940-1980. Epilepsia, 1991; 32 (3 Suppl.): 83-88.
31. Plouin P. Benign idiopathic neonatal convulsions. In: Roger J, Bureau M,

Dravet Ch, et al., editors. Epileptic Syndromes in infancy, childhood and adolescence. London: John Libbey, 1992: 3-13.
32. Aicardi J. Early myoclonic encephalopathy. In: Roger J, Dravet C, Bureau M, et al., editors. Epileptic syndromes of infancy, childhood and adolescence. London: John Libbey, 1985: 12-21.
33. Ohtahara S, Ohtsuka Y, Yamatogi Y. The early-infantile epileptic encephalopathy with suppression-burst: developmental aspects. Brain Dev., 1987; 9: 371-376.
34. Ohtahara S, Ohtsuka Y, Yamatogi Y, et al. Early-infantile epileptic encephalopathy with suppression-bursts. In: Roger J, Bureau M, Dravet Ch, et al., editors. Epileptic syndromes in infancy, childhood and adolescence. London: John Libbey, 1992: 25-35.
35. Consensus development Panel. Febrile seizures: long term management of children with fever associated seizures. Pediatrics, 1980; 66: 1009-1012.
36. Hauser WA. The natural history of febrile seizures. In: Nelson KB, Ellenberg J, editors. Febrile seizures. New York: Raven Press, 1982: 5-18.
37. Aicardi J. Epilepsy in children, second edition. New York: Raven Press, 1994: 18-44.
38. Dulac O, Plouin P, Jambaqué I, et al. Spasmes infantiles épileptiques bénins. Rev. EEG Neurophysiol. Clin., 1986; 16: 371-382.
39. Lacy JR, Penry JK. Infantile spasms. New York: Raven Press, 1976.
40. Riikonen R, Donner M. Incidence and aetiology of infantile spasms from 1960 to 1976. A population study in Finland. Dev. Med. Child Neurol., 1979; 21: 333-343.
41. Dravet C, Bureau M, Roger J. In: Roger J, Dravet C, Bureau M, et al., editors. Epileptic syndromes of infancy, childhood and adolescence. London: John Libbey, 1985: 58-67.
42. Dalla Bernardina B, Capovilla G, Gattoni MB, et al. Epilepsie myoclonique grave de la première année. Rev. EEG Neurophysiol., 1982; 12: 21-25.
43. Hurst DL. Epidemiology of severe myoclonic epilepsy of infancy. Epilepsia, 1990; 31: 397-400.
44. Olsson I. Epidemiology of absence epilepsy. I: Concept and incidence. Acta Paediatr. Scand., 1988; 77: 860-866.
45. Blom S, Heijbel J, Bergfors PG. Incidence of epilepsy in children: a follow up study three years after the first seizure. Epilepsia, 1978; 19: 343-350.
46. Livingstone S, Torres I, Pauli LL, et al. Petit mal epilepsy. Results of prolonged follow-up study of 117 patients. JAMA, 1965; 194: 113-118.
47. Lennox WG. The petit mal epilepsies. JAMA, 1945; 129: 1069-1073.
48. Roger J, Genton P, Bureau M, et al. Less common epileptic syndromes. In: Wyllie E, editor. The treatment of epilepsy. Philadelphia: Lea and Febiger, 1993: 624-635.
49. Rowan AJ, Meijer JWA, Binnie CD, et al. Sodium valproate and sodium valproate-ethosuximide combination therapy: Intensive monitoring studies. In: Johannessen SI, editor. Antiepileptic Therapy: Advances in Drug Monitoring. New York: Raven Press, 1980: 161-166.
50. Landau WM, Kleffner FR. Syndrome of acquired aphasia with convulsive disorder in children. Neurology, 1957; 7: 523-530.
51. Janz D. Die Epilepsien. Stuttgart: Thieme, 1969.
52. Asconapé J, Penry JK. Some clinical and EEG aspects of benign juvenile myoclonic epilepsy. Epilepsia, 1984; 25: 108-114.
53. Kurland LT. The incidence and prevalence of convulsive disorders in a small

urban community. Epilepsia, 1959; 1: 143-161.
54. Beaussart M, Faou R, Defayes J. Epidémiologie de l'épilepsie dans la région du Nord-Pas-de Calais (à propos de 12.290 cas). Lille Médical, 1980; 25: 183-191.
55. Viani F, Beghi E, Atza MG, et al. Classifications of epileptic syndromes: advantages and limitations for evaluation of childhood epileptic syndromes in clinical practice. Epilepsia, 1988; 29: 440-446.
56. Roger J, Dravet C, Bureau M, et al. Editors. Epileptic syndromes in infancy, childhood and adolescence. London, Paris: John Libbey, 1985: 335.
57. Roger J, Bureau M, Dravet C, et al. Editors. Epileptic syndromes in infancy, childhood and adolescence (second edition). London: John Libbey, 1992: 418.

3 Seizures in the newborn

WIL O. RENIER
Department of Neurology, University of Nijmegen, Nijmegen, The Netherlands

INTRODUCTION

Seizures are the most frequent neurological problem in neonates. Although they cause great concern to parents and nurses, their medical significance is controversial. Neonatal seizures are defined as seizures occurring in the first two weeks after birth, although some authors extend this period to one month. Although seizures may be difficult to recognize in the newborn and several nonepileptic phenomena may easily be confused with seizures, neonatal seizures are estimated to occur in 0.2-1.5% of live births.[1] Due to the immaturity of the brain, the epileptic manifestations are less pronounced, compared to older children or adults. In general the seizures tend to occur serially over a short period.[2] The diagnosis of neonatal seizures has traditionally been based on the clinical observation of paroxysmal stereotyped behaviour phenomena. By cinematographic and electroencephalographic registration of these activities, Dreyfuss-Brisac & Monod[3] were able to demonstrate the unique character of neonatal seizures. From their observations it appears that not all seizures are convulsive.

Two main patterns may be distinguished in true neonatal convulsions:
- tonic convulsions: which are characterized by extension of the trunk and upper and lower limbs, resembling decerebral posturing, often associated with apnoea and cyanosis, and sometimes deviation of the eyes;
- clonic convulsions: in which the rate of jerking is 1-3 per second, and which may be focal, multifocal, or generalized.[4]

The clinical classification of Volpe[5] includes also myoclonic and subtle seizures.
Subtle seizures are composed of:
- isolated abnormal eye movements, eyelid blinking or fluttering;
- sucking, lip smacking and salivation;
- rowing, pedalling or swimming movements;
- tonic posturing of a single extremity;
- apnoea spells, hyperpnoea, vasomotor phenomena.

The sudden jerking which can occur on awaking from sleep, the intense agitation and tremulous movement which may be caused by hunger, and the jittery movements which may be associated with tetany have to be differentiated from convulsions.

Jitters are nonconvulsive and can be distinguished from seizures because change of posture or restraint can ablate them. They are not accompanied by abnormal eye movements. Abnormal movements can also be induced by drug withdrawal, impaired cerebral inhibition, cerebral palsy or sleep. Recurrent apnoea is frequently due to pulmonary, cardiovascular or gastro-intestinal disturbances, but may also be convulsive.

In the newborn, classical generalized tonic-clonic seizures are seldom observed in contrast to multifocal, subcortical and localization related seizures. Elaboration of dendrites, formation of synapses, glia-neuron connection and interaction, and myelination are incomplete at birth. This immaturity of the central nervous system (CNS) and especially the incomplete intracortical connections may explain the incomplete spread and slow diffusion of electrical discharges and the more localized seizure manifestations in neonates.

INVESTIGATIONS

Clinical phenomena designated as seizures of the newborn have varied, depending on techniques and the points of view of the investigators. Modern neuroimaging techniques and encephalography (EEG) have contributed to better insight in the underlying mechanisms. Paroxysmal high-voltage slow waves, positive or negative spikes or alpha-like waves, and burst-suppression patterns are EEG characteristics associated with ictal phenomena. The use of EEG-polygraphic-video-monitoring systems has confirmed that paroxysmal stereotyped behaviour in neonates are not always epileptic in nature and that 'epileptic' phenomena in the EEG are not always accompanied by clinical manifestations (occult or subclinical seizures).

In the study of Mizrahi and Kellaway[6] focal clonic and focal tonic seizures are always associated with focal rhythmic sharp-wave discharges in the EEG, while none of the generalized tonic seizures with decerebrate or decorticate posturing are. Only some of the myoclonic or slow serial jerking movements of the extremities or axial muscles, and few of the subtle seizures (those characterized by tonic eye deviation with nystagmus) have a consistent electrographic correlate. Their study confirms the earlier observations of Kellaway & Hrachovy.[7] According to these authors many of these motor manifestations may be due to lack of cerebral inhibition and have to be considered as 'brainstem release phenomena'. These stimulus-sensitive reflex-like motor patterns are easily elicited in neonates with a diffuse brain damage because of the tonic inhibition, normally exerted by cortical structures is lacking. The fact that such seizure-like behaviour responds to anti-epileptic drugs (AED's) does not prove its epileptic nature because these drugs have a general depressing effect on the CNS. In the study of Scher et al.[8] only 10% of the clinically identifiable seizures have an EEG correlate. From their study the authors conclude that electro-graphic seizures are not without a prognostic significance. The diagnostic contribution of the EEG lies in the relation between the background abnormalities with the severity of neuropathological changes in the brain and the frequent association of brain disease with epileptic seizures. The interpretation of EEG's in the newborn needs some experience. Vigilance and sleepy states have to be taken into account. Transient flattenings in the EEG in sleeping newborns occurring near the onset of state 1 have no clinical significance beyond their importance as a normal variation; post-flattening bursting should not be misinterpreted as epileptic activity.[9] Interictal sharp waves can occur in infants who do not have epileptic seizures and perhaps are the electro-physiological expression of brain maturation. Electro-graphic seizures can, but do not necessarily, refer to an underlying brain disease. The problem of what constitutes a seizure in the neonatal period and the contribution of diagnostic tests and particularly the EEG is extensively described by Tuchman and Moshé.[10] In paralysed babies EEG monitoring gives the best indication of epileptic activity.

Neonatal seizures are, with the exceptions of benign familial autosomal dominant and idiopathic cases, the manifestation of an insult to the CNS. Most seizures occur within three days after birth. Convulsions due to brain damage tend to occur in the first three days of life or after the eight day, and convulsions due to metabolic disturbance between the fifth and eight days.[4] Both structural and metabolic disturbances can occur in association and have mutual unfavourable influences. The most common causes of neonatal seizures are the hypoxic-ischaemic encephalopathy, intracranial haemorrhage (especially in premature newborns), hypertensive encephalopathy, cerebral trauma, developmental abnormalities of CNS, congenital infections, meningo-encephalitis, disturbances of the metabolic homeostasis, inborn errors of metabolism, pyridoxine dependency, narcotic withdrawal and toxins. Recently a familial case with low activity of Na^+/K^+-ATPase in the brain has been described.[11] Careful observation and examination of the newborn shall guide further diagnostic investigations for underlying etiology (Table 1).

TREATMENT

The possible dissociation between clinical and EEG findings in neonates presenting paroxysmal stereotyped behaviour, raises the question of treatment. For some investigators neonatal seizures are simply epiphenomena of brain injury while for others seizures per se are injuries to the CNS.

Animal experiments have proven that prolonged seizures, electrically or chemically induced, in early life interfere with brain development by inhibiting mitotic cell activity. The impairment of energy supply in a situation with higher energy utilisation results in inhibition of protein synthesis, dissociation of polysomes, and inhibition of DNA synthesis. In the neonatal period the cell number and in a later period the cell size are reduced.

Table 1. Essential investigations in neonatal seizures

Clinical examination	tension, heart rate, respiration skin inspection fontanelle palpation neurodevelopmental examination funduscopy
Neuroimaging	cranial echography, CTscan, MRI
Neurophysiology	EEG, preferably EEG-videomonitoring
Laboratory	serum BUN, sodium, potassium, calcium, magnesium blood gases glucose, bilirubine, ammonia, liver enzymes
	urine glucose, protein, cell count
	CSF glucose, protein, cell count glycine
	cultures of blood, urine, CSF
	TORCH Titers
	inborn errors of metabolism

In addition, myelination is affected disproportionately to overall brain growth.[12-14] This brain growth defect cannot be compensated for later in life. In the presence of seizures, neurotoxic agents such as glutamic and aspartic acids accumulate.[15] Moreover, seizures per se can be damaging to the brain development by elevation in intracranial pressure and cerebral blood flow velocity.[16] The importance of this data to the human condition is uncertain. Nevertheless, based on all these experimental arguments and studies on prognosis in newborns with seizures, no controversy exists concerning the necessity of treatment of severe or prolonged neonatal seizures.

The proper approach to the treatment of neonatal seizures depends on the etiology. While awaiting the results of neuroimaging and laboratory investigations, a dextrose infusion can be started. For this purpose, 25% dextrose, 2-4 ml/kg by intravenous infusion, is recommended to maintain blood glucose above 40 mg/dL in association with a mild hypohydration.[17] Hypocalcemia should be treated by slow intravenous administration of calcium gluconate (up to 10 ml. of 10% solution) under cardiac rate monitoring. For hypomagnesemia therapy consists of intravenous injection of 2 to 6 ml. of 2-3% magnesium sulcate or intramuscular injection of 0.2 mL/kg of 50% magnesium sulcate. The diagnosis of pyridoxine deficiency and dependency is made by administration of 25 to 50 mg of pyridoxine intravenously, preferably under EEG control.

Table 2. Dose scheme of anti-epileptic drugs in neonates[32,33]

Pyridoxine	25-100 mg IV (under EEG monitoring)
Phenobarbital (PB)	loading doses of 20 mg/kg I.V. maintenance doses of 2-4 mg/kg/day I.V. serum level > 16-50 mg/L < PB binding appears to decrease in sick newborns, and can lead to cardiovascular instability especially decrease of heartrate when level > 50 mg/L
Phenytoin (PHT)	loading doses of 20 mg/kg I.V. (not exceeding a rate of 1 mg/kg/min) maintenance doses of 4-6 mg/kg/day I.V. serum level > 15-20 mg/L < PHT binding appears to decrease in sick premature infants, and, in the presence of hyperbilirubinaemia, can lead to cardiovascular dysrhythmias
Primidone (PMD)	loading doses of 20 mg/kg maintenance doses of 10-20 mg/kg/day serum level of PMD > 6 mg/L and PB component as for PB
Sodium Valproate (VPA)	loading doses of 15 mg/kg infusion of 1 mg/kg/hr
Diazepam (DZP)	loading doses 1 mg/kg over 2-5 min. infusion of 3-12 mg/kg/day or rates of 0.3 0.8 mg/kg/hour or 0.7-2.75 mg/hour (until seizure free) serumlevel > 10-35 µmol/L <
Lorazepam (LZP)	loading doses 0.05 mg/kg/over 2-5 min
Clonazepam (CZP)	loading doses 0.1 mg/kg/over 2-5 min beware: administration in jaundiced infants
Paraldehyde	loading dose of 200 mg/kg infusion of 15 mg/kg/hr or 200 mg/kg/hr for 2 hours or 150 mg/kg/hr for 3 hours serum level > 100-250 mg/L < can lead to hypotension and lipoid pneumonitis
Lidocaine	4 mg/kg/hr IV for the first 12 hours 3 mg/kg/hr IV for the next 24 hours 2 mg/kg/hr IV for another 24 hours serum level > 0.5-4 or even 3.4-10.5 mg/L < can lead to hypotension, arrhythmias, seizures

Determination of the etiology of hyponatremia is important because hyponatremia secondary to excessive water retention should be treated with fluid restriction, but excess sodium loss requires sodium replacement. Hypernatremia needs slow rehydration.[18] If seizures are not due to specific metabolic causes but to structural abnormalities of the brain AED's should be administered.

The choice of AED's in the treatment of neonatal seizures has been based on tradition rather than on the proven superiority of one agent over another.[19] Phenobarbital (PB) remains the drug most frequently chosen as the initial agent in treatment. Other anticonvulsants that have been used in the treatment of neonatal seizures include phenytoin (PHT), benzodiazepines (BZP), primidone (PRM), valproic acid (VPA), carbamazepine (CBZ), paraldehyde, lidocaine (Table 2).

The clearance of drugs is dependent of the free fraction in the blood and the intrinsic clearance capacity of the liver. Antiepileptic drugs as PHT, BZP, CBZ, VPA have an important protein binding capacity. By influencing this binding capacity clinically important changes in serum concentration of the free fraction and in clearance can occur. PB, PHT, and CBZ have an enzyme inducing effect, while VPA causes an intrinsic clearance reduction. The quick changes in the neonatal period in composition of body and brain compartments, protein content, bilirubin, free fatty acids, blood flow in different organs, gastrointestinal kinetic and resorption, blood-brain-barrier, and brain receptor development makes strict control of serum levels of AED's necessary. In contrast to older children more fractioning of the daily dose over 24 hours is advised.[20,21] The rapid changes of metabolism during the first weeks make adjustment of dosage and drug monitoring mandatory.

In general phenobarbital is started with a loading dose of at least 20 mg/kg intravenously until a serum level of 25 mg/L is obtained. A serum level above 16 mg/L is minimally required, while levels above 40-50 mg/L can cause cardiovascular suppression, causing a deleterious effect in neonates with hypoxic-ischaemic encephalopathy. Because newborns have relatively long phenobarbital half-lives, maintenance doses of 2-4 mg/kg/day suffice to maintain PB levels between normal reference values in the first week, but after approximately two weeks increase of dose to approximately 5 mg/kg/day may be necessary.[22] If seizures are not under control with PB eventually phenytoin is added. Loading dose for PHT is the same as for PB and, in general, a maintenance dose of 4-6 mg/kg/day IV is needed in order to obtain a serum level between 15-20 mg/L. When given orally higher doses of PHT (15-20 mg/kg/day) may be necessary in the first weeks because of slow absorption. Because PHT does not have a linear kinetic, careful monitoring is necessary once levels of 20 mg/L are obtained. Positive results have been described in refractory neonatal seizures with primidone as adjunctive oral medication with loading dose of 15-20 mg/kg and maintenance dose of 12-20 mg/kg/day.[23] The associated increase of PB makes conclusions about effectiveness unclear. For status epilepticus benzodiazepines are preferable.[24] Diazepam 0.3 mg/kg/hr is in most cases an effective dose. The combination of PB and BZP can be responsible for respiratory depression. Drug resistance is another problem in chronic use of benzodiazepines and therefore these drugs should be reserved for acute intervention strategies and not for maintenance therapy. The early ontogenetic appearance of benzodiazepine receptors in the brain and spinal cord predisposes the developing brain for increased affinity for benzodiazepines.

Critical evaluation of the effectiveness of AED's in the treatment of neonatal seizures by continuous EEG recording reveals that neonatal seizures are relatively resistant to AED's[25] By cassette recording it has become clear that PB only in a minority of cases achieves complete seizure control.[26-28] Bolus doses of diazepam seems largely ineffective[27,28], while infusions of benzodiazepines have eventual seizure control.[24] This conclusion is not surprising regarding the etiologies of neonatal seizures and the neurotransmitter receptor state of the immature neonatal brain.[29]

Critical questions remain regarding the influence of chronic use of AED's on brain growth, denditric arborization and receptor site development.[29,30] Therefore many authors are of the opinion that treatment with AED's has to be of short duration, but suggested time limits vary from 1 to 6 months. Volpe[5] advises discontinuation of PHT before discharge of the infant, with PB being maintained until all neurological signs have disappeared, and not to be withdrawn before 3 months of age.

CONCLUSION

Because seizures in neonates are the expression of an underlying disease, a causal therapy if possible has priority. In all other conditions, treatment with AED's should be started in order to prevent exacerbation of cerebral injury by preventing cerebral metabolic demands above the energy supply. However, one has to keep in mind that clinical features accompany electrically-recorded seizures on only about 50% of occasions and that the prognosis for epilepsy[31] or severe neurological disability or death[26] seems equally serious.

REFERENCES

1. Dehkharghani F, Sarnat HB. Neonatal seizures. In: editors. Topics in Neonatal Neurology. New York: Grune & Stratton, 1984: 209-238.
2. Aicardi J. Neonatal and infantile seizures. In: Morselli PL, Pippenger CE, Penry JK, editors. Antiepileptic Drug Therapy in Pediatrics. New York: Raven Press, 1983: 103-113.
3. Dreyfuss-Brisac C, Monod N. Electroclinical studies of status epilepticus and convulsions in the newborn. In: Kellaway P, Petersén I, editors. Neurological and Electroencephalographic Correlative Studies in Infancy. New York: Grune & Stratton, 1964: 250-272.
4. Brown JK, Cockburn F, Forfar JO. Clinical and chemical correlates in con-

vulsions of the newborn. Lancet, 1972; 1: 135-139.
5. Volpe JJ. Neonatal seizures. In: Volpe JJ, editor. Neurology of the Newborn. Major Problems in Clinical Pediatrics, vol. 22. New York: Saunders, 1987; 129-157.
6. Mizrahi EM, Kellaway P. Characterization and classification of neonatal seizures. Neurology, 1987; 37: 1837-1844.
7. Kellaway P, Hrachovy RA. Status epilepticus in newborns: A perspective on neonatal seizures. In: Delgado-Escueta AV, Wasterlain CG, Treiman DM, et al. editors. Status Epilepticus. Advances in Neurology, vol.34. New York: Raven Press, 1983: 93-95.
8. Scher MS, Painter MJ, Bergman I, et al. EEG diagnosis of neonatal seizures: clinical correlations and outcome. Pediatr. Neurol., 1989; 5: 17-24.
9. O'Brien MJ, Lems YL, Prechtl HFR. Transient flattenings in the EEG of newborns - a benign variation. Electroencephalography and Clin. Neurophys., 1987; 67: 16-26.
10. Tuchman RF, Moshé SL. Neonatal seizures: diagnostic and treatment controversies. In: Sillanpää M, Johannessen SI, Blennow G, et al. editors. Paediatric Epilepsy. Petersfield: Wrightson Biomedical Publishing, 1990; 57-64.
11. Renkawek K, Renier WO, de Pont JJHHM, et al. Neonatal status convulsivus, spongiform encephalopathy, and low activity of Na^+/K^+-ATPase in the brain. Epilepsia, 1992; 33: 58-64.
12. Wasterlain CG, Plum F. The vulnerability of developing rat brain to electroconvulsive seizures. Arch. Neurol., 1973; 19: 38-45.
13. Wasterlain CG. Effects of neonatal status epilepticus on rat brain. Neurology, 1976; 26: 975-986.
14. Wasterlain CG. Neonatal seizures and brain growth. Neuropädiatrie, 1978; 9: 213-228.
15. Olney JW. Glutamate, a neurotoxic transmitter. J. Child Neurol., 1989; 4: 218-226.
16. Perlman JM, Volpe J. Seizures in the preterm infant: effect of cerebral blood flow velocity, intracranial pressure, and arterial blood pressure. J. Pediatr., 1983; 102: 288-293.
17. Lombroso CT. Differentiation of seizures in newborns and in early infancy. In: Morselli PL, Pippenger CE, Penry JK, editors. Antiepileptic Drug Therapy in Pediatrics. New York: Raven Press, 1983; 85-102.
18. Freeman JM. Neonatal seizures - diagnosis and management. J. Pediatr., 1970; 77: 701-708.
19. Boer HR, Gal P. Neonatal seizures. Clin. Pediatr., 1982; 21: 453-457.
20. Morselli PL. Development of physiological variables important for drug kinetics. In: Morselli PL, Pippenger CE, Penry JK, editors. Antiepileptic Drug Therapy in Pediatrics. New York: Raven Press, 1983; 1-12.
21. White HS, Kemp JW, Woodbury DX. Effects of central nervous system maturation on drug responses. In: Morselli PL, Pippenger CE, Penry JK, editors. Antiepileptic Drug Therapy in Pediatrics. New York: Raven Press, 1983: 13-35.
22. Painter MJ, Bergman I, Crumrine P. Neonatal seizures. Pediatric Clinics of North America, 1986; 33: 91-109.
23. Powell C, Painter MJ, Pippenger C. Primidone therapy in refractory neonatal seizures. J. Pediatr., 1984; 105: 651-654.

24. Painter MJ, Gaus LM. Neonatal seizures: diagnosis and treatment. J. Child Neurol., 1991; 6: 101-108.
25. Hakeem VF, Wallace SJ. EEG monitoring of therapy for neonatal seizures. Dev. Med. Child Neurol., 1990; 32: 858-864.
26. Eyre JA, Oozeer RC, Wilkinson AR. Continuous electroencephalographic recording to detect seizures in paralyzed newborn babies. Br. Med. J., 1983; 286: 1017-1018.
27. Connell J, Oozeer R, DeVries L, Dubowitz LMS, Dubowitz V. Continuous EEG monitoring of neonatal seizures: diagnostic and prognostic considerations. Arch. Dis. Child., 1981; 64: 452-458.
28. Connell J, Oozeer R, DeVries L, Dubowitz LMS, Dubowitz V. Clinical and EEG response to anticonvulsants in neonatal seizures. Arch. Dis. Child., 1989; 64: 459-464.
29. Moshé SL, Sperber EF, Brown LL, et al. Experimental epilepsy: developmental aspects. Cleveland Clin. J. Med., 1989; 56 S: 92-99.
30. Yanai J, Bergman A. Neuronal deficits after neonatal exposure to phenobarbital. Exper. Neurol., 1981; 73: 199-208.
31. Legido A, Clancy RR, Berman PH. Recent advances in the diagnosis, treatment and prognosis of neonatal seizures. Pediatr. Neurol., 1988; 4: 79-86.
32. Painter MJ. Therapy of neonatal seizures. Cleveland Clin. J. Med., 1989; 56 S: 124-131.
33. Marlow N, Cooke RWI. Intravenous sodium valproate in the neonatal intensive care unit. In: Chadwick D, editor. IVth International Symposium on Sodium Valproate. London: Royal Society of Medicine, 1989: 208-210.

4 The malignant epilepsies of childhood and adolescence

WIL O. RENIER

Department of Neurology, University of Nijmegen, Nijmegen, The Netherlands

INTRODUCTION

The highest incidence of epilepsy occurs in infancy, young childhood and adolescence, typically in specific periods of brain maturation. Because many of the epilepsies in infancy and young childhood are difficult to treat and are frequently associated with mental retardation and/or mental regression, the term 'malignant' has been used in contrast to the 'benign' childhood epilepsies. In the last decennium, the increased interest in these malignant epilepsies of childhood has resulted in more or less well-defined individual epilepsy-syndromes (Table 1).

Under the age of 6, a series of age-related malignant epileptic encephalopathies can be identified, but not all of these epilepsy-syndromes are well delineated. From the well-known West syndrome and Lennox-Gastaut syndrome, a very early epilepsy with suppression-bursts on the electroencephalogram (EEG) and different severe myoclonic epilepsies have been separated as new syndromes. A special group of epilepsies is characterized by continuous spikes and waves during sleep and is associated with seizures, acquired aphasia and cognitive and behavioural disturbances. Epilepsia partialis continua is a focal status epilepticus characterized by prolonged rhythmically repeated twitching of a group of muscles for many days, weeks or even months, and on the EEG rhythmical epileptic paroxysms over the contralateral hemisphere. Different progressive diseases associated with brain pathology and epilepsy start in childhood. They constitute the group of progressive myoclonus epilepsies.

Table 1. The malignant epilepsies under the age of 6 year

Early myoclonic epilepsy
    (Aicardi-Goutières syndrome)
Early infantile epileptic encephalopathies with suppression bursts
    (Ohtahara syndrome)
Infantile spasms (West syndrome)
Severe myoclonic epilepsy of infancy (Dravet syndrome)
Lennox-Gastaut syndrome and related myoclonic epilepsies of childhood
Epilepsy with continuous spikes and waves during slow sleep/
    Landau-Kleffner syndrome
Epilepsia partialis continua type II
Progressive myoclonus epilepsies and related disorders

In this chapter clinical and electroencephalographic characteristics of these malignant epilepsy-syndromes and their treatment controversies will be discussed.

EARLY INFANTILE EPILEPTIC ENCEPHALOPATHIES
WITH SUPPRESSION-BURSTS

Aicardi and Goutières[1] and Ohtahara[2] have both described, in the same year (1978) but separately, epileptic syndromes with onset in the neonatal or early infantile period (20 to 28 days and 2 to 3 months). These babies have a severe encephalopathy, frequent seizures and an interictal suppression-bursts trace. The syndrome described by Aicardi and Goutières has been called 'early myoclonic epilepsy' (EME) and by Ohtahara 'early infantile epileptic encephalopathy' (EIEE). Both descriptions share many features in common but small differences can be distinguished (Table 2). In 1992 Tassinari[3] has proposed the term 'neonatal encephalopathy with suppression bursts', EME and EIEE being two subtypes, while Schlumberger et al.[4] have the opinion that no overlap exists between the two syndromes. Infants with a severe hypoxic-ischaemic encephalopathy, cerebral dysgenesis, neurometabolic or neurodegenerative diseases, or cerebral pathology of other origin can manifest this epileptic syndrome. It has a severe course and treatment is disappointing. The intractability contains a real danger for intoxicating regimens of medication are a real danger.

Early myoclonic encephalopathy (EME) is characterized by myoclonic and partial seizures. Fragmentary or erratic myoclonias can be almost continuous or only while awake, and massive myoclonic jerks can alternate with erratic myoclonias. Partial seizures are of a simple type, usually clonic or subtle, without relation to the jerks. Tonic spasms appear later in the course of the disease. Infants are in poor neurological condition with hypotonic, hypertonic or dystonic posturing. There is a lack of development associated with an acquired microcephaly in half of the cases and a lack of visual contact.[1,5]

The characteristic pattern of the EEG is that of suppression-bursts. It consists of 1-5 sec of an irregular generalized mix of spikes, sharp waves and spike-and-wave complexes, alternating with flat trace during 3-15 sec. From 3-4 months on the EEG pattern evoluates in atypical hypsarrhythmia. The suppression-bursts trace may be associated with localized discharges of sharp waves or rhythmical paroxysmal activity accompanying seizures.[1,5]

Many infants die before the age of one year. Approximately one third of the children become older than one year, but even then some do not survive. In one quarter of the cases more than one child with the disease is noted in the family, suggesting a genetic influence. The etiology is unknown. Hypoxic-ischaemic encephalopathy, cerebral spongiosis, poliodystrophy, progressive cerebral atrophy, glycine encephalopathy, and propionic acidopathy are some of the associated diagnoses.[1,5]

Early infantile epileptic encephalopathy (EIEE) has its onset at the same young age. The seizures are typically tonic spasms with later on transition to infantile spasms, and sporadic erratic myoclonias, especially in early onset cases. If the child survives, severe mental retardation and poor development complete the picture of severe encephalopathy. Prognosis is bad. The disease is fatal in approximately 1/3 of the cases.[2,6]

The EEG shows in both awaking and sleeping stages suppression-bursts. In cases of later onset, suppression-bursts trace is not present at onset but develops later during sleep. After 3-9 months the trace changes into hypsarrhythmia, initially in the awake state. However, some cases do not evolve into infantile spasms with hypsarrhythmia, but to an epilepsy with multifocal EEG discharges, indicating that EIEE is not always a very early form of West syndrome.

Most of the survivors show focal spikes indicating more localized cortical pathology. Multiple causes have been described but in many cases there is a severe cerebral malformation.[2,6]

Table 2. Common features and differences between EIEE and EME

	EIEE	EME
Onset	< 28 d - 25 m >	< 12 d - 3 m >
Seizure pattern	repetitive tonic spasms	frequent myoclonic seizures (massive, fragmentary, erratic)
	sporadic erratic myoclonias and partial seizures	late tonic spasms partial seizures (clonic, subtle)
EEG	periodic suppression-bursts	periodic suppression-bursts
Course	fatal or severe mental retardation, transition into West-syndrome or multifocal epilepsy	fatal or cerebral palsy with severe mental retardation acquired microcephaly
Etiology	unknown; cerebral malformation various etiologies	unknown; cerebral spongiosis or atrophy genetic/metabolic origin

In these very early epileptic encephalopathies the clinical picture shows a broad spectrum of tonic spasms, partial seizures and myoclonic seizures (massive, fragmentary, erratic). The expression in the individual case can be different, depending of age, vigilance state, neurological defects, degree of cerebral malformation, etiology, gastro-oesophageal reflux, general condition. In many cases it is the first stage of what can be called 'a malignant childhood epilepsy'.

Treatment is not very successful. Phenobarbital (3-5 mg/kg/d), valproate (25-50 mg/kg/d) and nitrazepam (0.3-1.0 mg/kg/d) or clonazepam (0.05-0.20 mg/kg/d) are drugs that have been used, but are in most cases not effective.[5,6] ACTH also has limited effect.[6] At present there is insufficient experience with vigabatrin and lamotrigin and felbamate in this type of symptomatic epilepsy.

Two other types of epilepsy can occur at this early stage of life.

Nonketotic hyperglycinaemia (glycine encephalopathy) starts on the first or second day after birth with erratic myoclonias, partial seizures and later on with infantile spasms. Seizures are refractory and stimulus-sensitive. In some cases no myoclonic jerks but tonic spasms and dystonic movements are described. The infants are deeply comatose between spells. Half of the children die before the end of the first month of life. EEG shows suppression-bursts and later atypical hypsarrhythmia.[7,8] The diagnosis can be confirmed by a high level of glycine in the cerebrospinal fluid.

Early partial epilepsy can present as generalized spasms with interictal focal abnormalities and diffuse ictal discharges on EEG. The difference with atypical West syndrome (West syndrome without hypsarrhythmia) has to be determined.

INFANTILE SPASMS

Infantile spasms have been recognized as an epileptic phenomenon since their early description by West in 1841. The triad of infantile spasms, arrest of mental development, and a typical interictal EEG pattern, hypsarrhythmia, has become known as West syndrome. The incidence of infantile spasms has been estimated to be 1 per 4000 to 1 per 6000 live births.[9] The onset is usually in the age group 4-7 months. The apparent association between infantile spasms and immunization seems coincidental.

Infantile spasms are generalized brief myoclonic contractions of muscles of the neck, trunk and extremities, occurring both as single seizures and as series of seizures. Spasms occur more or less bilateral symmetric even when focal cerebral lesions are present. They are the expression of a disturbed interaction between cortex and subcortical structures and have some resemblance with the Moro-reflex. Three main types of spasms can be recognized: flexor (35%), extensor (20%), or mixed (45%). Most infants with this disorder have more than one type of spasm, influenced by body position. They frequently occur upon arousal or when unexpected auditory or tactile stimuli occur.[10]

The interictal EEG is characterized by hypsarrhythmia, particularly in younger infants. Hypsarrhythmia is a mix of continuous irregular polymorphous theta and delta waves and spikes and sharp waves with variable localization. This typical pattern can be modified by state of wake or sleep, by age and by etiology. Fragmentation of hypsarrhythmia can be seen in NREM-sleep and disappearance in REM-sleep; in younger infants an increase of hypsarrhythmia in sleep can occur. Different ictal EEG patterns have been described by Kellaway et al.[10] Disappearance of the paroxysms by 0.5 mg/kg diazepam intravenously may signify a better prognosis. The most common EEG-event associated with spasms is a high-voltage, generalized, slow-wave transient, followed by a general attenuation of background activity that may last from one to many seconds. Longer periods can be associated with the arrest phenomenon.

While infantile spasms are age-related phenomena and tend to disappear six to twelve months after onset of the disease, mental development after a first decline may stabilize at a lower level. Depending of the etiology, mental development may normalize after arrest or control of spasms in 10 to 30% of the cases. But, in all other cases severe mental retardation with dyspraxia and dysgnosia, dysphasia and autistic behaviour are the dramatic outcome.

Syndromes of infantile spasms can be classified according to etiology into idiopathic (5%), cryptogenic (15%) and symptomatic (80%). Idiopathic refers to a pure functional cerebral dysfunction, while cryptogenic and symptomatic refer to an encephalopathy with respectively unknown and known etiology. The characteristics and prognosis of these three forms of infantile spasms are slightly different (Table 3). Prognosis is best in idiopathic cases, and in cases with prematurity, porencephaly, postnatal disorders, Down syndrome or neurofibromatosis. Response to treatment is moderate in normal term babies with respiratory distress or with focal cerebral malformations. Protracted spasms are the rule in extensive clastic lesions (hydranencephaly, multicystic leucomalacia) or cortical dysplasias (lissencephaly, Aicardi syndrome, tuberous sclerosis).

The issue of therapy is still under debate (Table 4). In 1958, Sorel and Dusaucy-Bauloye[11] have reported excellent results of a therapy with 4-10 units of ACTH-retard during 10-25 days, depending of the age of the child (5 weeks-27 months). Subsequently, many publications on hormonal therapy have appeared, reporting different results while using a variety of products and dosages. Some authors proclaim high doses of ACTH for several months while others prescribe low doses of ACTH for a few weeks. In a retrospective study, Renier and Le Coultre[12] have demonstrated that, independent of the type of therapy (conventional antiepileptics, low-dose ACTH or high-dose ACTH), there is good result in proximately 10% of the cases. Severe or fatal complications occurred only in the high-dose ACTH group. One of the best studies is the doubleblind and 24-hour polygraphic/video monitoring study of Hrachovy et al.[13] This study demonstrated that the response to hormonal therapy is all or none, that only a short course (2 weeks) of therapy is required and that there is no major difference in therapeutic efficiency between ACTH (20-30 units/day) and prednisone (2 mg/kg/day). In their opinion etiology and treatment delay are not useful predictors of response and normalization of the EEG is not an absolute criterion for clinical evaluation or prognosis. Of the conventional AED's, valproate (25-50 mg/kg/d) and nitrazepam (0.3-1.0 mg/kg/d) have been reported to be effective. Liver enzymes and thrombocytes have to be controlled carefully in infants with mental retardation under treatment with valproate. In mitochondriopathies valproate should be avoided.

In symptomatic types of infantile spasms, especially tuberous sclerosis, vigabatrin (40-80 mg/kg/d) has been promoted as a drug of first choice, but more longterm experience is needed.

Table 3. Characteristics of different types of infantile spasms

Characteristics of infantile spasms	idiopathic	cryptogenic	symptomatic
Development prior to onset of spasms	normal	± normal	abnormal
Seizures:			
independent spasms during clusters	+	+	±
Symmetry of spasms	+	±	±/-
Mental deterioration	-/±	+	+
EEG			
symmetry of hypsarrhythmia	symmetry	asymmetry focal abnorm.	asymmetry focal abnorm.
effect of IV-benzodiazepine	+	±	-
Therapy:			
low dose of ACTH	+	+/±	-
valproate/nitrazepam or vigabatrin	+	+	+
focal surgery	-	+	?
success or seizure control	± 90%	± 15%	± 15%
Long term outcome: normal or only mild impairment	± 90%	± 30%	± 5%

Occasionally a high dose of pyridoxine (200-400 mg/d) produces clinical and EEG improvement. Beneficial results have been described with intravenous immunoglobulins (IV-Ig), with thyrotropin-releasing hormone (TRH) and with methysergide and alpha-methyl-paratyrosine. Most studies however suffer from methodological shortcomings. The study of Chugani et al.[14] demonstrates that in infants with cryptogenic spasms, positron emission tomography can identify those due to unsuspected focal cortical dysplasia, for which resective surgery may offer a better prognosis.

Table 4.   Different schemes of treatment of infantile spasms

*   after investigations for etiological diagnosis
    (neuroradiological, metabolic, virological, chromosomal

*   conventional treatment

1. idiopathic,   ACTH
    cryptogenic   week 1: 40 IU/d
                                          week 2: 30 IU/d
                                          week 3: 20 IU/d
                                          week 4: 10 IU/d
                     or
                                          week 1: 10 IU/d   or week 1: 20 IU/d
                                          week 2: 10 IU/d   or week 2: 20 IU/d
                                          week 3: 10 IU/d

                     corticosteroids:
                     dexamethason   week 1: 2.0 mg
                                          week 2: 1.5 mg
                                          week 3: 1.0 mg
                                          week 4: 0.5 mg

                     prednison   week 1: 2.0 mg/kg/d
                                          week 2: 2.0 mg/kg/d

2. symptomatic   nitrazepam: 0.3-1.0 mg/kg/d
                     valproate: 25-50 mg/kg/d
                     vigabatrin: 40-80 mg/kg/d

*   experimental   pyridoxine: 200-400 mg/d
    treatment   IV-Ig: 400 mg/kg/d during 5 days,
                                          then every 2 weeks for 3 months

                     TRH: 0.5-1.0 mg IV or IM for 1 or 4 weeks
                     methysergide/
                     alpha-methyl-
                     paratyrosine

*   resective surgery

MYOCLONIC EPILEPSY OF INFANCY (MEI)

Myoclonic epilepsy of infancy (MEI) can be classified in a benign and a severe type.[15,16]

The *benign myoclonic epilepsy of infancy (BMEI)* occurs typically in normal male infants, 4 months to 3 years of age. Ictal manifestations consist of massive symmetrical myoclonic jerks when drowsy or when he/she is falling asleep. Occasionally atonic seizure with head dropping, myoclonic seizures and febrile convulsions are possible. Tonic seizures have never been noted. The epilepsy is not associated with mental retardation.

The background activity of the EEG is normal in the awake state, but can be abnormal in sleep phase 1 and 2. During the myoclonic manifestations fast generalized spike-waves and polyspike-waves are generated. Sometimes there is photosensitivity.

After nearly one year, the epileptic activity disappears but rare tonic-clonic seizures in adolescence are mentioned in some cases. In 30% the family history is positive for epilepsy. Treatment with valproate and/or benzodiazepine is successful in all cases correlating with the benign character of the disease.[15]

The *severe myoclonic epilepsy of infancy (SMEI)* has an early onset at 5-6 months of age with generalized or unilateral clonic seizures, associated with fever. Ictal phenomena consist of alternating hemiconvulsions, vibratory tonic seizures, a typical partial seizures and myoclonias (massive, segmental, epileptic or non-epileptic). Mental retardation appears in most cases after the age of one year. In the series of Dravet et al.[16] 14% of the children died. There is a high percentage of family history positive for epilepsy or convulsions.

The EEG is fairly abnormal at onset, but after nearly one year coexistence of focal and diffuse epileptic discharges is seen; later on, after a period of lateralisation and focalisation of the epileptic activity, disappearance of all epileptic discharges is possible.[16]

Neuroradiology only seldomly contributes to treatment and diagnosis. Until now only one autopsy has been published[17] illustrating developmental abnormalities of the brain and brainstem. An important characteristic of SMEI is its intractability. Valproate, benzodiazepines, high-dose phenobarbital, ACTH/ corticosteroids, and bromide have been tried with poor or temporary results.

Difficulties in the identification of SMEI can be due to the polymorphic character of the seizures. Moreover, myoclonias in SMEI are not always prominent nor typical, especially at onset. Other questions concern the transitory character of the epilepsy in some cases, the mechanism of mental deterioration, and the underlying pathology. The differential diagnosis with febrile convulsions and with early Lennox-Gastaut syndrome can be difficult at onset.

LENNOX-GASTAUT SYNDROME AND RELATED MYOCLONIC EPILEPSIES OF CHILDHOOD

In the first 2 to 8 years of life, a group of epilepsies occur, characterized clinically by myoclonic and drop attacks, cognitive impairment, arrest of mental development or even mental regression, and intractability. Drop attack is a sudden (<1 sec) incapacity to maintain posture and therefore can be missed if the patient is lying down. It can be caused by tonic extensor or flexor spasms, myoclonic and atonic seizures. The EEG is characterized by interictal slow (<2.5 Hz) or fast (>2.5 Hz) spike-wave complexes, slow background activity, and focal or multifocal spikes or sharp waves. Although many of the clinical signs overlap, a classification of myoclonic epilepsies of childhood into more or less well defined syndromes has been proposed: true myoclonic epilepsies, Lennox-Gastaut syndrome, myoclonic variant of Lennox-Gastaut syndrome and myoclonic-astatic epilepsy. A more recent approach tends to use both clinical and EEG criteria for subdividing the broad group of myoclonic epilepsies of infants and young children.[18,19] However, in the individual patient a precise classification may be difficult.

The *classical Lennox-Gastaut syndrome (LGS)* is characterized by a triad of characteristic multiple and complex seizures, diffuse slow spike and wave discharges in the interictal EEG and cognitive dysfunctions with personality disorders and behaviour problems.[20] There is a male: female ratio of 1.5:1. Several seizure types may brief tonic spasms, accentuated by phase 1 sleep, atonic-astatic attacks, atypical absences, myoclonic seizures, and frequent status epilepticus (absence, tonic, myoclonic). Short tonic, atonic and myoclonic seizures may lead to frequent drop attacks. The clinical characteristics of LGS varies with age of onset and neurological defects. Tonic fits are more frequent in symptomatic cases, whereas myoclonic attacks and atypical absences are more frequent in cryptogenic cases. Infantile spasms can precede LGS in 20-30% of the cases.[18,20]

Interictal EEG shows slow background activity, slow (1-2.5 Hz) spike-wave complexes while awake and polyspikes while asleep, and fast rhythms of 10 c/sec during sleep; in REM-sleep the paroxysmal activity decreases markedly.

Behaviour problems in LGS are nonspecific. Hyperkinetic motor behaviour, short attention span, slow mental speed, perseveration, stereotype behaviour and progressive mental retardation are frequently found. Mental retardation is present before seizure onset in approximately 60% of the cases.

A poor dendritic arborization of the inner layers of the cerebral cortex has been found in idiopathic cases.[21] Diffuse or more localized hypometabolism on the interictal PETscan can be demonstrated.

The distinction between *myoclonic epilepsies of infancy and early childhood* and LGS is in many cases difficult to make.[19] Prolonged observation and EEG recording are needed. The myoclonic epilepsies are characterized by predominantly repetitive myoclonic jerks with frequently a saccadic character in the neck and shoulder and external ocular and eyelid muscles. They are provoked by awakening and by photic stimulation. Except for the pure myoclonic epilepsy, myoclonic seizures are seldom the only type of ictal phenomena and clonic, tonic and atonic elements may be associated. Mental retardation is less common than in LGS. As in LGS, infantile spasms can precede the myoclonic epilepsy. The interictal EEG has a normal or slow background activity and shows brief irregular polyspike-wave paroxysms.

Myoclonic epilepsy of childhood has its onset between 2 and 4 years, and is considered cryptogenic. A subtype with favourable and one with unfavourable course and mental deterioration have been mentioned.[22] At onset, the children present generalized clonic, tonic, or tonic-clonic seizures sometimes in series. A status epilepticus with massive or erratic myoclonias is a frequently occurring clinical problem. In the favourable group, seizures are more of the massive myoclonic type while in the unfavourable group erratic myoclonics predominate. In the unfavourable group status consists in most cases of erratic myoclonias in the face, tongue and hands or of vibratory tonic seizures. Drop attacks or a status of atonic absences can be seen. The distinction between myoclonic epilepsy of childhood and Lennox-Gastaut syndrome can be very difficult.

The EEG is characterized by generalized slow waves. If mental deterioration is present, EEG becomes slow with an encephalitis-like pattern. Benzodiazepines intravenously make spikes disappear but slow waves remain.

Carbamazepine, phenobarbital and vigabatrin can provoke seizures at onset especially in the unfavourable group.[22]

The differential diagnosis with *epilepsy with myoclonic astatic seizures (EMAS),* or *Doose syndrome* is also difficult. The patients described by Doose[23] have a more cryptogenic type of epilepsy because clinically and electroencephalographically they are normal at onset of the disease (Table 5). A genetic predisposition for this type of epilepsy is supposed because of frequent bilateral theta-rhythms and photosensitivity in the EEG, and early febrile convulsions.

Seizures are predominantly generalized myoclonic, astatic and myoclonic-astatic, absences with tonic and/or clonic component, occasionally tonic-clonic or late tonic. In contrast to LGS, no tonic seizures during sleep occur. EMAS has in general a somewhat better outcome than LGS.

The myoclonic epilepsies of infancy and early childhood are often therapy resistant. Sodium valproate and benzodiazepines (clonazepam, nitrazepam, clobazam) are the antiepileptic drugs most often used in these epilepsy-syndromes; valproate being the drug of first choice that should be tried. The best results are observed in true or cryptogenic myoclonic epilepsy and less effects are seen in symptomatic cases. Hyperactivity and dysphoria can occur as side-effects of the medication. Sometimes a combination of valproate and benzodiazepines proves to be more effective than a monotherapy of either drug. In that case benzodiazepines should be added to the initial valproate treatment. The dosage of both drugs is best evaluated by clinical observation of seizure frequency, alertness, and side-effects. Plasma levels are important when high doses of the drugs are prescribed or when they are combined with other antiepileptic drugs.

Table 5. Common features and differences between LGS and EMAS

	LGS	EMAS
Etiology	more symptomatic	more cryptogenic
Family history	negative	genetic predisposition
Development before onset of seizures	(ab)normal	normal early febrile convulsions
Onset epilepsy	8 m - 8 y	7 m - 6 y
Seizures	tonic-clonic (T-C), tonic tonic in sleep + + atypical absence with T or C (myo)clonic/atonic frequent status	tonic-clonic, late tonic, no tonic in sleep absences w. myocl/atonia myoclonic astatic + + frequent status
EEG at onset	(normal)/slow/background slow spike-waves multifocal abnormalities	normal background bilateral theta rhythms rapid spike-waves rare focal abnormalities
In sleep	10 c/sec rhythms	no 10 c/sec rhythms
Prognosis	bad	variable

Benzodiazepines commonly show an initial and quick reduction of seizure frequency ('honeymoon period'), followed, after 2 or 3 months, by tolerance. Change of the product or alternate day medication can be tried in such cases, however without guarantee of success. Starting with a low dose and a higher dose at bedtime is to be advised. Drowsiness or paradoxical hyperactivity, lack of concentration, problems with short memory functions, hypotonia, increased bronchial secretion and drooling, limit the use of benzodiazepines in young children. Because alertness and active wakefulness are usually associated with fewer seizures, a reduction of sedative drugs can give better results. Exceptionally, a tonic status epilepticus can be provoked by benzodiazepines. For all these reasons benzodiazepines should be used only transiently if possible, for intravenous or rectal use in acute situations with no respiratory distress. Phenobarbital, primidone and phenytoin are usually ineffective. They have side-effects on memory and behaviour during chronic use. Nevertheless primidone and ethosuximide can be used if other drugs fail. Carbamazepine can be helpful in older children who have more generalized tonic-clonic and complex partial seizures, associated with mood disturbances. Occasionally carbamazepine can provoke non-convulsive and/or myoclonic seizures.[24] Lamotrigine is a promising new drug, but more longterm experience is needed.

Because many of these epilepsy-syndromes have a waxing and

waning course ACTH or corticosteroids can be given in a period of deterioration which is frequently a period of physical growth. At our department we use dexamethasone 1 or 2 mg in the morning, following the physiological day and night rhythm of corticosteroid production. This rhythm is frequently disturbed in these children. The add-on treatment with corticosteroids is given for 5 or 10 days. The best results are observed in children under the age of four years, and in cryptogenic cases.

Because treatment with conventional AED's in many cases is disappointing, alternative treatment regimens have been tried. There is no proof that treatment with special diet (ketogenic diet, oligo-antigenic diet, food without additives) is of help in children with an allergic constitution or with hyperactivity and behaviour problems. Thyrotropin- leasing hormone (0.5-1.0 mg i.v. or i.m. for 1 or 4 weeks) is reported to have beneficial but transient effect in half of the cases of West or Lennox epilepsies.[25] High dose of intravenous immunoglobulins have been used by clinical researchers. The rationale for immunoglobulins in therapy resistent epilepsies is based on the hypothesis that immunological mechanisms may play a role in the pathogenesis of these epileptic encephalopathies.[26,27] A review of the literature reveals that immunoglobulins given in different ways and different doses to a heterogeneous group of patients with therapy resistent epilepsies have a beneficial effect in approximately 50% of the cases. A prospective clinical study at our department in a homogeneous group of children with a cryptogenic type of West or Lennox epilepsy could demonstrate a seizure reduction of more than 50% and a reduction of epileptic discharges in the EEG of more than 30% in nearly all cases. Moreover behaviour and social contact improved, sometimes even before seizure reduction.

Considering most of the clinical trials in malignant childhood epilepsies it must be stated that most studies suffer from methodological shortcomings. On the other side, because many of these epilepsy-syndromes are intractable, small results have their value and all new treatment should be fully exploited.

THE LANDAU-KLEFFNER SYNDROME AND THE EPILEPSIA PARTIALIS CONTINUA

In the *Landau - Kleffner syndrome* (LKS) all children suffer from acquired receptive and expressive aphasia and EEG abnormalities, especially during sleep. In 75 % of the cases seizures occur. Neuropsychological examination reveals a verbal auditory agnosia. The EEG background activity is usually normal, but theta waves in the posterior derivations are possible. Spikes, sharp waves, or spike-wave complexes are generally found with a predominance over the occipito-temporal and parietal regions, but without a stable lateralisation. The variability of the EEG abnormalities

in time and intensity is frequently parallelled by a variable clinical expression of the aphasia. The electrical status epilepticus during slow sleep is a striking feature of the disease. Therefore the question arises if LKS is not to be considered a subtype of the epilepsy with continuous spikes and waves during slow sleep as discussed by Tassinari et al.[28]
Antiepileptic drugs have been of limited value for the aphasic problems. Benzodiazepines, carbamazepine, ACTH/corticosteroids can give partial and temporary results. The different seizure types can require different medication. Our experience with four children with the syndrome has learned that linguistic communicative processing may be possible, provided the input is through the visual modality and the output by gesture; sometimes writing in telegram style is possible. If a good nonverbal communication with the children can be elaborated, the behaviour problems are less pronounced. Children with onset of the disease at 6 year or later have a greater chance to recover, in contrast with younger children. Recovery occur more frequently in girls than in boys.

Epilepsia partialis continua (EPC) has been classified by Bancaud et al.[29] into two groups. One corresponds to the classic description of the Kojewnikow syndrome, due to focal, nonprogressive pathology of the central motor area. The second group presents with completely different characteristics: early onset of seizures, presence of other seizure types, relatively short delay to onset of myoclonic jerking, localization of myoclonus over large parts of the body, and progressive evolution of the motor signs and cognitive disturbances. Elevated levels of IgG and lactate in cerebrospinal fluid and neuromorphological findings in such cases[30] are well explained by a chronic inflammatory process as postulated by Rasmussen et al.[31]

THE PROGRESSIVE MYOCLONUS EPILEPSIES

Generalized epilepsies, secondary to static or progressive encephalopathy constitute approximately 9% of the chronic epilepsies in childhood. These epilepsies are characterized by a variety of seizure types and by cerebral pathology of acquired or genetic origin. *The progressive myoclonus epilepsies* are part of this group of epilepsies.[32] Under the major causes are Baltic myoclonus, Lafora body epilepsy, neuronal ceroid lipofuscinosis, sialidosis and mitochondrial encephalomyopathy with ragged-red fibers.[33] Antiepileptics are helpful in the initial stages of the diseases, but due to the progressive character of the disease vital supportive measures are more required.

CONCLUSION

Treatment and prognosis of 'spike and wave syndromes in childhood' are determined by the respectively benign or malignant character of the epilepsy. Benign epilepsy with pure functional disturbances of the brain, as expressed by few seizures and a normal intelligence, has a relatively good prognosis. It generally needs only small doses of antiepileptic drugs. In contrast, in the malignant epilepsies prognosis and efficacy of treatment are dependent of the known or unknown underlying cerebral pathology, which can be progressive in some cases. Most of these patients are on polytherapy and are potential candidates for drug intoxication. When confronted with a history of seizures in children, the diagnostic procedure must always include an extensive search for etiology. Before prescribing medication we must realize that 'epilepsy is more than just having fits'. Treatment of these malignant epilepsy syndromes must include educational and psychosocial advise for the parents. Selfhelp groups for parents with a child with a therapy resistant epilepsy have proven to be of great emotional help.[34]

REFERENCES

1. Aicardi J, Goutières F. Encéphalopathie myoclonique néonatale. Rev. EEG Neurophysiol., 1978; 8: 99-101.
2. Ohtahara S. Clinico-electrical delineation of epileptic encephalopathies in childhood. Asian Med. J., 1978; 21: 499-509.
3. Tassinari CA. Tentative for a consensus. Communication at the 4th International Workshop on Childhood Epilepsies: Epilepsies and generalized epileptic syndromes before 6 years. Marseilles, June 23-24, 1992.
4. Schlumberger E, Dulac O, Plouin P. Early infantile epileptic syndrome(s) with suppression-burst: nosological considerations. In: Roger J, Bureau M, Dravet C, et al. editors. Epileptic Syndromes in Infancy, Childhood and Adolescence. 2nd edition. London: John Libbey & Company Ltd, 1992: 35-42.
5. Aicardi J. Early myoclonic encephalopathy (neonatal myoclonic encephalopathy). In: Roger J, Bureau M, Dravet C, et al. editors. Epileptic Syndromes in Infancy, Childhood and Adolescence. 2nd edition. London: John Libbey & Company Ltd., 1992: 13-23.
6. Ohtahara S, Ohtsuka Y, Yamatogi Y, et al. Early-infantile epileptic encephalopathy with suppression-bursts. In: Roger J, Bureau M, Dravet C, et al. editors. Epileptic Syndromes in Infancy, Childhood and Adolescence. 2nd edition. London: John Libbey & Company Ltd., 1992: 25-34.
7. Dalla Bernardina B, Aicardi J, Goutières F, et al. Glycine encephalopathy. Neuropädiatrie 1979; 10: 20--25.
8. Seppäläinen AM, Similä S. Electroencephalographic findings in three patients with nonketotic hyperglycinemia. Epilepsia, 1971; 12: 101-107.
9. Van den Berg BJ, Yerushalmy J. Studies on convulsive disorders in young children. I. Incidence of febrile and nonfebrile convulsions by age and other

factors. Pediatr. Res., 1969; 3: 298-304.
10. Kellaway P, Hrachovy RA, Frost JD, et al. Precise characterization and quantification of infantile spasms. Ann. Neurol., 1979; 6: 214-218.
11. Sorel L, Dusaucy-Bauloye H. A propos de 21 cas d'hypsarhythmia de Gibbs. Son traitement spectaculaire par L'ACTH. Acta Neurol. Belg., 1958; 58: 130-141.
12. Renier WO, Le Coultre R. ACTH-kuur en ketogeen dieet: een kritische evaluatie (ACTH and ketogenic diet: a critical analysis). T. v. Kindergeneesk., 1989; 57: 81-86.
13. Hrachovy RA, Frost JD, Kellaway R, et al. Double-blind study of ACTH vs. prednisone therapy in infantile spasms. J. Pediatr., 1983; 103: 641-645.
14. Chugani HT, Shields WD, Shewmon DA, et al. Infantile spasms: I. PET identifies focal cortical dysgenesis in cryptogenic cases for surgical treatment. Ann. Neurol., 1990; 27: 406-413.
15. Dravet C, Bureau M, Roger J. Benign myoclonic epilepsy in infants. In: Roger J, Bureau M, Dravet C, et al. editors. Epileptic Syndromes in Infancy, Childhood and Adolescence. 2nd edition. London: John Libbey & Company Ltd., 1992: 67-74.
16. Dravet C, Bureau M, Roger J. Severe myoclonic epilepsy in infants. In: Roger J, Bureau M, Dravet C, et al. editors. Epileptic Syndromes in Infancy, Childhood and Adolescence. 2nd edition. London: John Libbey & Company Ltd., 1992: 78-88.
17. Renier WO, Renkawek K. Clinical and neuropathologic findings in a case of severe myoclonic epilepsy of infancy. Epilepsia, 1990; 31: 287-291.
18. Aicardi J. Lennox-Gastaut syndrome. In: Aicardi J. editor. Epilepsy in Children. 2nd edition. New York: Raven Press, 1994: 44-66.
19. Aicardi J. Myoclonic epilepsies of infancy and early childhood. In: Aicardi J. editor. Epilepsy in Children. 2nd edition. New York: Raven Press, 1994: 67-79.
20. Beaumanoir A. Les limites nosologiques du syndrome de Lennox-Gastaut. Rev. EEG Neurophysiol., 1981; 11: 468-473.
21. Renier WO, Gabreëls FJM, Jaspar HHJ. Morphological and biochemical anal'-is of a brain biopsy in a case of idiopathic Lennox-Gastaut syndrome. Epilepsia, 1988; 29: 644-649.
22. Dulac O. Cryptogenic myoclonic epilepsy of childhood. Communication at the 4th International Workshop on Childhood Epilepsies: Epilepsies and generalized epileptic syndromes before 6 years. Marseille, June 23-26, 1992.
23. Doose H. Myoclonic astatic epilepsy of early childhood. In: Roger J, Bureau M, Dravet C, et al. editors. Epileptic Syndromes in Infancy, Childhood and Adolescence. 2 nd. edition. London: John Libbey & Company Ltd., 1992: 103-114.
24. Brett EM. The Lennox-Gastaut syndrome: therapeutic aspects. In: Niedermeyer E, Degen R, editors. The Lennox-Gastaut Syndrome. New York: Alan R Liss, Inc., 1988: 329-339.
25. Matsumoto A, Kumagai T, Takeuchi T, et al. Clinical effects of thyrotropin-releasing hormone for severe epilepsy in childhood. A comparative study with ACTH therapy. Epilepsia, 1987; 28: 49-55.
26. Haraldsson A, van Engelen BGM, Renier WO, et al. Light chain ratios and concentrations of serum immunoglobulins in children with epilepsy. Epilepsy Res., 1992; 13: 255-260.
27. van Engelen BGM, de Waal LP, Weemaes CMR, Renier WO. Serologic

HLA typing in cryptogenic Lennox-Gastaut syndrome. Epilepsy Res., 1994; 17: 43-47.
28. Tassinari CA, Bureau M, Dravet C, et al. Epilepsy with continuous spikes and waves during slow sleep - otherwise described as ESES (epilepsy with electrical status epilepticus during slow sleep). In: Roger J, Bureau M, Dravet C, et al. editors. Epileptic Syndromes in Infancy, Childhood and Adolescence. 2 nd. edition. London: John Libbey & Company Ltd., 1992: 245-256.
29. Bancaud J. Kojewnikow's syndrome (epilepsia partialis continua) in children. In: Roger J, Bureau M, Dravet C, et al. editors. Epileptic Syndromes in Infancy, Childhood and Adolescence. 2 nd. edition. London: John Libbey & Company Ltd., 1992: 363-379.
30. Verhagen WIM, Renier WO, ter Laak H, et al. Anomalies of the cerebral cortex in a case of epilepsia partialis continua. Epilepsia, 1988; 29: 57-62
31. Rasmussen T, Olszewski J, Lloyd-Smith D. Focal seizures due to chronic localized encephalitis. Neurology, 1958; 8: 435-445.
32. Marseille Consensus Group. Classification of progressive myoclonus epilepsies and related disorders. Ann. Neurol., 1990; 28: 113-116.
33. Berkovic SF, Andermann F, Carpenter S, Wolfe LS. Progressive myoclonus epilepsies: specific causes and diagnosis. N. Engl. J. Med., 1986; 315: 296-305.
34. Renier WO. Self-help groups, second opinion and quality of life. (Abstract). Epilepsy Europe Conference 1992, Glasgow, 1992.

5   The benign epilepsies of childhood and adolescence

JUDITH MANELIS
MICHAEL DUBLIN
Western Galilee Regional Hospital, Nahariya, Israel

BENIGN PARTIAL EPILEPSIES

Identification and classification of benign partial epilepsies has been an important development in paediatric epilepsy in recent years. Diagnostic criteria proposed include:
- The absence of neurological or intellectual deficit
- A family history of epilepsy, especially of benign types
- An onset of seizures after the age of two
- Brief seizures which are stereotyped in clinical manifestation
- Frequent nocturnal occurrence
- Spontaneous remission in adolescence
- Favourable response to antiepileptic therapy

The absence of neurological or intellectual deficit not only signifies a good prognosis, but it also constitutes a specific part of the definition of all idiopathic partial epilepsies of childhood (IPEC). A high percentage (up to 40%) of close relatives have been found to have a history of febrile convulsions, partial or generalized seizures, or epileptic discharges in the EEG of focal or generalized nature. In most cases relatives of children with IPEC had a benign type of epilepsy with seizures recovering during adolescence. In all the types of IPEC described, the mean age of onset ranges between 4 and 8 years, very rare cases showing the first seizure before the age of two. The majority of cases have simple partial motor seizures or seizures with sensory symptomatology. About 25% of the cases have complex partial seizures. Seizure frequency is high, early in the course of idiopathic partial epilepsies of childhood. Most of the seizures are nocturnal. Status epilepticus of partial seizures has been observed in about 8% of the IPEC cases, mostly with the unilateral clonic seizure, lasting more than 30 minutes. Seizures get rare during the later evolution of IPEC and spontaneous remission occurs for the majority of the children in adolescence.[1,2]

The International Classification of the Epilepsies and Epileptic Syndromes recognizes the following idiopathic localization related (focal, local, partial) epileptic syndromes :
- Benign partial epilepsy with centro-temporal spikes (Rolandic Epilepsy).
- Benign epilepsy of childhood with occipital paroxysms.
- Benign epilepsy with affective symptoms (benign psychomotor epilepsy).

- Benign partial epilepsy with extreme somato-sensory evoked potentials.
- Benign frontal epilepsy
- Benign epilepsy associated with multiple spike foci
- Benign partial epilepsy of adolescence
- Reading epilepsy.
- Other unclassified syndromes

By definition, the prognosis of these syndromes, in contrast to the malignant partial seizures of childhood, is favourable. Genetic influences in partial seizures have previously been considered to be marginal when compared to generalized seizure disorders. However, with further evaluation of the family histories of children with benign partial epilepsy, it became clear that inheritance plays a major role in the pathophysiology of the disorder. It is suggested that the inheritance of the EEG pattern is transmitted by an autosomal dominant gene with a low penetrance and marked age dependency.

Benign partial epilepsy with centro-temporal spikes (Rolandic Epilepsy)

This syndrome was originally described by Marinus Rulandus in 1597, but only during the past 30 years, the interesting features of benign partial epilepsy with centro-temporal spikes (BECT) have become fully recognized. Gastaut, in 1952 was the first to describe the electroencephalographic features, already noting that these pre-rolandic discharges are not related to focal lesions. BECT occurs only children in and a major genetic factor is involved. Other characteristics are the nocturnal generalized seizures of probable focal onset and diurnal partial seizures, arising from the lower Rolandic area and an EEG pattern consisting of a mid temporal-central spike foci.[3] The failure to recognize this rather common syndrome is probably due to the nocturnal activity and the lack of continuous EEG monitoring. Moreover, the absence of identified underlying cerebral pathology also played an important role. The diagnosis of this syndrome, which depends on history and EEG features, allows the clinician to offer the patient and parents a rational plan for treatment, prognosis and genetic counselling. In a Swedish study of 52,252 children, followed from birth to the age of 15 years, an incidence rate of epilepsy was reported to be 134/100,000 and BECT incidence rate of 21/100,000 (16% of all epileptic seizures).[4]

Clinical Characteristics
Typically this syndrome has its onset during childhood (3-13 years, with a peak between the 8th and 9th years of life). In fact, this form of epilepsy appears in the first decade, to disappear in the second. The syndrome is

labelled '*rolandic*', because of the characteristic feature of partial seizures involving the region around the lower portion of the central gyrus of Rolando. It is one of the commonest forms of epilepsy encountered in childhood. The sex prevalence is approximately 60% males and 40% females. Seizure frequency varies: single seizure 10-13%, seizure recurrence once every 2-12 months 66-70%, multiple daily seizures (clusters are common) 20%. There is no correlation between the severity of EEG abnormalities and seizure frequency. The duration of seizures ranges from several seconds to 2 minutes. Most of the seizures (65%) occur only during sleep. The nocturnal seizures tend to be longer and more severe and most of them occur in early morning hours. Eighteen per cent of children suffering from BECT had febrile convulsions compared to 8% in controls. In family history we can find 50% of close relatives who had febrile seizures compared to 21% in controls. Migraine-like manifestations are frequent. Studies dealing with BECT emphasize the fact that we deal with neurologically intact and mentally normal children.

Seizures description
The characteristic features of BECT are:[5,6]
1. Somatosensory onset with unilateral paresthesias involving tongue, lips, gums, and inner cheeks
2. Unilateral, tonic, clonic or tonic-clonic convulsions involving the face, lips, tongue, pharyngeal and laryngeal muscles causing 3. and 4.
3. Speech arrest or anarthria
4. Drooling due to saliva pooling
5. Preservation of consciousness

In a typical seizure, occurring either upon waking or out of sleep, the child will come to his parents terrified with fear, unable to speak but fully conscious, pointing to his mouth which is drawn to one side, drooling from one corner of his mouth. Sometimes hemifacial twitching accompany these features. The whole episode lasts from a few seconds to 2 minutes, there is no postictal confusion. Most of the time the child describes that the seizure started with a feeling of electrical twitching in his gums, cheeks or tongue. This clinical story, told by the patient and his parents is sufficient to diagnose BECT even without EEG. In case of nocturnal seizures, the parents hear noises coming from the child's bed. They find him with grunting, gurgling sounds emanating from his mouth, which is drawn to one side and drooling. At this stage the seizure may end or it may develop into generalized convulsion.

The nocturnal seizure might be of 3 different types:
1. Typical brief hemifacial seizure, with speech arrest, drooling in a conscious state. Usually the child is awakened from sleep by the somatosensory component.
2. Like those above, but with loss of consciousness, associated with

gurgling, vomiting like noises.
3. Secondary generalized convulsions lasting from few minutes to half an hour.[7] Todd's paralysis may follow prolonged seizures.

The somatosensory aura is quite common but usually is missed due to the young age of the patients. Pure sensory seizures are rare and can delay diagnosis for a long time. In rare cases, atypical seizures may occur with ictal abdominal pains, blindness, vertigo, flashing lights and absence seizures.

EEG findings
The disorder is characterized by a very distinctive EEG pattern. The characteristic interictal EEG correlate is distinct high amplitude, usually diphasic spike with a prominent following slow wave. The spikes (<70 msec) or sharp waves (<200 msec) appear singly or in groups in the midtemporal and central (Rolandic) region. When unilateral, they are always synchronous in the central and midtemporal areas, although they are sometimes of different amplitudes. When bilaterally asynchronous spikes occur, both the rate of firing and amplitude varies from side to side. In a minority of patients, amplitude tends to be more prominent during sleep. In 60% of the patients the focus is unilateral. In the remaining 40%, we can find bilateral foci which shift from side to side. Most of the time no other focal abnormalities, such as focal slowing, occurs. However, when the spikes occur frequently there may appear to be focal slowing in the region of the spikes. This 'pseudo slowing' is secondary to the slow waves accompanying the spikes. Topographic EEG investigations with instant voltage mapping showed maximal negativity of Rolandic spikes over the central or midtemporal electrodes with spread to parietal or upper frontal areas with a horizontal dipole formation (centro-temporal negativity, frontal positivity). Spike amplitude or duration was not correlated with spread adjacent areas.[8] The generator of the spikes is likely situated in the lower rolandic region where the zero potential zone existed, between the centro-temporal negativity and frontal positivity. Some records show generalized spike-wave discharges without any concomitant clinical sings of absence seizures.[9] There is a weak correlation between the frequency or location of the spikes and frequency, length and duration of the seizures. While multiple spikes are unusual, some patients with the typical clinical course of BECT have multiple spike discharges. Conversely, some patients with typical Rolandic spikes have medically intractable seizures. Spike frequency varies significantly from patient to patient and may vary, in the same patient, from record to record. Not all patients with Rolandic spikes or sharp waves will have seizures. Centro-temporal spikes may occur also in: Rett syndrome, Gilles de la Tourette syndrome and Fragile - X syndrome. It is, however, not difficult to differentiate between these syndrome and BECT.

Various authors have used different terms to describe the EEG findings. American authors have labelled these spike discharges as 'midtemporal' while the French preferred the term 'Rolandic spikes'. The most appropriate name centro-temporal spikes was proposed by Blom et al. in 1972[10], since discharges occur in both areas and the site of origin is probably the lower rolandic cortex. Furthermore, it avoids confusion with temporal lobe epilepsy, in which spike discharges are recorded in the anterior temporal area and are of different configuration.

Pathophysiology
The clinical symptomatology of BECT strongly suggest that the source of disturbance lies in the lower rolandic cortex representing the face and oropharynx. The alteration in lateral predominance of the foci in the evolution of BECT, the coexistence or alternation with occipital spike and /or generalized discharges, as well as their evanescence within several years, coupled with the lack of evidence for neurological deficits in these children, all this lends further support to the concept of an idiopathic type of partial epilepsy lacking an anatomical substrate, much in contrast to the majority of the partial epilepsies seen in adults. Hormones influence brain function throughout life and may also affect the seizure threshold by altering neuronal excitability.[11] It was proven experimentally that estrogen enhances and progesterone diminishes neuronal excitability. Hormonal effects on the CNS depend on brain region in which the hormones acts. Steroid hormones act on the hypothalamus and limbic cortex, providing a mechanism for modulating behaviour and endocrine function. Seizure patterns may change at certain life stages. Perhaps as a result of alternations in hormones at puberty BECT may remit.

Genetic factors
While a family history of epilepsy is higher than expected in BECT, the reported incidence varies considerably, i.e. from 9% to 59%.[2,10] In addition, a high percentage of relatives of patients with BECT will have abnormal EEGs. There is some evidence that centro-temporal spike discharges are transmitted by an autosomal dominant gene with a low penetrance and marked age dependency and very low penetrance after 20 years of age. Degen and Degen[12] performed waking and sleep EEGs in 69 siblings of 43 patients. Five siblings had history of seizures. Epileptiform activity was recorded in 38% of all siblings. In children between 5-12 years epileptiform activity was observed in 54%. compared to 33% in younger children and 23% in older. A patient carrying a gene for BECT may never a have clinical seizure but can have an abnormal EEG.

Diagnosis
A correct diagnosis is essential for correct management. As the condition is benign, it is possible to avoid unnecessary diagnostic procedures. In the rare cases in which structural pathology is found, it is not causally related to the seizure disorder and hence is non-contributory to management or prognosis of the epilepsy. It is easy to differentiate BECT from complex

partial seizures (temporal lobe epilepsy). There is no aura, consciousness usually is preserved and there is no evidence of automatism or psychic phenomena. In case of secondary generalized convulsions, typical EEG findings may help to make the correct diagnosis. Sequential topographic mapping EEG may differentiate epileptic from non-epileptic rolandic spikes. The topographic pattern of stationary potential fields is morphologically represented by a single spike and wave complex whereas that of non-stationary potential fields is morphologically represented by a double spike and wave complex.[13] BECT may coexist with benign occipital epilepsy in the same child.[2] The main clinical features include: early onset (2.5-6 years), occurrence of several types of seizures such as partial motor fits, atypical absence and myoclonic-astatic seizures. Waking EEG shows focal discharges but sleep tracing shows an almost continuous diffuse slow spike-wave activity. Prognosis of this atypical combined syndrome is good and the course is favourable.

Drug therapy
Although BECT results in spontaneous remission with or without treatment, some patients have many seizures for several years prior to resolution. Antiepileptic medication is advisable after the second seizure. Focal seizures, a short interval between the first and the second seizure and an early onset appear to forecast repeated seizures and a necessity to treat. The drug of choice to this syndrome is carbamazepine which has excellent efficacy combined with minimal side effects. Valproic acid also can be effective. Discontinuation of medication can be considered after a 1-2 years seizure-free period, even before EEG normalizes. Most cases respond well to low doses of medication. Prognosis is good even in children who do not respond well to low levels of monotherapy. As BECT does not exist in patients older than 16 years of age it is not necessary to continue treatment after this age.

Prognosis
The prognosis of BECT is excellent with the vast majority of children going into remission by the teenage years.[14,15] School and family problems occur in some patients during the acute stage of the disease, the social adaptability of such patients is excellent. Loiseau et al.[5] found that 165 of 168 patients with BECT followed from 7 - 30 years after remission were seizure free. Only 3 of their patients suffered from generalized tonic-clonic seizures after recovery from BECT.

Benign epilepsy of childhood with occipital paroxysms

Forty years ago (1950) benign epilepsy with occipital paroxysms (BEOP) was suggested by Gastaut, who described an epilepsy with ictal symptoms and interictal occipital rhythmic spike waves, appearing only after eye closure. This epilepsy is age related and no occipital cortical lesion can be found. Gibbs and Gibbs in 1952 described seizure foci in occipital lobes which tend to disappear with age in young epileptic children. This syndrome has not been acknowledged before. This is probably due to the visual phenomena, that may be poorly described by younger children; by the disappearance of the occipital spike-wave at eye opening preventing detection in routine EEG recording where eye opening may occur once or twice for a few seconds only. The incidence of BEOP is still to be determined. Of 418 children with onset of epilepsy before age 13 years 4.3% had BEOP.[15] Girls are more often affected than boys.

Clinical features
All characteristics of the syndrome can be found in about 55% of the patients with BEOP. In 25%, the patients either lack the ictal or visual symptoms or the interictal occipital EEG abnormalities.[17,18] The age of onset ranges from 15 month to 17 years with a peak age of onset between 5 - 7 years.[16] Both the clinical manifestations and the seizure frequency are variable, depending on whether they are nocturnal or diurnal.[1]

In nocturnal seizures motor symptoms predominate whereas in diurnal seizures visual symptoms are most common. Nocturnal seizures are more common in younger age and have a good prognosis.

Seizures starting after the age of 8 years are more likely to be frequent and diurnal and continue for longer periods of time. The visual symptoms consist of amaurosis, elementary visual hallucinations - i.e. phosphenes, complex visual hallucinations, or visual illusions including micropsia, metamorphopsia or palinopsia. There is a wide discrepancy in the reports on the visual phenomena. Gastaut[17,18] describes ictal visual symptomatology as common, while Beaumanoir[19,20] found that only 44% of the patients had visual symptoms. Panayiotopoulos[16] found that visual symptoms are rare in children with age onset below 8 years. It is possible that these discrepancies are due to 'under-reports' of visual phenomenon in younger age groups. Motor activity in this disorder includes hemiclonic seizures (in 43% of the cases); complex partial seizures with automatisms (in 14% of the cases); generalized tonic clonic seizures (in 13% of the cases); dysphasia, dysesthesia, adversive seizures (in 25% of the cases). In 15% of all new cases the onset was stormy and alarming. The first seizure was characterized by prolonged loss of consciousness lasting up to 12 hours. Typically children with BEOP have a normal neurological and developmental history.

EEG findings
The interictal EEG is characterized by normal background and very distinct occipital discharges. The occipital discharges are typically high in voltage (200-300 micV), diphasic, with a main negative peak followed by a relatively small positive peak and negative slow wave. The spike component is higher in amplitude than the slow wave, often exceeding 100

micV and a duration of less than 70 msec. The amplitude is highest over the occipital and postero-temporal regions and can be unilateral or bilateral. When occurring bilaterally the spikes are frequently asymmetrical.[16] Location of the spikes may change over time and migrate from the occipital to the centrotemporal region. The spikes usually appear rhythmically at a frequency from 1-3 Hz in bursts or trains. In most of the cases, the spikes disappear with eye opening, the spikes reappear 1-20 sec after eye closure. Occipital spikes can be activated by darkness as opposed to eye closure per se.

Genetics
In Gastaut's series[18] a family history of epilepsy was established in 37% of all cases, migraine in 16%, febrile seizure in 14%. 40% of the patients may also develop migraine. In other reports no positive family history was found.[16] Kuzniecky and Rosenblat[21] studied 25 relatives of a proband of 4 siblings; 3 had BEOP and the fourth had occipital spike-waves in the EEG. Occipital spike-wave was found in 26% of EEGs performed. All of the occipital spikes were found in children. Based on these findings the authors suggest an autosomal dominant pattern for the EEG abnormalities with age dependent expression and variable penetrance of the seizure disorder.

Pathophysiology
The anatomical connections of the occipital lobe with the temporal lobe and central regions may explain why some of the children develop generalized tonic-clonic, partial motor seizures or complex partial seizures following the visual phenomenon. In many reports migraine was found to precede the seizures or to act as a stressor, leading to seizures. It is possible that in some patients the visual aura of migraine triggers the epileptic discharge which than continues autonomously.[1]

Therapy and prognosis
Prognosis is generally favourable with complete seizure control achieved in 60% of patients.[17,18] Children with an early onset (< 8 years), had an excellent prognosis.[16] BEOP remits in most of the children by the age of 16 years. Gastaut[17,18] found no patients with typical seizures persisting past adolescence. Other seizure types occur in adulthood in only 5%. Drug therapy is similar to BECT.

Benign partial epilepsy with affective symptoms (Benign Psychomotor Epilepsy)

Epileptic seizures with predominant affective symptoms, particularly an expression of terror, are complex partial seizures (CPS) usually attributed to temporal lobe epilepsies. The onset of partial epilepsy with CPS in a child generally suggests a poor prognosis: mental retardation, behavioral disturbances are common in this form of epilepsy. A lesional etiology is suspected in most cases and refractoriness to drug therapy is considered a frequent event. In 1980 Dalla Bernadina et al. reported a group of children with complex partial seizures in whom affective symptoms, predominately fear, were the major clinical features. The authors reported 26 children, ranging from 7 - 17 years with the syndrome. As with other benign partial seizures, the children had normal neurological examinations and CT scans. Other researchers reported on partial epilepsy with CPS that had favourable prognosis: single type seizures, early onset, age related disappearance of all clinical features, disappearance of all EEG abnormalities, the existence of familial epileptic antecedents and the lack of any etiology. In 1980 this clinical syndrome was named Benign Psychomotor Epilepsy (BPE).

Clinical features
The predominant feature of the seizure is sudden fright or terror. The terror is expressed by the child in screaming or calling his mother, clinging on her or anyone nearby, or hiding in the corner of the room. These terrorized expressions are sometimes associated with either chewing or swallowing movements, distressed laugh, arrest of speech with glottal noises, moans and salivation, or some kind of autonomic manifestation such as pallor, sweating or abdominal pain. During the seizure some degree of impaired consciousness is noted. The average duration of the seizure is 1-2 minutes (maximum 10 min.). No post-ictal deficit was ever observed. The seizure is frequently followed by sleep. In about 50% of the patients the seizure becomes frequent soon after onset. In the majority of the children, the seizures occur both during waking and sleep with similar semiology. Generalized tonic-clonic seizures were not observed.

Genetics
A family history of epilepsy was elicited in 40% of the patients, febrile seizures occurred in 19% of the children.

EEG findings
The background activity is normal. The most frequent interictal abnormality is a spike followed by slow wave in the fronto-temporal or parieto-temporal area of one or both hemispheres. In some patients we can find rhythmic sharp waves in the fronto-temporal or the parietal or temporal area of one hemisphere. These abnormalities fluctuate during the course of the disorder and are activated by sleep. In about 60% of the cases brief bursts of generalized spike-wave, alone or in association with focal abnormalities, can be seen. The ictal pattern is relatively stereotyped during awakeness or sleep.

Treatment and prognosis
As in other partial epilepsies the course of BPE is favourable. About 80% of the children respond well to anti-epileptic treatment. In most of them remission is achieved during adolescence. The most effective therapy is monotherapy with carbamazepine. It is still very difficult to distinguish between CPS and BPE at onset, but during follow-up BPE has a more favourable prognosis.

Benign partial epilepsy with extreme somato-sensory evoked potentials

In 1% of children without neurological pathology, tapping the feet could elicit high voltage evoked potentials. Such extreme somato-sensory evoked potentials (ESEP) were evident on the EEG involving the parietal and parasagital regions. These spikes occur in children between 1-13 years, with a peak age between 4-6 years of age. A strong male predominance is found (3:1). About 25-30% of these children develop non-febrile seizures. None of these children suffer from organic brain lesion, they have normal development and normal IQ values. History of simple febrile seizure is noted in about 40% of these children.

ESEP and epilepsy
Patients with this syndrome show four distinct periods:
- First period - between the age of 2.5 and 5.5. The only abnormal EEG finding is the presence of ESEP by tapping one or both feet.
- Second period - appearance of spontaneous focal EEG abnormalities during sleep.
- Third period - appearance of spontaneous focal abnormalities in waking. The focal discharges are similar in morphology and topography to ESEP. The time interval between ESEP onset and onset of focal discharges varies between 9 months to 4 years, with a mean interval of 2 years.
- Fourth period - characterized by appearance of electroclinical seizures. Seizures occur with a delay of 5-24 months from the third period, with mean interval of 1 year.

Clinical data
Partial motor seizures occur in about 60% of the children, most of them consist of motor symptoms involving head and upper limbs. Consciousness is impaired in most of the cases. 40% suffer from generalized tonic-clonic seizures. Seizures are rare, between 2-6 times a year. In 90%, seizures occur during waking. In 10% seizures occur both during night and day. In the minority of the patients partial motor status epilepticus is observed. Seizures usually persist for about 1 year and then disappear. The focal interictal discharges and ESEP may persist for years, after the seizures have stopped. In most of the cases, about 3 years after the disappearance

of seizures ESEP are no longer observed.

Treatment
Antiepileptic drug treatment consists mainly of carbamazepine. ACTH is the drug of choice in treatment of partial status epilepticus of this rare syndrome.

Comment
It seems that we are dealing with a particular form of benign epilepsy, not yet fully investigated, which is expressed primarily in partial motor seizures, preceded by ESEPs and associated with interictal focal EEG abnormalities involving mainly the parietal regions. The appearance of ESEPs before the possible occurrence of seizures constitutes an interesting and important EEG-sign, which is worth further exploration. In the absence of cerebral organic lesions, the ESEPs suggest that we deal with an age-related functional phenomenon, with a maximum expression around 4 years of age, with greater prevalence in boys.

Benign partial seizures of adolescence

Epileptic seizures are frequent in adolescence. The average annual incidence rate of epilepsy in adolescents is about 25/100,000 between the ages 10-14 years and 19/100,000 between the ages 15-19 years. These data show that about 25% of all epilepsies occur between the ages 10-19 years. Within this age span, 45% of all epilepsies have a focal onset and about 80% of them are idiopathic. The syndrome of idiopathic benign partial seizure in teenagers was described in 1978.[22]

Clinical data
This syndrome represents 25% of partial seizures with onset between the ages 12-18 years. Most patients (71%) are males. A family history of epilepsy is not very rare (3%). The syndrome peaks between 12-15 years (50% of all cases), and it is very exceptional to find it after the age of 20 years or before the age of 10. Combined signs and symptoms are noted in about 60% of all patients:
- Motor symptoms with or without march (in 95% of the patients)
- Sensory symptoms (80%)
- Jacksonian seizure (7%)
- Somato-sensory symptoms (40%)
- Visual symptoms (20%)
- Psychic symptoms (5%)
- Autonomic symptoms (17%)

Most of the seizures (90%) are diurnal The interictal neurologic and mental status are normal in almost all patients.

Seizure classification
- Simple partial seizures (in 15% of the patients)
- Complex partial seizures (28%)
- Simple partial seizures with secondary generalization (34%)
- Complex partial seizures with secondary generalization (7%)
- Simple partial seizures with evolvement to complex partial seizures with secondary generalization (16%)

EEG findings
A normal interictal EEG or non-specific generalized abnormalities in EEG are mandatory to the diagnosis of this syndrome. Bilateral posterior or diffuse slow waves may be found when the EEG is recorded a few days after a seizure. No typical spike-wave complex is seen in any case, nor any focal abnormality.

Prognosis
Single seizures or clusters of seizures are observed in 80% of the patients. No recurrence is mentioned in the reported cases after 20 years of follow-up.

Reading Epilepsy

The syndrome was first described by Bickford et al. in 1956. They reported two groups of adolescents in whom reading precipitated epileptic seizures. To describe this phenomenon they named it primary reading epilepsy. The commission on classification and terminology of the international league against epilepsy, in 1989, found the syndrome sufficiently established as an idiopathic localization related epilepsy.

Epidemiology
Reading epilepsy (RE) is a rare syndrome. Most publications concern case reports. The epidemiology is unknown. The largest series of 11 cases was collected over a period of 20 years. This syndrome is the only epileptic syndrome which depends on a 'social factor', i.e. reading. The syndrome occurs only in literate individuals and societies. Male predominance is striking. Of the 111 patients reported, 104 of them were males.

Genetics
There seems to be strong genetic component. In about 45% of patient's families, there was at least one relative with epilepsy. Most of them were first degree relatives. In 10% of all the cases there was another family member with RE (including two known cases of identical twins).

Clinical data
The age of onset is a most interesting feature because clinical

manifestation does not occur in early school age when reading skills are acquired but in adolescence. The mean age of onset is 17.7 years with an age range of 12-25 years. These observations indicate that the mechanism of RE is not related primarily to reading but rather to the biological maturation of a subsystem which becomes epileptogenic and happens to be involved in reading. The most striking feature of RE is its clinical uniformity which leaves no doubt that we are dealing with a syndrome. The most constant symptoms of the seizures consist of:

- Abnormal sensations or movements which develop in clear consciousness after a certain amount of reading which involve tongue, throat, jaws, lips, and face. These sensations include stiffness, numbness, tightness, clicking sensation, stammering. Most of the movements observed were myoclonic and rarely tonic.
- Speech motor symptomatology can be followed by generalized tonic-clonic seizures (GTCS). Most of patients avoid having secondary generalization by learning to avoid reading when partial seizure sensations starts. Some of the patients gave up reading because GTCS would develop too rapidly.
- A minority of the patients describe some kind of visual or ocular manifestations (including hallucinations) in addition to the motor manifestations.
- Feelings of discomfort and anxiety are ictal sensations described by many patients.

The length of the material to read and its level of complexity are important factors precipitating seizures. In many of these patients, it had been pointed out that the provocation of epileptic events was directly connected with words or passages where difficulties of transforming or transcoding script into speech became manifest. Typically, paroxysmal EEG discharges and clinical seizures are connected with misreading.

It appears that there is some kind of feedback loop between misreading that provoke seizure and spike discharge during reading causing misreading. So far, CT scans and MRI have never revealed any findings suggesting a symptomatic origin.

EEG findings
Routine interictal EEG will show paroxysmal findings only in 5% of the patients. Photic stimulation is negative in 91% of the patients. Reading provocation may be negative in 16% of the patients. In the ictal EEG we can find bilateral symmetric activity in 32% of the patients, bilateral nonsymmetric activity in 38% of the patients and focal paroxysmal activity in 30% of the patients.

Therapy and prognosis
Most of the known patients decide not to take any antiepileptic drug but to prevent GTCS by reacting to the partial seizure by stopping to read. This

is possible when there is no rapid progress to a convulsive seizure. In most of the patient this warning mechanism is sufficient for everyday reading. Behavioral therapy may be successful in some patients, but it does not stop the seizures in all patients. Due to minimal side effects and good results valproate is considered to be the drug of choice in the treatment of RE. In cases unresponding to valproate, clonazepam is reported to be effective. Prognosis is good. In about 30% of the patients, the seizures last for about 10 years but none of them had spontaneous seizures. In the majority (70%) remission is reported to be achieved in adulthood.

IDIOPATHIC GENERALIZED EPILEPSIES

According to the ICE, generalized epilepsies and syndromes are epileptic disorders with generalized seizures in which the first clinical changes indicate initial involvement of both hemispheres. The ictal electroencephalographic patterns initially are bilateral. Idiopathic generalized epilepsies are forms of generalized epilepsies in which all seizures are initially generalized, with an EEG expression that shows generalized, bilateral, symmetrical discharges. The patient usually has a normal interictal state, without neurological or neuroradiological pathology. In general, interictal EEGs show normal background activity and generalized discharges, such as spikes, polyspikes, spike-waves and polyspike waves. The discharges are increased by slow wave sleep. The various syndromes of idiopathic generalized epilepsies differ mainly in age of onset. The International Classification of the Epilepsies and Epileptic Syndromes recognizes the following idiopathic generalized epilepsies:
- Benign neonatal familial convulsions
- Benign neonatal convulsions
- Benign myoclonic epilepsy in infancy
- Childhood absence epilepsy (pyknolepsy)
- Juvenile absence epilepsy
- Juvenile myoclonic epilepsy
- Epilepsy with GTCS on awakening
- Other generalized idiopathic epilepsies not defined above

Childhood Absence Epilepsy

The first descriptions of absence epilepsy (AE) was given by Poupart in 1707 and, later, by Tissot in 1770. The term 'absence' was quoted for the first time by Calmeil (1824). In 1838 Esquirol suggested the term 'petit mal', in 1916 Sauer introduced the term pyknolepsy. In 1935 Gibbs et al. described petit mal as 'brief interruptions of consciousness, associated with rhythmic 3 c/s discharges of regular spike and wave complexes on the EEG. Childhood absence epilepsy (CAE) was recognized as a separate epileptic syndrome by the Commission on Classification of the ILAE in 1989.

Definition
The term CAE is restricted to epilepsy with the following characteristics:
- Onset before puberty.
- Occurring in previously normal children.
- Absence seizures (AS) as the initial type of seizure.
- Very frequent absence seizures of any kind except for myoclonic absences
- Absence seizures associated in the EEG with bilateral, symmetrical and synchronous discharges of regular 3 c/s spike and wave complexes on a normal background activity. Less regular spike-wave activity is possible, when compatible with a clinical diagnosis of typical absence.

Epidemiology
Childhood absence epilepsy accounts for 2-8% of patients with epilepsy. The variation probably depends largely on the mode and source of case collection (hospitalized patients or population based studies, and the mean age of the selected cases). The annual incidence of CAE has been estimated at 6.3/100,000.[23] CAE is clearly more frequent in girls than in boys, 60 to 76% of the affected children being girls. CAE typically has its onset between the ages of 4 and 8, but the age of onset is not strictly limited. It is rare to find a typical absence before the age of 3 years, although a case report of a 6 month old baby with typical AS has been described. About 20% of all patients have febrile seizures prior to the onset of AS.

Genetics
A positive family history of epilepsy is found in about 30% of the patients. About 10% of brothers and sisters to children having AS will suffer from epilepsy. Febrile seizures are frequent in siblings of these children. The risk of epilepsy in offspring of subjects having CAE was estimated to be 6 - 8%.[24] The responsible gene of CAE is probably located on the short arm of chromosome 6.[25,26] It was shown that CAE can be acquired: Perinatal complications, head trauma and cerebral inflammatory disease are found in the case histories of 7 - 30% of the patients. The initiation of CAE may thus depend on a combination of an autosomal dominant gene with age dependent penetrance with multifactorial environmental factors.

Seizure description
AS are characterized by the following features: short duration, abrupt onset and termination, impairment of consciousness and a high frequency. Absence duration ranges from 5-30 sec. As a rule, the onset of AS is sudden. In very rare cases a warning sensation is felt before the absence.

A retrograde amnesia lasts from 4 to 15 sec while the patient is fully awake and has a normal EEG. This amnesia is partly reversible.

The seizure ends as abruptly as it begins and the patient carries on with his activity as if nothing had happened. It takes however few seconds before returning to normal behaviour. The child is often unaware of the seizure. A stimulus such as pain, or a call can shorten the seizure. The essential features of AS are loss of awareness and responsiveness with cessation of on-going activities. The patient stops talking, eating, walking, he remains motionless, with vacant eyes, staring straight ahead or upwards. Breathing continues normally or slows - in long lasting attacks. The International Classification of Epileptic Seizures in 1981 distinguished between 6 variants of AS:

- Simple absence. With only impairment in consciousness. This type occurs in only 10% of all cases.
- AS with a mild clonic component. Typically, clonic movements are restricted to eyelids, blinking at a rhythm of 3 c/s, rarely it involves lips and chin. Jerks of the head, arm and shoulders have also been described as a clonic component. The clonic component is very common (in over 50% of the patients).
- AS with atonic components. A diminution of muscle tone resulting in a gradual lowering of the head and/or arms, often rhythmic, mixed with clonic jerks. In these cases, the patient may drop what he is holding in his hands. Absences with a sudden fall are usually atypical absences. The atonic component is present in about 20% of AS. In most cases it is combined with other components.
- AS with a tonic component. An increase in extensor muscle tone is the characteristic feature in this seizure type. In most cases this is limited to the eyes, which rotate upward, while the head may draw backward. Tonic components can be found mixed with mild clonic component.
- AS with automatisms. Two categories of ictal automatism exists:
    a) preservative. The patient persists in what he is doing (eating, walking). These activities can be correctly done or distorted (pouring water into full glass).
    b) de novo - in the majority of the cases they can be simple as: lip licking, swallowing, face rubbing, scratching, or they can be more complex: catching objects, grunting, mumbling, humming.
  Automatisms as a sole component or associated with other components are very frequent, occurring in more than 60% of the patients. The frequency of these automatic components, is related to the duration of the AS. The frequency is ranging from 22% in a 3 sec seizures to 95% in seizure lasting more than 16 sec.
- AS with autonomic components - These components are easily recognized and include urinary incontinence, frequent pupilary dilatation, pallor, flushing, tachycardia, change in blood pressure.

The seizures are usually very frequent during a day. In most of the cases the parents note short episodes of immobility or eye rolling, but pass it off as daydreaming or emotional reactions. In time, however, the blackout periods increase in frequency or in duration and cannot be disregarded any longer. Because of the minimal clinical manifestations, the evaluation of the real frequency of the seizures, prolonged EEG monitoring is needed. AS can occur spontaneously, but can also be precipitated by emotions (anger, sorrow, fear), metabolic disturbance (hypoglycaemia, hyperventilation) or time of the day (evening, on awakening). Absence status (petit mal status) is very rare under the age of 10 years and can be found in about 10 to 15% of the patients with CAE. The event starts with acute confusional state and mild to moderate disturbances in daily functioning. Clonic components or automatisms can accompany the confusional state. The status may last from hours to months.

EEG findings
AS are associated with a bilaterally synchronous and symmetrical discharges of rhythmic spike and wave complexes. The paroxysmal activity occurs bilaterally, begins abruptly and synchronously in both hemispheres. The termination of the spike and wave sequences are less abrupt than the onset. It takes few seconds to the slow waves to decelerate and decrease in amplitude before merging with the normal background. The frequency of the spike-wave complexes is 3 c/s at the beginning of the event and slows down to 2 - 2.5 c/s towards the end. More irregular spike-wave discharges as polyspike waves do not change the diagnosis of CAE, as long as the other components of the syndrome exist. The interictal EEG background activity is usually normal, although mild background abnormalities can occur.

Treatment
Both ethosuximide and valproic acid suppress AS in 80% of the cases. As it is important to preserve cognitive skills, valproic acid is the drug of choice in the treatment of AS because of its low neurotoxicity. Valproic acid is also active against tonic-clonic seizures which some of the patients develop, shortly after AS onset. When valproic acid does not control AS, Ethosuximide is the drug of second choice. If it fails as sole drug treatment, the combination of Valproic acid and Ethosuximide is recommended.

Prognosis
A wide range of remission rates have been given, ranging from 33% to 80%. It seems that a longer follow-up renders a lower percentage of controlled patients. This is partly due to the late onset of tonic-clonic seizures which may occur many years after the remission of absence seizures. Thus control of AS is not a guaranty for good prognosis. Only in 6% of the patients, AS persists over the age of 14 years. The seizures tend

to be short and less frequent. Tonic clonic seizures (GTCS) occur late in the course of CAE in about 40% of the patients. The seizures are frequent and begin in most cases between 10 and 15 years of age. GTCS may occur in some patients 10 to 20 years after the cessation of AS. Behavioral problems occur in about 30% of the patients with CAE and may be a consequence of frequent seizures. Psychomotor slowing was found in chronic CAE. About half of the children suffering from AS will have also GTCS. This illustrates that the prognosis of CAE partly depends on the further occurrence of GTCS. Patients with GTCS will have epileptic seizures during a longer period of their life; beyond childhood and adolescence.

Juvenile Absence Epilepsy

After CAE was established as an epileptic syndrome, more attention was given to patients who had absence seizures with the same clinical appearance as in CAE, with identical EEG, but with low seizure frequency.

Epidemiology and clinical data
The epidemiology of this disorder has not been well studied, but most studies report that about 20% have an onset after the age of 10 years. It is claimed that JAE is still underdiagnosed, most probably due to tonic-clonic seizures that cause AS to be overlooked. No sex prevalence was found between males and females. The age of onset ranges from 10-17 years with a peak of onset near the age of 12 years.

Seizures description
Absences occur -per definition- in all cases. Unlike the multiple clustered pattern in CAE, which may involve dozens or hundreds of seizures per day, the absences in JAE occur relatively infrequent, with only one or a few episodes daily. AS in JAE are similar to those in CAE, except that consciousness is not as severely impaired as in the latter. Tonic-clonic seizures are more frequent in JAE than in CAE, occurring in about 80% of the patients and in most cases are the presenting symptoms of this syndrome. Tonic clonic seizures may occur shortly after waking or they may appear randomly during the day. Absence status is common and can be found in about 40% of the patients. Without specific history of JAE this kind of status can be easily overlooked. Myoclonic seizures occur in 15% of the patients. The clinical overlap of JAE and JME suggests that these two phenotypic syndromes may share the same genetic determinants.

Treatment and prognosis
Both ethosuximide and valproic acid control AS in 80% of the cases. The long term prognosis of JAE has not been properly evaluated. Education

about the disorder, attention to drug compliance, avoidance of sleep deprivation and abstinence from alcohol and drugs are essential components of the management in the adolescent patient.

Juvenile Myoclonic Epilepsy

Juvenile myoclonic epilepsy (JME) is a syndrome of idiopathic generalized epilepsy with age related onset. It is characterized by seizures with bilateral, single or repetitive arrhythmic, irregular myoclonic jerks, predominantly in the arms. By the second century AD, Galen had recognized the sudden, brisk muscular contractions that characterize some epileptic syndromes. In 1822, Pritchard called myoclonias 'motor auras' and assigned them a spinal cord origin. The first extensive report of a patient with the syndrome was by Herpin in 1867, while Janz and Christian gave the first detailed description that was based on the study of a substantial group of patients and suggested the term 'impulsive petit mal'. In the 1970s, Delgado-Escueta et al. named such familial epilepsies 'benign myoclonic and tonic-clonic seizures in adolescence and late childhood'. In 1980, these authors found that this syndrome and Janz's 'impulsive petit mal' were the same disease. Delgado-Escueta et al. described the excellent response of seizures to valproic acid and the lifetime persistance of seizures. In appreciation of Janz's earlier work, Delgado-Escueta called the syndrome 'juvenile myoclonic epilepsy of Janz'.

Prevalence
JME is estimated to account for about 10% of all epilepsies (7-9%). JME may be an underdiagnosed syndrome. In many patients, myoclonias are recognized only after repeated interviews or documentation of seizures on closed-circuit television and CCTV - EEG recordings. Many of the patients are referred to the clinics due to primary tonic-clonic seizures. Sex distribution for JME is about equal.[27] Obeid and Panayiotopopulos[27] reported an average age at onset of 15.5 years, with 76% manifesting between the ages of 12 and 18 years. Similar findings were reported in other studies.

Clinical features
The characteristic features of the syndrome are the myoclonic jerks that are bilateral, mostly but not always symmetric, single or repetitive, rapid and of variable amplitude. They are often restricted to the arms, where they seem to affect the extensor muscles. When they are multiple, they are usually arrhythmic. In some cases jerks are reported to be extended to the legs, causing the patient to fall down. In most off these cases falling down was well remembered by the patients and consciousness was retained during the seizures. The seizures are precipitated by sleep withdrawal and mostly occur shortly after awakening. Reports show that about 50% of the

seizures occur after morning awakening, 20% after nocturnal awakening, 15% in the evening relaxation and the rest at sleep onset. Some patients (5-10% of the cases), report seizure precipitation by intermittent light stimuli. Physical examination fails to reveal gross pathological findings. The most consistent findings are a significant increase in skull diameter and frequent calvarial hyperosteoses. Major mental defects are uncommon, but many of these patients suffer from a rather immature personality that often results in inadequate social adjustments. Sleep habits contribute in many patients to seizure precipitation, and dissimulation may prove an important obstacle to compliance and successful therapy. Most of the patients who come for treatment (95%), also have generalized tonic-clonic seizures (GTCS). GTCS of these patients are often preceded by series of jerks. The other seizure type which may be associated with JME is the absence seizure. About 10-20% of the patients have absence seizures of the juvenile type with infrequent recurrence, about 5% of the patients have AS combined with the CAE syndrome.

EEG findings
The EEG that correlates with the myoclonic jerks is a burst of medium to high amplitude 10-16 Hz polyspikes followed by irregular 1-3 Hz slow waves of different amplitude. The bilateral, symmetric polyspikes have a maximum amplitude in the frontocentral region and are much less synchronized and regular than the classic 3 Hz spike and wave complexes seen in childhood absence epilepsy. A diffuse discharge of irregular 2-5 Hz spike and wave complexes may precede the multispike discharge. The spikes have negative polarity and may reach amplitudes of 150-350 micV. The slow waves have an amplitude of 200-300 micV. Intermittent photic stimulation may induce the myoclonic jerks. In clonic-tonic-clonic seizures, myoclonic jerks occur during 10-16 Hz rapid spikes.

Genetics
JME became the first of the major epilepsy syndromes where a gene could be located. The first to realize that JME is a genetic based disorder was Janz who found that 25% of his patients had a positive family history of epilepsy. The seizure type of affected relatives are generalized tonic-clonic in 85% of the cases. The most important progress in JME genetics, however, was brought about by linkage studies. The provisional localization of a gene involved in the syndrome to the p 21.3 area of chromosome 6 is meanwhile confirmed.[25,28,29] At present, two important questions concerning the genetics of JME are still unanswered. First, the mode of inheritance remains unsolved. Whereas some present data seem to favor autosomal dominant inheritance, two other possibilities are suggested: a recessive mode of inheritance and a two-locus model with the dominant gene on 6p and another, unidentified, recessive gene. One important problem for these investigations is the question whether individuals without seizures but presenting typical epileptiform EEG

findings should be considered as affected or not. One other question still open is whether the gene located on 6p is specific for JME or a common gene in idiopathic generalized epilepsies. There is still a possibility that the four generalized syndromes: CAE, JME, epilepsy with GTCS on awakening, benign familial neonatal convulsions, have the 6p locus in common. It remains open whether the syndromatic differences are due to a variance in gene expressivity.

Treatment and prognosis
Management of the patient with JME include not only pharmacological treatment of epileptic seizures, but also control of precipitating factors and prevention of accidents and injuries during GTCS. Patients with JME must be made aware of precipitating factors such as alcohol, sleep deprivation and drug intake. A regular sleep cycle of 8-10 hours is advised and should be maintained regularly. Due to reports of seizure relapse during pregnancy in this syndrome, female patients should be advised during pregnancy about the type of drug treatment and about labour management. Valproic acid is the drug of choice and is effective in approximately 90% of the patients. Seizures invariably return after valproic acid withdrawal. Lifetime treatment is usually required. The reason in most cases for seizure relapse is sleep deprivation, non compliance with the drug regimen, stress and alcohol consumption. If monotherapy valproic acid is not effective, carbamazepine should be added. The prognosis is favorable in the majority of patients.

Epilepsy with Grand Mal on Awakening

The relationship between circadian rhythm and grand mal seizures (generalized tonic clonic seizures: GTCS) was already noticed by Gowers in 1885. Janz in 1953 published the first paper that was dedicated to what he called 'awakening epilepsy'. He found that many of these patients had a second seizure peak in the late afternoon or evening hours on relaxation. In a small group, the evening peak was the main or the exclusive seizure peak. He found that patients with long lasting epilepsy tend to start having seizures during sleep. The patients studied by Janz had not been investigated with EEG. Loiseau, in 1964, concluded that seizures of different kinds could be related to the awakening situation but there was still one syndrome of awakening epilepsy, grouped around GTCS of this biorhythmic peculiarity, sometimes associated with myoclonic seizures absences, and with generalized epileptic EEG discharges.

Epidemiology
The reported prevalence rates vary considerably. The reasons for the differences between the reports are not yet clear but they may have been caused by the changes in diagnostic methods. In all the reports a slight male preponderance was found. The age range of clinical manifestation of

this syndrome is wider than in other epileptic syndromes. In about 80% of the cases GMA became manifest between the ages 5-25 years.
In all studies, there is a clear peak age of onset around puberty.

Clinical manifestations
GTCS on awakening is the characteristic feature of this syndrome. Many patients have, in addition, minor generalized seizures, either absence or myoclonic or both. About 40% of the patients have GTCS only, 20% have GTCS + JME, 30% have GTCS + AS and 10% have GTCS + Focal seizures. It is not uncommon that the individual GTCS seizure is preceded by series of absence or a volley of bilateral jerks in clear consciousness. New technologies as 24 hours video monitoring combined with continuous EEG recording assisted to verify that epilepsy with GTCS's without preceding absence or myoclonic seizures are rare.

Clinical data
The definition of GMA requires that, for this diagnosis, a clear majority of seizures must occur either in the first two hours after awakening (regardless of the time of day), or in the second seizure peak, during the relaxation phase. In order to firmly establish the chronobiology, a certain number of seizures must have been observed. With the advances of antiepileptic therapy and the growing awareness of the importance of rapid seizure control, it is rare to see the natural history of any convulsive disorder. Most of the patients will receive effective therapy early enough to have their seizures controlled before chronobiology can definitely be determined. As in other generalized idiopathic epilepsies, the presence of gross physical or mental abnormalities is rare. In some reports patients with GMA were taller than the control groups. These patients tend to suffer from vasomotor lability (acrocyanosis, intense dermographism) and a relatively early onset of puberty. These patients are also reported to have a typical personality. They are unstable, unreliable and always ready to follow the slightest temptation even against their better judgment. Often they have little ability to suppress, to contradict and renounce. Conflicts, tensions and disinclinations are usually momentarily disposed of by denial. Many of these patients have little or no discipline which is often an obstacle to successful therapy. Seizures, as in JME, are precipitated by sleep withdrawal and factors such as excessive alcohol intake.

EEG findings
The most frequent findings in GMA are increased slow waves, disorganized background activity with sleep transients and generalized spike-wave activity. In the minority of the patients spike-wave develop during hyperventilation. Focal abnormalities were extremely rare. GMA is one of the epileptic syndromes that are related to photosensitivity. In about 60% of patients suffering from GMA photosensitivity is very prominent in EEG. In women this phenomenon is twice as common than in males.

Therapy and prognosis

As in JME, avoidance of precipitating factors is as important for therapy as antiepileptic medications. Valproic acid is very effective. In resistant cases combination therapy with carbamazepine is effective. The risk for relapse after drug withdrawal is very high, even in seizure-free patients (85%).

REFERENCES

1. Fejerman N, Di-Blasi AM. Status epilepticus of benign partial epilepsies in children: report of two cases. Epilepsia, 1987; 28(4): 351-355.
2. Lerman P, Kivity, S. The benign partial nonrolandic epilepsies. J. Clin. Neurophysiol., 1991; 8(3): 275-287.
3. Holmes GL. Benign focal epilepsies of childhood. Epilepsia, 1993; 34(S3): S49-61.
4. Heijbel, et al. Benign epilepsy of children with centro-temporal EEG foci. A study of incidence rate in outpatient care. Epilepsia, 1975; 16: 657-664.
5. Loiseau P, Duche, B. Rolandic paroxysmal epilepsy or partial benign epilepsy in children. Rev. Prat., 1988; 38(18): 1194-1196.
6. Loiseau P, Duche, B. Benign rolandic epilepsy. Adv. Neurol., 1992; 57: 411-417.
7. Aicardi J. Epileptic syndromes in childhood. Epilepsia, 1988; 29(S3): S1-S5.
8. Lischka A, Graf M. Benign rolandic epilepsy of childhood: topographic EEG analysis. ustria. Epilepsy Res., 1992; 6: 53-58.
9. Dalla Bernardina B, Sgro V, Caraballo R, et al. Sleep and benign partial epilepsies of childhood: EEG and evoked potentials study. Epilepsy Res., 1991; 2: 83-96.
10. Blom, et al. Benign epilepsy of children with centro-temporal EEG foci: prevalence and follow-up study of 40 patients. Epilepsia, 1972; 13: 609-619.
11. Morrell MJ. Hormones and epilepsy through the lifetime. Epilepsia, 1992; 33 Suppl. 4: S49-S61.
12. Degen R, Degen HE. Some genetic aspects of rolandic epilepsy: waking and sleep EEGs in sibling. Epilepsia, 1990; 31(6): 795-801.
13. Van der Meij W, Van Huffelen AC, Willemse J, et al. Rolandic spikes in the inter-ictal EEG of children: contribution to diagnosis, classification and prognosis of epilepsy. Dev. Med. Child Neurol., 1992; 34(10): 893-903.
14. Deonna T, Ziegler AL, Despland PA, et al. Partial epilepsy in neurologically normal children: clinical syndromes and prognosis. Epilepsia, 1986; 27(3): 241-247.
15. Shinnar S, Berg AT, Moshe SL, et al. Discontinuing antiepileptic drugs in children with epilepsy: a prospective study. Ann. Neurol., 1994; 35(5): 534-545.
16. Panayiotopoulos CP. Benign nocturnal childhood occipital epilepsy: a new syndrome with nocturnal seizures, tonic deviation of the eyes, and vomiting. J. Child Neurol., 1989; 4(1): 43-49.
17. Gastaut H. A new type of epilepsy: benign partial epilepsy of childhood with occipital spike-waves. Clin. Electroencephalogr., 1982; 13(1): 13-22.
18. Gastaut H. Benign spike-wave occipital epilepsy in children Rev. Electroencephalogr. Neurophysiol. Clin., 1982; 12(3): 179-201.

19. Beaumanoir A. Infantile epilepsy with occipital focus and good prognosis. Eur. Neurol. 1983; 22(1): 43-52.
20. Beaumanoir A, Thomas P. Benign epilepsy of childhood with occipital paroxysms. Epilepsy Res., 1992; 6: 105-109.
21. Kuzniecky R, Rosenblatt B. Benign occipital epilepsy: a family study. Epilepsia, 1987; 28(4): 346-350.
22. Loiseau P, Pestre M. Benign epilepsies in adolescence. Presse Med., 1987; 16(37): 1823-1824.
23. Olsson I, Hagberg G. Epidemiology of absence epilepsy; clinical aspects. Acta Paediatr. Scand., 1991; 80(11): 1066-1072.
24. Janz D, Beck Mannagetta G, Sander T. Do idiopathic generalized epilepsies share a common susceptibility gene? Neurology, 1992; 42 Suppl. 5: 48-55.
25. Greenberg DA, Delgado-Escueta AV. The chromosome 6p epilepsy locus: exploring mode of inheritance and heterogeneity through linkage analysis. Epilepsia, 1993; 34 Suppl. 3: S12-S8.
26. Whitehouse WP, Rees M, Curtis D, et al. Linkage analysis of idiopathic generalized epilepsy (IGE) and marker loci on chromosome 6p in families of patients with juvenile myoclonic epilepsy: no evidence for an epilepsy locus in the HLA region. Am. J. Hum. Genet., 1993; 53(3): 652-662.
27. Obeid T, Panayiotopoulos CP. Clonazepam in juvenile myoclonic epilepsy. Epilepsia, 1989; 30(5): 603-606.
28. Weissbecker KA, Durner M, Janz D, et al. Confirmation of linkage between juvenile myoclonic epilepsy locus and the HLA region of chromosome 6. Am. J. Med. Genet., 1991; 38(1): 32-36.
29. Durner M, Janz D, Zingsem J, et al. Possible association of juvenile myoclonic epilepsy with HLA-DRw6. Epilepsia. 1992; 33(5): 814-816.

6 Neurophysiological aspects of epilepsy in children and adolescents

ANTÓNIO MARTINS DA SILVA

*Department of Neurophysiology - Hospital Santo António - Porto
and Unit of Human Physiology - Biomedical Institute "Abel Salazar" - University of Porto - Porto, Portugal*

The different epilepsy syndromes are often characterized by typical EEG patterns that can help with the diagnosis, with seizure classification and with exploring the relationship between seizures, age and sleep-wake cycle. In this review the EEG characteristics of different age-related and state-related epilepsies will be described with respect to their clinical-EEG characteristics.

INTRODUCTION

Adequate neurophysiological qualification and quantification of seizure events, of interictal epileptiform activity and of background characteristics provide relevant information to a correct diagnosis and accurate seizure classification. This could be used to characterize seizure types, to quantify therapeutical benefits and to specify epilepsy prognosis.

Epileptiform EEG events are events meeting IFSECN standards:[1] spikes, sharp waves, spike or polyspike and waves, runs of slow waves and extreme EEG amplitude changes occurring in patients having seizures or with an history of seizures. They can be recorded at rest or during activation procedures and they do not have clinical involvement. On the other hand, the same events during seizures and myoclonic jerks, occurring and exceeding the photic stimuli are considered ictal events.

EEG PATTERNS IN AGE-RELATED GENERALIZED EPILEPSIES

The EEG in idiopathic generalized epilepsies

In these epilepsies the background EEG is normal and interictal discharges have a common pattern. In both forms of benign neonatal convulsions the EEG is non-specific.[2] The interictal EEG activity may also show an increase of theta (4-5 Hz) frequencies and of sharp waves. Apneic seizures can occur in idiopathic generalized epilepsies, showing impressive EEG patterns that can be followed, later on, by normal maturational EEG patterns. On the contrary, when apneic seizures persist as a consequence of severe encephalopathy, the prognosis is less favourable. Typical examples of the generalized idiopathic age-related epilepsies are most of the myoclonic epilepsies and pyknolepsy. These epilepsies have EEG patterns, activated by non-deep NREM sleep or drowsiness.[3] Three of these epilepsies are less state-dependent: the benign myoclonic epilepsy in infancy, childhood absence epilepsy (pyknolepsy) and juvenile absence epilepsy. The others: juvenile myoclonic epilepsy (JME - or impulsive petit mal) and epilepsy with grand mal seizures (GTCS) on awakening are strongly state-dependent. These epilepsies have a number of interictal EEG events in common: bursts of generalized discharges, spikes, spike-wave (SW) and polyspike-wave (PSW), lasting up to 3 sec. The frequency, duration, and location of the EEG discharges are significantly influenced by sleep. The interictal EEG events have been used to predict the prognosis. The association of transient focal abnormalities, the persistence of posterior delta activity and of brief episodes of SW bursts without clinical expression are frequently associated with an increased risk of generalized tonic-clonic seizures later in life.[4]

Ictal events are successive discharges of SW in absences (figure 1, end of chapter) and/or PSW in myoclonic attacks (figure 2, end of chapter). Typically they have to last more than 3 seconds to give behavioral (ictal) changes. However by using appropriated computerized tests, runs of discharges with less duration can give transitory cognitive impairment.[5,6] The EEG changes in generalized epilepsies, with seizures precipitated by specific modes of activation are well documented. Photosensitivity occurs in about 5% of the patients with epilepsy.[7] The typical pattern is the appearance of generalized bursts of PSW or SW lasting some hundreds of milliseconds after the stimuli. The interictal and ictal EEG during myoclonic jerks have the same pattern.

The EEG in cryptogenic and symptomatic generalized epilepsies

Most of these syndromes have SW patterns. The stability and characteristics of these patterns differ among the syndromes. The spike component changes more than the slow wave component. The syndrome of West shows characteristic EEG features. The malignant form is characterized by complete disorganization of electrogenesis with a total lack of basic rhythms in awake, mixed with hypsarrhythmia[8] and with sharp and slow waves with multiple localizations (figure 3, end of chapter). The hypsarrhythmia can precede or follow the clinical seizure. The paroxysmal activity is clearly influenced by sleep state with a strong reduction of hypsarrhythmia during REM. During NREM the diffuse bursts of PSW and SW and sharp waves, separated by periods of low amplitude background activity, occur more frequently. The ictal EEG events vary. High amplitude slow waves can be followed by diffuse high frequency activity or by a 'flattening' of the EEG activity that follows tonic or myoclonic seizures (see figure 2). Partial seizures commonly show asymmetric interictal background activity in the EEG without return to hypsarrhythmia between the spasms of a cluster.[9] The EEG can be helpful in determining the prognosis in the syndrome of West. The best prognosis is related to a decrease of interictal EEG abnormalities after benzodiazepine administration and reappearance of hypsarrhythmia between consecutive spasms of a cluster.[10] Patients with the Lennox-Gastaut syndrome also have complex and multiple types of seizures with characteristic EEG features. Interictal recordings show abnormal background activity with slowing of EEG rhythms and bursts of generalized SW at low frequency (less than 3 Hz). Slow and sharp waves and multifocal abnormalities are also common (figure 4, end of chapter). Ictal events, characterized by electro-clinical correlation[11] are: bilateral 10-13 Hz rhythm and generalized SW at 3Hz with tonic seizures; 2.5-3 Hz SW in myoclonic-atonic seizures; PSW and slow waves and diffuse spike-waves and fast rhythms with anterior predominance in short-lasting tonic spasm.[11] Atypical absences show different EEG abnormalities: diffuse or irregular SW with or without fragmentation, irregular and diffuse 10-13 Hz fast activity and fast rhythms with increasing amplitude followed by synchronous spike-wave discharges. Consciousness is regained at the end of EEG discharges even when fragmentation or low frequency (<2 Hz) SW discharges occur.[11] Seizures in patients with the Lennox-Gastaut syndrome decrease during REM sleep and rhythmic spike discharges spread to the anterior areas in NREM sleep.[12] Status epilepticus was reported in more than two thirds of patients. During status, the EEG of the Lennox-Gastaut syndrome is 'almost hypsarrhythmic' and is typically the only documented circumstance of Lennox-Gastaut influence on sleep structures. The interictal EEG can show normal background activity during the first period after the onset of myoclonic-astatic seizures although this can change to runs of generalized SW during clinical development.[13,14] Photic stimulation discharges follow the characteristic 4-7 Hz rhythms in parietal areas[14,15] and persist in cases with severe prognosis. Because of the frequent EEG changes in LGS stable epileptiform true foci or foci maintained in a specific region are rare, although spikes lateralization occur frequently. The ictal EEG depends on the seizure type: 2-3 Hz SW generalized and irregular high amplitude spikes disrupted by large amplitude slow waves occur in astatic or myoclonic astatic seizures; generalized PSW or irregular spikes occur in myoclonic seizures. Bursts of 10-15 Hz sharp waves or spikes take place in nocturnal tonic seizures. Status epilepticus is frequent in up to 36% of the patients[14] and is characterized by 2-3 Hz SW or rare PSW and irregular and polymorphic activity resembling hypsarrhythmia. EEG abnormalities in most of the symptomatic and cryptogenic generalized epilepsies are related to the nature and the development of the aetiology, especially the encephalopathy. The EEG pattern is less rhythmical and with more burst suppression low frequency SW and generalized fast rhythms than in idiopathic forms.

EEG PATTERNS IN LOCALIZATION-RELATED EPILEPSIES

The EEG in idiopathic localization-related epilepsies

The EEG has typical features: persistence of normal background activity, normal sleep organization, spikes demonstrating specific localization and morphology: focal (sometimes 'rolandic' sometimes in different places) spikes or sharp waves.

The EEG in the benign epilepsy with centrotemporal spikes (BECTS)
In about 60% of the patients, the typical feature is the presence of focal interictal spikes in the midtemporal or rolandic areas. In the remaining patients the focus is bilateral.[16] SW discharges are generalized (15%) or foci are independent (10%). These variations are without relevance for clinical outcome.[17] Migration to temporal regions (from occipital to midtemporal) can occur though the opposite should not be expected (temporal to occipital region).[18-21] Non-deep NREM sleep is a strong interictal activator (figure 5, end of chapter). 'Rolandic' spikes are high voltage diphasic events recorded in centrotemporal areas. The interictal spike dipole can be positive in surface recording. Atypical forms of BECTS have a more complex distribution of dipoles and a single dipole fits better on typical forms.[22] Identification of this dipole configuration may be useful to differentiate BECTS from focal epilepsy with other etiologies.[23] The relationship between spike morphology and the nature of BECTS is complex, though. Van der Meij et al.[24] reported that the morphological features of rolandic spikes, studied with EEG, CT-scan and MRI in five clinical groups of patients with BECTS appeared to be identical, but not predictive for etiology or prognosis and they do not provide a clue for 'epileptogenicity'. The EEG background activity analysis seems to provide relevant information in these groups of patients, because of the probability to find a cerebral lesion and to define prognosis. The ictal spike in BECTS is a stereotypic dipole with a negative pole in the frontal region and a positive pole in the centrotemporal region.[23] When the seizures persist, the spikes occur more frequently, last longer and progress from mid-temporal areas to the whole ipsilateral hemisphere, bilateralizing thereafter. High frequency and low amplitude spikes characterize seizure onset. Most seizures have been recorded during sleep. Sequences of spikes during the tonic phase and spike-waves during the clonic phase are reported with day-time seizure recording.[16]

The EEG in idiopathic non-rolandic epilepsies
The non-rolandic types of idiopathic localization-related epilepsies (NRE) are also characterized by normal background EEG activity and normal sleep structures.[25] The Landau-Kleffner syndrome and the epilepsy with continuous spike-waves during sleep (ECSWS) are exceptions because of the risk for permanent deficits after recovering. The benign childhood

epilepsy with occipital spikes[26] can also be considered as an exception as the seizures can remain until late adolescence. Interictal epileptiform EEG events are characterized by spikes or spike-waves, more frequently occurring during NREM sleep. Changes in the morphology of the discharges will depend on the type of epilepsy. In the benign occipital epilepsies, the sleep pattern discharges change to polyspike waves. The waking EEG of occipital epilepsies shows high amplitude spike or sharp-waves. PSW recurring rhythmically in posterior scalp areas spreads to the central and the temporal regions during seizures. Changes in EEG pattern during sleep, modification of EEG background and a high prevalence of focal seizures are indicators of a symptomatic origin of occipital seizures,[27] excluding the diagnosis of benign occipital epilepsy. The clinical and EEG similarity that is apparent in BECTS is also present in other non-rolandic epilepsies: a) idiopathic temporal lobe epilepsy with strong affective components, with typical high voltage temporal spikes with frontal or parietal extension, strongly activated by sleep;[28,29] b) benign epilepsy with extreme somatosensory evoked potentials (ESEPs) with rare and short duration seizures, less involving facial muscles, more frequent in waking, with a focus parasagital and parietal. High amplitude (200 µV) somatosensory potentials with decremental voltage evoked by repetitive stimulation of tapping one or both feet at low frequencies (less than 3-5 Hz) are present.[30,31] The Landau-Kleffner Syndrome and epilepsy with continuous spike-waves during slow wave sleep (ECSWS) are now classified among syndromes with both focal and generalized seizures.[32] Due to the clinical and graphical affinity with NRE or with BECTS both syndromes are described in this paragraph. Both have neurological deficits (neuropsychological deficits in ECSWS and aphasia and auditory agnosia in LKS) and a typical strong sleep activation of the seizures and of the EEG discharges. Interictal discharges of SW are at 2-3 Hz frequencies, generalized or focal and have been reported in ECSWS (figure 6, end of chapter). Focal temporal or bitemporal discharges are common in the Landau Kleffner syndrome. Focal continuous SW pattern (focal electrical status) have been recorded in LKS during REM sleep.[33] Non-rolandic common EEG features were also found in epilepsy with continuous focal discharges during sleep[34] and in the atypical benign epilepsy of childhood. Both have a picture similar to BECTS and 'almost continuous spike and wave' discharges during sleep.

The EEG in symptomatic localization-related epilepsies

Chronic progressive epilepsia partialis continua of childhood is characterized by persistent normal EEG background activity and non-specific focal paroxysmal activity: bursts of slow waves or spike/sharp-wave and spikes that are of increasing frequency on ictal events until constant discharges. The epilepsies in which the seizures are precipitated by specific stimuli are grouped under the denomination of 'reflex epilepsies'.[35,36] The EEG bursts of generalized or focal SW are elicited by specific stimuli. If the stimuli remain, ictal changes will occur and partial seizures can progress into other types. Symptomatic localization-related epilepsies also include syndromes with multiple individual clinical and EEG expressions. These syndromes are not strictly age-related. Localization is often based on careful description of the symptoms, rather than on the EEG signal (even if depth EEG recording or neuroimaging are used). The anatomical localization and functional structures involved, explain the seizure characteristics. Syndromes related to anatomic localizations are: a) temporal lobe epilepsies (TLE) mostly originating in amygdalo-hippocampal and temporo-lateral areas; b) frontal lobe epilepsies, originating from varies areas: supplementary, motor, cingulate, frontopolar, orbitofrontal, dorsolateral and opercular cortex; c) parietal lobe epilepsies; d) occipital lobe epilepsies. The characteristics of these syndromes are similar to those in adults, although the highly organized gestural or behavioral sequences are rare in young children.[37,38] These localization-related epilepsies are linked to an organic origin and ictal and interictal events are influenced by sleep cycles. NREM increases the frequency and propagation of interictal epileptiform events, whereas REM decreases the frequency and maximizes focalization.[39] Interictal and ictal spikes of frontal lobe epilepsies originating from the intermediate, fronto-orbital, or operculo-insular cortex can not be detected in the scalp EEG recording and can thus lead to misinterpretations.[40,41] The temporal lobe is the common place to onset of partial seizures. Interictal EEG recordings can be normal or, with few modifications show slow waves, spikes or sharp waves. Long-term EEG monitoring shows that the ictal EEG is characterized by runs of anterior temporal rhythmic delta activity. Mesial anterior spikes in depth recordings can be found in seizures from mesiobasal origin. Unilateral or bilateral temporal and posterior temporal spikes predominantly in the lateral regions are common in latero-temporal seizures. Seizures originating from the parietal lobe are more frequently characterized by sensitive phenomena with or without secondary generalization. Interictal and ictal events (spikes, sharp waves and runs of slow activity) are restricted to focal parietal areas or may spread to the neighbouring areas mimicking other types. Over 80% of individuals with posterior temporal and occipital SW paroxysms have clinical seizures. The ictal paroxysms are discharges with a longer duration than 6 sec.[42] The spikes are temporo-occipital located in about 50% of the patients. Discharges with surface negative spikes on O1 or O2 locations have been observed in only 18% of the patients.[43] The discharges can spread very fast to the contralateral occipital lobe, to the temporal areas (leading to TLE manifestations), to the convexity when the focus is supracalcarine, or to the parietal and frontal areas. All these different symptomatic epilepsies have common clinical and EEG features. The EEG discharges can originate from different structures but may nevertheless show common symptoms. The initial focus is frequently difficult to be identified. The background EEG gives little information (irregular runs of slow or sharp waves) with this respect. The use of techniques of automatic analysis of EEG may help us to surpass some of these difficulties.[44,45]

THE EEG IN STATE-RELATED EPILEPSIES

The EEG- and seizure manifestations of many epilepsies are influenced by the sleep-wake cycle. Hopkins[46] and Janz[47] have reported that seizures predominantly occur during sleep or during lowering of arousal. The circadian distribution of interictal EEG activity also shows a stable pattern from one day to another with a consistent peak of discharges related to sleep.[48] The EEG characteristics of JME are an excellent example. They occur predominantly during transition phases, strictly related to the sleep-wake cycle (awakening, falling asleep, or afternoon relaxation period). The clinical events of JME are, however, deactivated during sleep.[49] In most state-related epilepsies, the interictal EEG will not show any abnormalities, except for the transition phases. For adequate seizure identification, the EEG must thus be recorded at convenient times or situations. For example, if the EEG of a patient with a suspected form of epilepsy, related to sleep or awakening is normal, then an additional EEG recording will be necessary in awakening state, to confirm the epilepsy diagnosis.[50] The most critical time is 10 minutes to two hours after awakening,[47,51] more frequently from nocturnal sleep than from day-time naps. Epileptiform interictal events and seizure manifestations, including absences, occur in conjunction with prolonged periods of drowsiness.

EEG in awakening epilepsies

'Awakening epilepsies' frequently belong to the primary generalised epilepsies and are often characterized as 'developmental disorders' as they are also age-dependent.[39] Examples of awakening epilepsies are the juvenile myoclonic epilepsy (JME) and epilepsy with grand mal or GTCS on awakening.

The EEG in juvenile myoclonic epilepsy
The interictal EEG shows bursts of generalized 3-5 Hz SW patterns with an intradischarge frequency of 3-5 Hz. SWs and PSWs are bilateral symmetrical, or frontocentral accentuated or restricted to the frontal leads. Paroxysmal activity is very frequent (up to 80% of the records) and background activity is mostly normal. Episodic slowing of alpha rhythm and runs of theta have been recorded. Individual complexes varied from 2-10 Hz[52] and are of multiple types: double-spike-wave, polyspike-wave, spike-wave at different frequencies.[53] Poly-spikes at 10-16 Hz frequency occurs simultaneously with jerks. Slow waves of various amplitude or

frequency (2-5 Hz) follow the interictal or ictal discharges. The spikes have negative polarity and may reach amplitudes up to 300 μV and are related to the duration of the jerks. The PSW complex lasts for 2-10 sec, exceeding the jerking. The slow waves have a mean amplitude of 200-400 μV. The PSW pattern does not display the characteristic synchronisms of pyknoleptic discharges.[53-55] Focal abnormalities and photoconvulsive responses also occur frequently (respectively up to 30 and 27%).[50] The JME seizures are typically a manifestation of transition phases (awakening, falling asleep, day-time naps) and JME manifestations are deactivated during sleep.

The role of arousal as a trigger is supported by provocation of evoked seizures. Provocation of arousal manually[56] and by photic stimulation[57] is concomitant to increase seizure frequency and of interictal events. The interictal discharges are often entrained to sleep spindles and k-complexes. These phasic events characterise non-deep NREM sleep stages and they reflect aborted arousal attempts[58] and support the link between interictal discharges, seizures and arousal levels.

The EEG in Epilepsy with grand mal seizures on awakening
The seizures manifest exclusively or predominantly short after awakening with a second peak at relaxation period. The seizures are clearly related to external factors such as lack of sleep, early rising, or stress.[47,59,60] The EEG background is normal. Interictal EEG may show a disorganisation of background, generalized bursts of SW (2.5-4 Hz) and photosensitivity. In this epilepsy the sleep is unstable and easily modified by external factors.[60]

The EEG in focal seizures, related to sleep

Focal seizures with secondary generalisation are common in sleep-related epilepsies.[61,62] The paroxysmal generalised EEG activity is prominent in NREM. The extreme example is ECSWS, a generalised seizure disorder in children manifested only during NREM sleep. Secondary generalised seizures with a frontal or temporal origin are commonly sleep-related. The majority of the patients with these epilepsies have seizures in sleep transitions: into and at the end of NREM or REM and during deep NREM.[47,63-65] The sleep stage in which a first seizure occurs also provides relevant information about the recurrence risk.[66] In children, the risk of recurrence was >50% if the first seizure occurred during sleep and the subsequent seizure occurred during the same sleep state of the initial seizure in 73% of the cases.[67]

CONCLUSION

EEG parameters change with age and with the development of the epilepsy and they can predict epilepsy outcome. In idiopathic forms of age-related epilepsies, both with generalized or with partial seizures, the background EEG and the type of paroxysmal activity are typical: interictal or ictal discharges (bursts of SW, PSW, focal or generalized), strongly activated by NREM sleep. If focal discharges occur, their origin is mostly better identified during REM sleep. The EEG changes tend to disappear over time, even in the extreme cases of continuous sleep discharges. Symptomatic and cryptogenic forms of age-related epilepsies have a more permanent impact on the interictal EEG background. Ictal recordings show complex EEG-characteristics. Symptomatic forms of localization-related epilepsies are more often state-related than cryptogenic forms. Generalized cryptogenic forms frequently have prominent EEG abnormalities: hypsarrhythmia and burst suppression, which are factors related to bad prognosis. State-related epilepsies can be divided into awakening and sleep epilepsies. Although the awakening epilepsies are strongly activated by NREM-sleep, the stable sleep state is less activator than transitions or change of arousal-level; arousals within sleep (K complexes) or from sleep are the most activating factors. The interictal EEG pattern is typical: generalized SW bursts, normal background EEG. Symptomatic forms of state-related epilepsies are often related to sleep: NREM increases the frequency of EEG-discharge with a tendency to generalize.

Acknowledgements
We thank Dr. MS Santos for providing computer help with the bibliography, Dr. F. Barbot for reviewing the text, the secretariat of the Neurophysiology Department in Porto for technical assistance, and my family who gave up valuable leisure time.

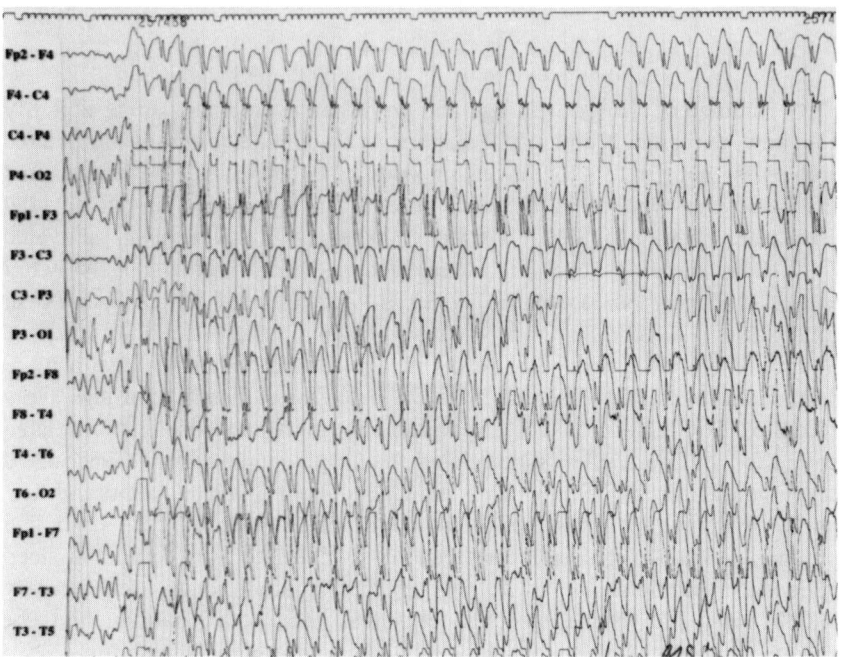

Figure 1. Typical absence seizure. Absence recorded during hyperventilation, with mild motor components. Girl, 12 years old and absences since the age of 8.

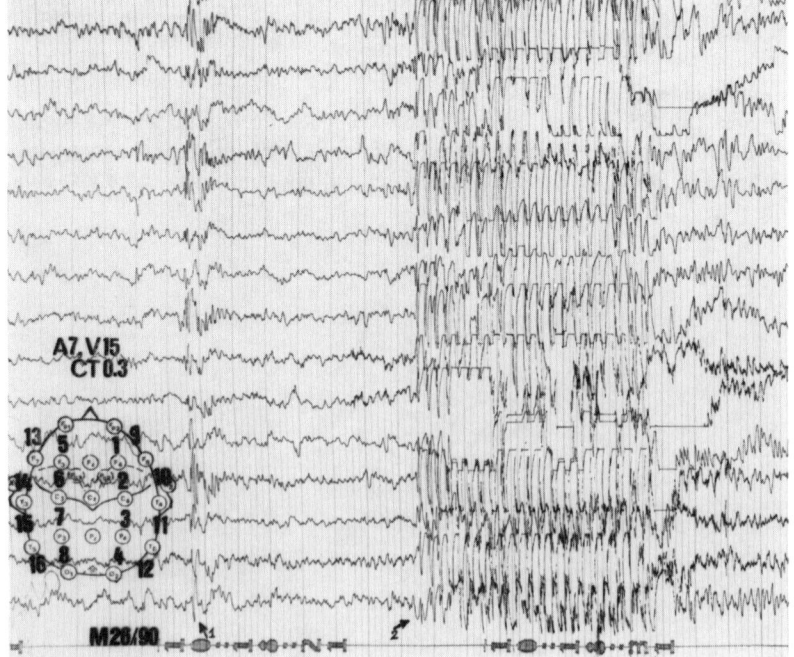

Figure 2. Myoclonic seizure. Girl, 14 years old. Drowsiness elicits a PSW discharge (Left arrow). Note the prolonged discharge (right arrow) with myoclonus at the transition (see EEG background after seizure).

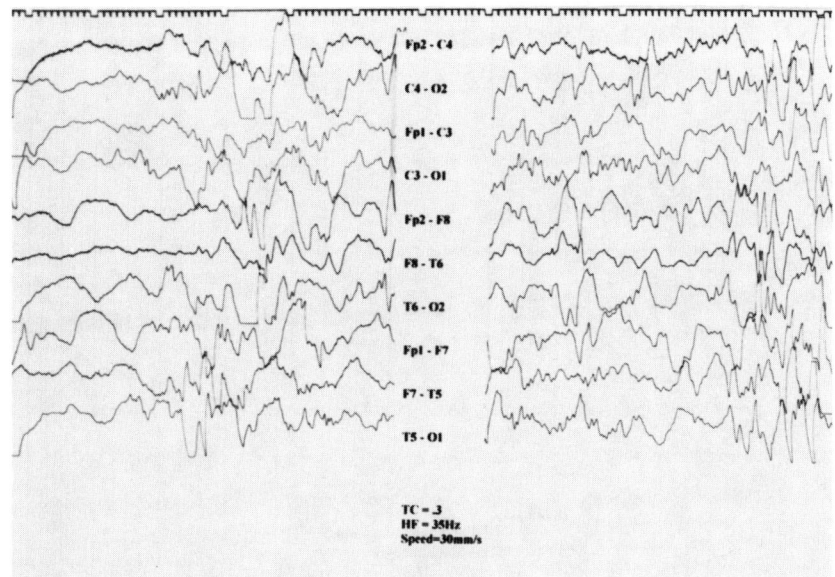

Figure 3. West Syndrome. Child, 5 months. Massive spasm(L), hypsarrhythmia(R).

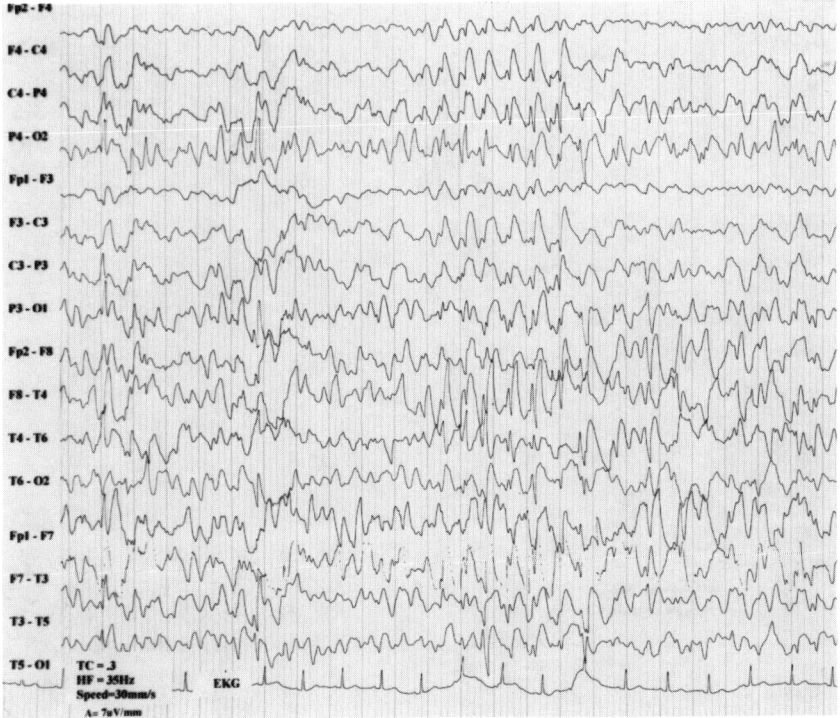

Figure 4. Lennox Gastaut syndrome. Boy, 16 yrs. Slow SW bursts generalized or with irregular focalization.

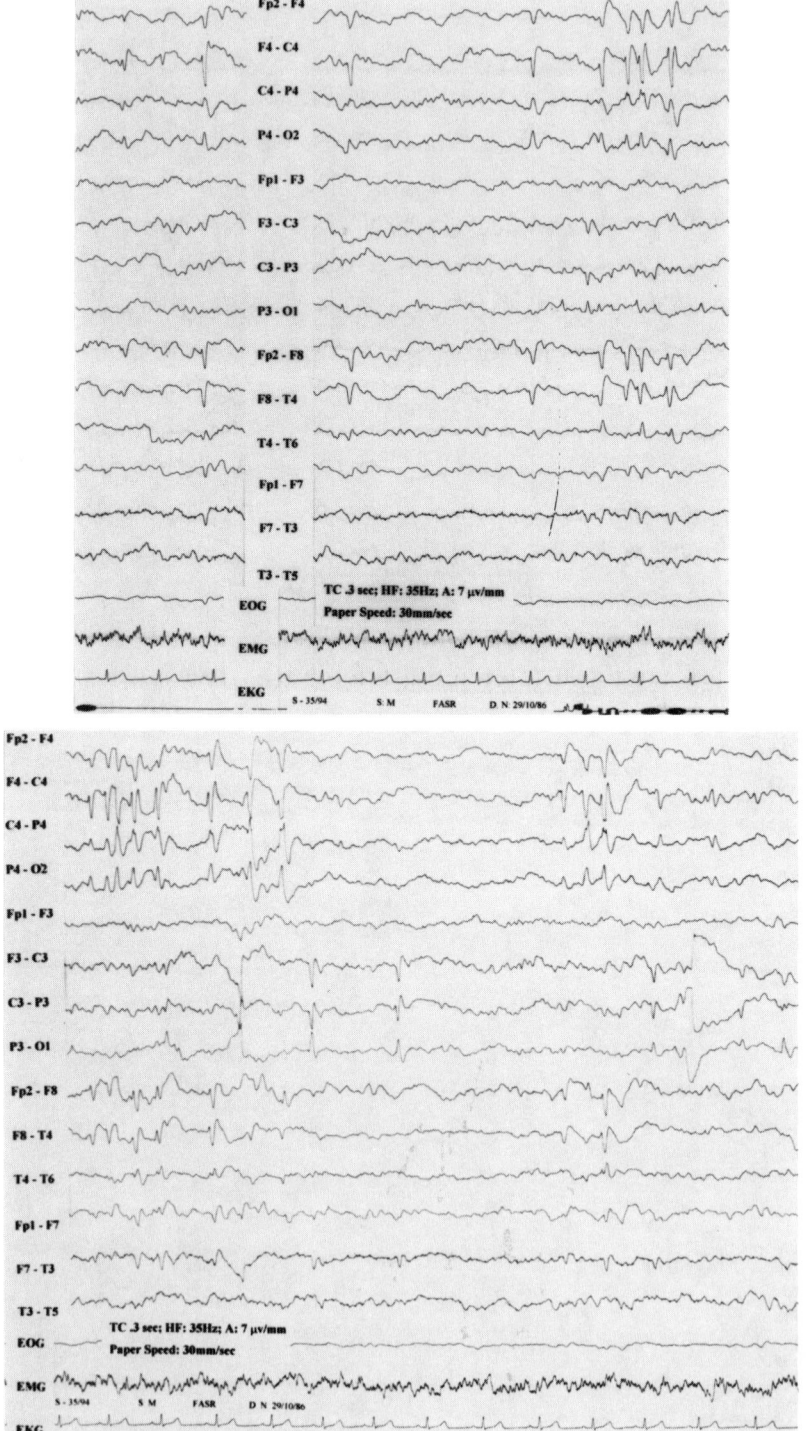

Figure 5. Boy, 8 yrs. with BECTS. Sleep recording. Typical right rolandic focus (top). In NREM Sleep increased bilateralization and contralateral focus appear (bottom).

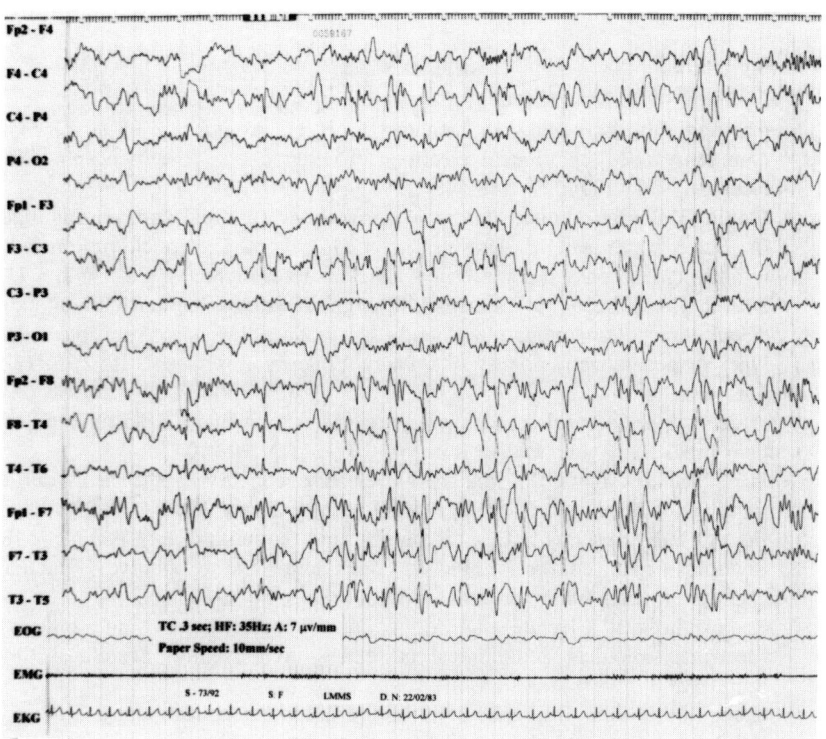

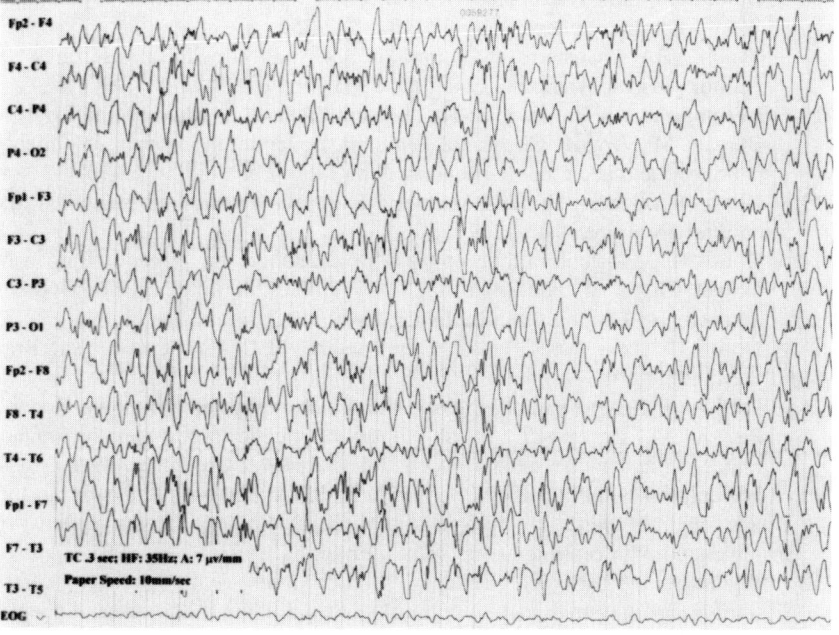

Figure 6. Girl, 9 yrs, ECSWS. NREM sleep. Focus or predominance on left temporal region (top) is loosed in deep sleep (bottom).

REFERENCES

1. Chatrian GE, Bergamini L, Dondey M, et al. A glossary of terms most commonly used by clinical electroencephalographers. Electroenceph. Clin. Neurophysiol., 1974; 37: 538-548.
2. Plouin P. Benign idiopatic neonatal convulsions (familial and non-familial). In: Roger J, Bureau M, Dravet Ch, et al, editors. Epileptic syndromes in infancy, childhood and adolescence. London: John Libbey & Co, 1992: 3-11.
3. Dravet Ch, Bureau M, Genton P. Benign Myoclonic epilepsy in infancy: electroclinical symtpomatology and differential diagnosis from the other types of generalized epilepsy of infancy. Epilepsy Res. Suppl. 1992; 6: 131-135.
4. Hedstrom A, Olsson I. Epidemiology of absence epilepsy: EEG findings and their predictive value. Pediatr. Neurol., 1991; 7(2): 100-104.
5. Aarts JHP, Binnie CD, Smit AM. Selective cognitive impairment during focal and generalised epileptiform EEG activity. Brain, 1984; 107: 239-308.
6. Binnie CD. Seizures EEG discharges and cognition. In: Trimble MR, Reynolds EH, editors. Epilepsy, Behaviour and Cognitive Function. Chichester: John Willey & Sons, 1987: 45-49.
7. Kasteleijn-Nolst Trenité DGA, editor. Photosensitivity in epilepsy. Electrophysiological and clinical correlates. Acta Neurol. Scandin., 1989; Suppl 125.
8. Gibbs FA, Gibbs EL, editors. Atlas of Electroencephalography. Epilepsy. Vol 2. Massachussets: Addison-Wesley Press Inc, 1952.
9. Plouin P, Dulac O, Jalin C, et al. Twenty-four-hour ambulatory EEG monitoring in infantile spasms. Epilepsia, 1993; 34(4): 686-691.
10. Dulac O, Plouin P, Jambaque I. Predicting favourbale outcome in idiopathic West syndrome. Epilepsia, 1993; 34(4): 747-756.
11. Yaqub BA, Genton P, Lipinski CG. Electroclinical seizures in Lennox-Gastaut syndrome. Epilepsia, 1993; 34(1): 120-127.
12. Beaumanoir A, Dravet Ch. The Lennox Gastaut Syndrome. In: Roger J, Bureau M, Dravet Ch, et al, editors. Epileptic syndromes in infancy, childhood and adolescence. London: John Libbey & Co., 1992: 115-132
13. Doose H, Gerken H, Hein-Voelpel KF, et al. Genetics of photossensitive epilepsy. Neuropaediat., 1969; 1: 56-73.
14. Doose H. Myoclonic astatic epilepsy of childhood. In: Roger J, Bureau M, Dravet Ch, et al, editors. Epileptic syndromes in infancy, childhood and adolescence. London: John Libbey & Co, 1992: 103-114.
15. Doose H, Baier WK. Theta rhythms in the EEG - a Genetic trait. Brain Dev., 1988; 10: 347-354.
16. Lerman P. Benign partial epilepsy with centro-temporal spikes. In: Roger J, Bureau M, Dravet Ch, et al, editors. Epileptic syndromes in infancy, childhood and adolescence. London: John Libbey & Co, 1992: 189-200.
17. Beydoun A, Garofalo EA, Drury I. Generalized spike-waves, multiple loci, and clinical course in children with EEG features of benign epilepsy of childhood with centrotemporal spikes. Epilepsia, 1992; 33(6): 1091-1096.
18. Beaussart M. Benign epilepsy of children with rolandic (centro-temporal)paroxysmal foci. A clinical entity. Study of 221 cases. Epilepsia, 1972; 13: 795-811.
19. Luders H, Lesser RP, Dinner DS. Benign focal epilepsy of childhood. In: H

Luders, RP Lesser, editors. Epilepsy: Electroclinical Syndromes. Berlin, Heidelberg: Spinger-Verlag, 1987: 303-346.
20. Gibbs EL, Gillen HW, Gibbs FA. Disappearence and migration of epileptic foci in childhood. Am. J. Dis. Child, 1954; 88: 596-603.
21. Roger J, Bureau M, Genton P. Idiopathic partial epilepsies. In Dam M, Gram L, editors. Comprehensive Epileptology. New York: Raven Press, 1990: 155-170.
22. Weinberg H, Wong PKH, Crisp D, et al. Use of multiple dipole analysis for the classification of Benign Rolandic Epilepsy. Brain Topogr., 1990; 3(1): 183-190.
23. Gregory DL, Wong PK. Topographical analysis of the centrotemporal discharge in benign rolandic epilepsies of childhood. Epilepsia, 1984; 25(6): 705-711.
24. Van der Meij W, Wieneke GH, Van Huffelen AC, et al. Identical morphology of the rolandic spike-and-wave complex in different clinical entities. Epilepsia, 1993; 34 (3): 540-550.
25. Lerman P, Kivity S. The benign partial non rolandic epilepsies. J. Clin. Neurophysiol., 1991; 8(3): 275-287.
26. Beaumanoir A. Infantile epilepsy with occipital focus and good prognosis. Eur. Neurol., 1983; 22: 43-52.
27. Dalla Bernardina B, Fontana E, Cappellaro O, et al. The partial occipital epilepsies in childhood. In: Andermann F, Beaumanoir A, Mira L, et al, editors. Occipital Seizures and epilepsies in children. London: John Libbey, 1993; 173-181.
28. Dalla Bernardina B, Bureau M, Dravet C, et al. Epilepsie bénigne de l'enfant avec crises à séméiologie afective. Rev. EEG Neurophysiol., 1980; 10: 8-18.
29. Dalla Bernardina B, Colamaria V, Chiamenti C, et al. Benign partial epilepsy with affective symptoms ('Benign psychomotor epilepsy'). In: Roger J, Bureau M, Dravet Ch, et al, editors. Epileptic syndromes in infancy, childhood and adolescence. London: John Libbey & Co, 1992; 219-223.
30. De Marco P, Negrin P. Parietal focal spikes evoked by contralateral tactile somatotopic stimulations in four non-epileptic subjects. Electroenceph. Clin. Neurophysiol., 1973; 34: 308-312.
31. Tassinari CA, De Marco P. Benign partial epilepsy with extreme somatosensory evoked potentials. In: Roger J, Bureau M, Dravet Ch, et al, editors. Epileptic syndromes in infancy, childhood and adolescence. London: John Libbey & Co, 1992: 225-229.
32. Comission on Classification and Terminology of the International League Against Epilepsy - Proposal for Revised Classification of Epilepsies and Epileptic Syndromes. Epilepsia, 1989; 30 (4): 389-399.
33. Genton P, Maton B, Ogihara M, et al. Continuous focal spikes during REM sleep in a case of acquired aphasia (Landau-Kleffner syndrome). Sleep, 1992; 15(5): 454-460.
34. Galletti F, Sturniolo MG, Giannotti F, et al. Décharges EEG localisées continues pendant le sommeil chez des enfants sans troubles neuropsychiques. Neurophysiol. Clin., 1992; 22(6): 447-457.
35. Beaumanoir A, Gastaut H, Naquet R, editors. Reflex Seizures and Reflex Epilepsies. Genève: Médicine et Hygienne, 1989.
36. Wolf P, editor. Epileptic seizures and syndromes. London: John Libbey & Co, 1994.

37. Duchowny M. The syndrome of partial seizures in infancy. J. Child Neurol. 1992; 7(1): 66-69.
38. Wyllie E, Chee M, Granstrom M-L et al. Temporal lobe epilepsy in early childhood - Epilepsia, 1993; 34(5): 859-868.
39. Shouse MN, Martins da Silva A, Sammaritano M. Sleep and Epilepsy. In: Engel JP, Pedley TA, editors. Epilepsy: A comprehensive textbook. Vol 1. New York: Raven Press, 1995; In press.
40. Bancaud J, Talairach J. Clinical Semiology of Frontal Lobe Seizures. In: Chauvel P, Delgado-Escueta AV, Halgren E, et al, editors. Frontal Lobe Seizures and Epilepsies. Advances in Neurology. Vol 57. New York: Raven Press, 1992: 3-58.
41. Chauvel P, Delgado-Escueta AV, Halgren E, et al, editors. Frontal Lobe Seizures and Epilepsies. Advances in Neurology. Vol 57. New York: Raven Press, 1992.
42. Talwar D, Rask CA, Torres F. Clinical manifestations in children with occipital spike-wave paroxysms. Epilepsia, 1992; 33: 667-674.
43. Salanova V, Andermann F, Olivier A, et al. Occipital lobe epilepsy: electroclinical manifestations, electrocorticogrphy, cortical stimulation and outcome in 42 patients treated between 1932 and 1991. Brain 1992; 115(6): 1655-1680.
44. Ebersole JS. Non-invasive localization of the epileptogenic focus by EEG dipole modeling. Acta Neurol. Scand., 1994; Suppl. 152: 20-28.
45. Martins da Silva A, Cunha JP, Guedes de Oliveira P. Scalp EEG recording: Interictal/ictal location and spreading of epileptiform events. Acta Neurol. Scandin., 1994; Suppl. 152: 17-19.
46. Hopkins H. The time of appearance of epileptic seizures in relation to age, duration and type of syndrome. J. Nerv. Ment. Dis., 1933; 77: 153-162.
47. Janz D. The grand mal epilepsies and the sleeping-waking cycle. Epilepsia, 1962; 3: 69-109.
48. Martins da Silva A, Aarts JHP, Binnie CD, et al. The circadian distribution of interictal epileptiform EEG activity. Electroenc. Clin. Neurophysiol., 1984; 58: 1-13.
49. Gigli GL, Calia E, Marciani MG, et al. Sleep microstructure and EEG epileptiform activity in patients with juvenile myoclonic epilepsy. Epilepsia 1992, 33/5: 799-804.
50. Panayiatopoulos CP. Juvenile myoclonic epilepsy. An underdiagnosed syndrome. In: P Wolf, editor. Epileptic seizures and syndromes. London: John Libbey Co, 1994: 221-230.
51. Beyer L, Jovanovic UJ. Elektrecephalographishe und klinische korrelate bei aufwachepileptikern mit besonderer berucksichtigung der therapeutischen probleme. Nervenzart, 1966; 37: 333-336.
52. Panayiatopoulos CP, Obeid T, Tahan AR. Juvenile Myoclonic Epilepsy: A 5-year prospective study. Epilepsia, 1994; 35 (2): 285-296.
53. Janz D. Juvenile Myoclonic Epilepsy. In: Mogens Dam and Lennart Gram, editors. Comprehensive Epileptology. New York: Raven Press, 1991: 171-185.
54. Delgado Escueta AV, Enrile-Bacsal F. Juvenile Myoclonic Epilepsy of Janz. Neurology, 1984; 34: 285-294.
55. Janz D, Christian W. Impulsiv petit mal. Dtsch. Z. Nervenheilk. 1957; 176: 348-386.
56. Touchon J. Effect of awakening on epileptic activity in primary generalized

myoclonic epilepsy. In: Sterman MB, Shouse MN, Passouant P, editors. Sleep and Epilepsy. New York: Academic Press, 1982: 239-248.
57. Meier-Ewert K, Broughton RJ. Photomyoclonic response of epileptic and non-epileptic subjects during awakefulness, sleep and arousal. Electroenceph. Clin. Neurophysiol., 1967; 23: 142-151.
58. Halasz P. Generalized epilepsy with spike-wave pattern (GESW) and intermediate states of sleep. In: Sterman MB, Shouse MN, Passouant P, editors. Sleep and Epilepsy. New York: Academic Press, 1982: 219-238.
59. Janz D. Pitfalls in the diagnosis of grand mal on awakening. In: P Wolf, editor. Epileptic Seizures and Syndromes. London: John Libbey & Co, 1994: 213-220.
60. Wolf P. Epilepsy with grand mal on awakening. In: Roger J, Bureau M, Dravet Ch, et al, editors. Epileptic syndromes in infancy, childhood and adolescence. London: John Libbey & Co, 1992: 329-341.
61. Billiard M. Epilepsies and the Sleep-Wake cycle. In: Sterman MB, Shouse MN, Passouant P, editors. Sleep and Epilepsy. New York: Academic Press, 1982: 269-286.
62. Niedermeyer, E. Epileptic Seizure Disorders. In: Niedermeyer E, Lopes da Silva F, editors. Electroencephalography. Basic principles, Clinical Applications and Related Fields. 3rd edition. Baltimore: Williams & Wilkins, 1993: 461-564.
63. Cadillac J. Complex partial seizures and REM sleep. In: Sterman MB, Shouse MN, Passouant P, editors. Sleep and Epilepsy. New York: Academic Press, 1982: 315-324.
64. Stevens JR, Kodama H, Lonsbury B, et al. Ultradian characteristics of spontaneous seizure discharges recorded by radio telemetry in man. Electroenceph. Clin. Neurophysiol., 1971; 31: 313-325.
65. Stevens JR, Lonsbury BL, Goel SL. Seizure occurrence and interspike interval: telemetered electroencephalogram studies. Arch. Neurol. (Chic), 1972; 26: 409-419.
66. Meierkord H. Epilepsy and Sleep. Curr. Opin. Neurol., 1994; 7: 107-112.
67. Shinnar S, Berg AT, Ptachewich Y, et al. Sleep state and the rsik of seizure recurrence following a first unprovoked seizure in childhood. Neurology, 1993; 43: 701-706.

7   Established and new antiepileptic drugs: an overview

JEAN A. HANNAH
GRAEME J. SILLS
MARTIN J. BRODIE

Epilepsy Research Unit, University Department of Medicine and Therapeutics, Western Infirmary, Glasgow, Scotland

The current attitude towards antiepileptic drugs emphasizes the use of monotherapy with serum concentration measurement coupled with standard, weight-adjusted starting and maintenance regimens to guide initial therapy and subsequent dosage titration. The established first-line anticonvulsants are carbamazepine, sodium valproate and phenytoin. The pharmacokinetic interpretation of serum concentration of sodium valproate is limited by its wide therapeutic index, large fluctuations in the concentration-time profile and concentration-dependent protein binding.

Pharmacokinetic principles are more readily applied to carbamazepine as it has a flatter concentration-time profile, a more clearly defined target range, it undergoes autoinduction of metabolism and it commonly interacts with other drugs. Phenytoin has required the greatest pharmacokinetic input due to its non-linear pharmacokinetics and narrow therapeutic index. Dosage alterations may be needed for specific patient groups. In particular, children generally require higher dosages on a weight-for-weight basis than adults, while equivalently lower dosages should be given to neonates. There are few data available on the pharmacokinetics of the newer drugs, felbamate, gabapentin, lamotrigine, oxcarbazepine and vigabatrin in children. This review highlights the pharmacokinetic properties, mechanisms of action and adverse drug interactions of the most commonly used established and new antiepileptic drugs.

INTRODUCTION

Epilepsy in childhood and its pharmacological management possesses several characteristics not found in affected adults. The Commission on Antiepileptic Drugs of the International League Against Epilepsy[1] has described these variations in terms of the disorder itself, cognitive function and behaviour, care-givers and ethics, and the pharmacokinetics, metabolism, and tolerability of antiepileptic drugs. Pharmacokinetics present specific problems as they vary in children due to age, growth, comedication, and disease. This is especially true for the newborn. In older age groups there are considerable differences in drug handling between children of the same age.[2] In order to maintain stable concentrations of antiepileptic drugs, their metabolites and active co-factors at receptor sites over extended periods, the pharmacokinetic properties of the antiepileptic drugs and issues influencing absorption, distribution, metabolism and excretion need to be understood.[3]

Due to faster clearance rates and consequently reduced elimination half-lives, children generally require higher drug doses on a weight-for-weight basis than adults, while equivalently lower dosages are indicated in neonates.[4] Some children need almost twice the recommended adult dosage, particularly if combination therapy with an enzyme inducer is employed.[5] In the case of phenytoin, the differences in dosage requirements between children and adults may be due to differences in the ratio of liver to body weight and not to qualitative or quantitative variations in the activities of the enzymes concerned with its metabolism.[6]

In general, shorter half-lives result in increased frequency of dosing. Despite this, wide variations in peak-to-trough concentrations can result in breakthrough seizures when the concentration is low, toxicity with high peak levels, and difficulty in interpreting a measured drug concentration.[4] The formulation also affects these issues due to varying absorption rates and subsequent changes in peak-to-trough serum levels. Liquid formulations are often used in children, but can contain potentially damaging ingredients such as sucrose, have short shelf-lives which lead to more frequent renewal, and result in fluctuating concentrations because of the faster rate of absorption of drug molecules in suspension.[2] Solid oral preparations overcome this problem by providing a longer absorption phase that reduces the extremes of concentrations, as do crushable tablets which are also preferable to liquids. In the newborn, low bioavailability may be caused by absorption immaturity. In addition, drug absorption may be altered by formula or milk products.[1]

Sustained release preparations of carbamazepine and valproate have been assessed recently in children. Reductions in adverse events[7] and in seizure frequency[8] have been reported with sustained release carbamazepine. Controlled release valproate in children with epilepsy[9] showed no significant difference to conventional valproate with respect to mean diurnal trough and peak plasma levels, and to mean fluctuation. Another study of a slow release valproic acid,[10] however, suggested more stable pharmacokinetic features, a larger age-related area under the concentration-time curve, and improved clinical effect, when compared with a conventional preparation. Rapid gastrointestinal transit times in childhood, however, may reduce bioavailability if the product dissolution and absorption rates are too prolonged.[11]

ESTABLISHED ANTIEPILEPTIC DRUGS

Tables 1a and 1b summarise the pharmacokinetics of a number of established anticonvulsant drugs. These will now be discussed individually in alphabetical order.

Established and new antiepileptic drugs: an overview 103

Table 1a. Phamacokinetics of established antiepileptic drugs

Drug	Absorption (bioavailability)	Distribution volume (L/Kg)	Protein binding (% bound)
CARBAMAZEPINE	Slow absorption (75 - 85%)	0.8 - 1.6	70 - 80
CLOBAZAM	Rapid absorption (90 - 100%)	0.7 - 1.6	87 - 90
CLONAZEPAM	Rapid absorption (80 - 90%)	2.1 - 4.3	80 - 90
ETHOSUXIMIDE	Rapid absorption (90 - 95%)	0.6 - 0.9	0
PHENOBARBITAL	Slow absorption (90 - 100%)	0.51 - 0.57	48 - 54
PHENYTOIN	Slow absorption (85 - 95%)	0.5 - 0.7	90 - 93
PRIMIDONE	Rapid absorption (90 - 100%)	0.4 - 0.8	20 - 30
SODIUM VALPROATE	Rapid absorption (100%)	0.09 - 0.17	88 - 92

Table 1b. Phamacokinetics of established antiepileptic drugs

Drug	Elimination half-life (hours)	Route(s) of elimination	Comments
CARBAMAZEPINE	24 - 45 (single) 8 - 24 (chronic)	Hepatic metabolism Active metabolite	Enzyme inducer Autoinduction of metabolism
CLOBAZAM	10 - 30	Hepatic metabolism Active metabolite	Tolerance and rebound
CLONAZEPAM	30 - 40	Hepatic metabolism	Sedative Tolerance and rebound
ETHOSUXIMIDE	20 - 60	Hepatic metabolism 25% excreted unchanged	More rapid clearance in children
PHENOBARBITAL	72 - 144	Hepatic metabolism 25% excreted unchanged	Enzyme inducer Sedative Tolerance and rebound
PHENYTOIN	9 - 40	Saturable hepatic metabolism	Enzyme inducer Elimination half-life concentration-dependent
PRIMIDONE	4 - 12	Hepatic metabolism Active metabolites 40% excreted unchanged	Sedative Phenobarbital a metabolite Tolerance and rebound
SODIUM VALPROATE	7 - 17	Hepatic metabolism Active metabolites	Enzyme inhibitor Concentration-dependent protein binding

Carbamazepine

Pharmacokinetics
Carbamazepine is an iminostilbene derivative, which behaves as a neutral lipophilic substance. It was synthesised in 1953 by Schindler at Geigy in an effort to compete with the newly introduced antipsychotic agent, chlorpromazine. Clinical trials in epilepsy were begun in 1963.

Carbamazepine is absorbed slowly on oral administration due to its limited water solubility,[12] with peak plasma levels occurring usually 4-8 hours after ingestion. Absorption rates depend on the preparation used, but have been reported to be slower after an evening dose and faster if carbamazepine is taken with food. Carbamazepine is 70-80% protein bound, has a distribution volume of 0.8-1.6 L/kg and a bioavailability of 75-85%. The elimination half-life is 24-45 hours following a single dose and 8-24 hours on chronic dosing due to auto-induction of metabolism.[13]

Carbamazepine is subject to extensive biotransformation in the liver. Approximately 40% follows the major metabolic pathway, which results in the formation of the active metabolite carbamazepine 10,11 epoxide itself metabolised to the dihydrodiol.[14] Variable autoinduction of hepatic monooxygenase activity by carbamazepine results (figure 1) in substantial interindividual differences in concentration found with the same dose.[13]

Controlled-release formulations reduce this variation by about 50%,[15] and may be given once[16] or twice[17] daily, thereby ameliorating high peak concentrations and associated side-effects.[18] Following a dose of carbamazepine, approximately 2% is ultimately excreted unchanged in the urine.

Impaired renal function has not been found to alter the bioavailability or clearance rates of carbamazepine.[12] Similarly, moderate impairment of liver function does not significantly affect its pharmacokinetics, although such changes do occur in severe cirrhosis, in hepatitis and in biliary obstructive conditions. Cardiac failure may result in impairment of absorption, metabolism and clearance of carbamazepine due to the reduction in cardiac output and associated gut wall and hepatic congestion. Malabsorption syndromes may result in lower bioavailability, while pyrexia and pulmonary disease increase the drug's catabolism. In postoperative states, the alteration in carbamazepine pharmacokinetics has been attributed to a fall in plasma protein binding.

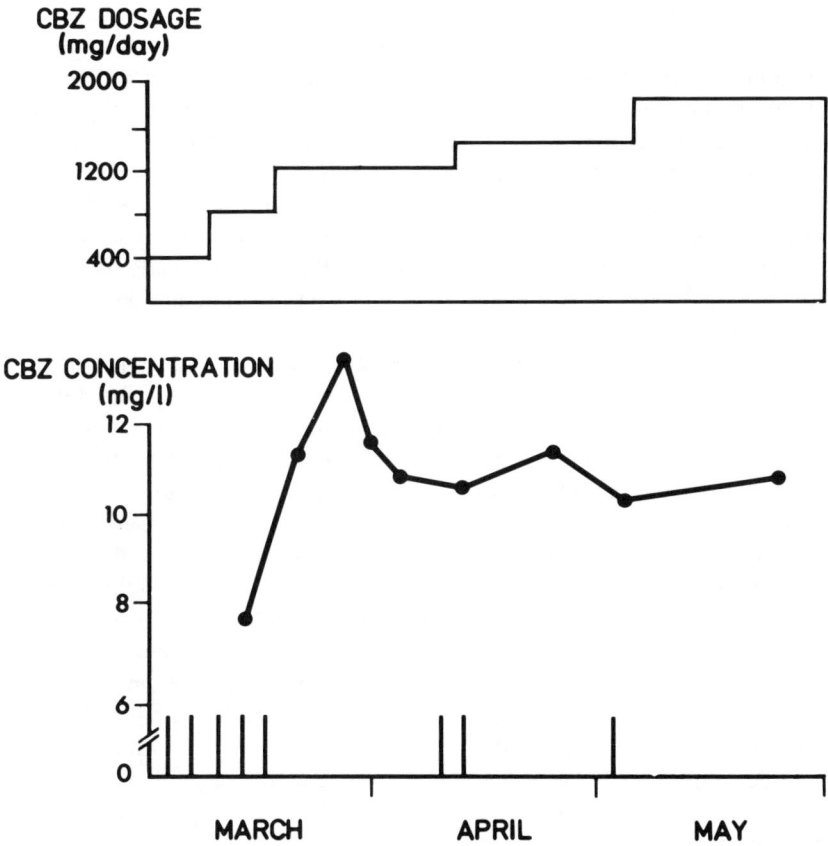

Figure 1. Effect of autoinduction of carbamazepine metabolism in a patient taken the drug as monotherapy. Vertical lines at foot are tonic-clonic seizures.

Mode of action
Carbamazepine has been proposed to exert its antiepileptic effects by interacting with the voltage-activated sodium channels responsible for the action potential upstroke in a highly specific voltage- and frequency-dependent manner.[19] It does not appear to reduce the amplitude or duration of single action potentials, but impairs the ability of neurones to fire trains of action potentials at high frequency.[20] This process is believed to be mediated by a shift of the sodium channel to an inactive state similar to that occurring normally but with slower recovery.[21] Other proposed mechanisms include an inhibition of voltage-dependent non-T-type calcium influx resulting in a reduction in excitatory neurotransmitter release[22] and an antagonistic action at brain adenosine receptors.[23]

Interactions
Carbamazepine is a powerful enzyme inducer which interacts with a wide range of drugs with important clinical implications.[24] Its metabolism too is particularly susceptible to enzyme inhibition. Drugs, such as erythromycin, cimetidine, dextropropoxyphene and isoniazid, increase plasma carbamazepine levels, while corticosteroid, cyclosporin, thyroxine, theophylline, doxycycline and anticoagulant (nicoumalone and warfarin) biodegradation is accelerated, the latter resulting in a reduced anticoagulant effect. Oral contraceptives are similarly affected leading to the need for higher oestrogen doses. Plasma concentrations of carbamazepine can be increased by the antidepressants fluoxetine, fluvoxamine and viloxazine. Mianserin and tricyclic antidepressant biotransformation are induced by carbamazepine.

A complex interaction results when phenytoin and carbamazepine are given together. The metabolism of the latter is induced by the former, whereas the latter inhibits the elimination of the former by competing for binding sites on the same population of metabolic enzymes.[25] This results in a rise in phenytoin and a fall in carbamazepine concentrations when one or other is introduced.[26] The opposite occurs when either drug is withdrawn. Interactions involving carbamazepine are further complicated by the presence of the metabolite, carbamazepine 10,11 epoxide, which possesses anticonvulsant activity and has been implicated in some of the neurotoxic side-effects associated with the drug.[27] Higher epoxide levels occur in epileptic patients taking carbamazepine in combination with phenytoin, sodium valproate and phenobarbital[28] compared with monotherapy. The former two drugs induce its transformation,[28] while valproate inhibits its breakdown[29] in an unpredictable way.[30]

Children
The absorption of carbamazepine in children is variable with peak plasma levels occurring at 4-8 hours after dosing. At doses over 25 mg/kg absorption is impaired with an apparent inverse relationship between dose and absorption rate.[31] The clearance of carbamazepine is dependent upon age, the fastest rates occurring during infancy and childhood and adult rates being reached at age 15-17 years.[12] Frequent dosing with larger weight-related total doses are necessary in children due to the rapid clearance rates in order to avoid wide variations in plasma levels. Eeg-Olofsson et al.[7] suggested an advantage in using of a controlled-release formulation of carbamazepine in children in terms of a reduction in side effects. A further study of a controlled release formulation compared with conventional carbamazepine, however, did not support any benefits in terms of improved efficacy or fewer side effects.[32]

Clobazam

Pharmacokinetics
Clobazam is a 1,5-benzodiazepine. It is a highly lipophilic compound, which is absorbed quickly and completely on oral administration.[33] The absorption rate may be increased or decreased by food.[34] Clobazam has a bioavailability of 90-100% with maximum serum concentrations reached at 1-4 hours after dosing. It is rapidly distributed with a volume of distribution of 0.7-1.6 L/kg. 87-90% of circulating clobazam is protein bound. Degradation in the liver results in the production of a number of metabolites including the active N-desmethylclobazam.[35] The elimination half-life of clobazam is 10-30 hours, while that of its active metabolite is longer at 36-42 hours.[33] In severe renal impairment, benzodiazepines should be introduced at low dosage due to increased cerebral sensitivity. Caution must also be taken in the presence of liver disease as benzodiazepines can precipitate coma.

Mode of action
Clobazam is believed to exert its pharmacological effects via an action at the $GABA_A$ receptor.[36] The benzodiazepines reliably enhance responses to GABA at this site without affecting those to other inhibitory amino acids that are also mediated by chloride selective channels.[37] These drugs have been shown to increase the frequency of chloride channel opening, without influencing the single channel conductance or the open time.[38] Other mechanisms suggested to play a part in the antiepileptic properties of the benzodiazepines include an inhibition of adenosine uptake and a blockade of neuronal voltage-dependent sodium and calcium channels.[39]

Interactions
Concomitant antacid use may reduce the absorption of clobazam. Plasma levels may be significantly increased by alcohol.[40] All enzyme inducers accelerate the breakdown of clobazam.[41] Interactions with other drugs are unusual, but may occur as a result of variations in patient susceptibility.[33]

Children
Unlike adults, children do not show a linear relationship between plasma levels and dose. They also appear to metabolise clobazam faster than adults.[33]

Clonazepam

Pharmacokinetics
Clonazepam is a chlorinated derivative of nitrazepam, which was first approved for use as an antiepileptic drug in 1975. It is absorbed rapidly with maximum plasma levels occurring at 1-4 hours after oral administration.[42] It has a bioavailability of 80-90% and volume of

distribution of 2.1 - 4.3 L/kg. Distribution occurs quickly due to its high lipid solubility, which is also the reason for the ease with which it crosses the blood brain barrier. Clonazepam is 80-90% protein bound. It is metabolised in the liver and has an elimination half life of 30-40 hours. The main metabolic process is reduction of the nitro group which produces inactive 7-amino derivatives. Less than 1% of a given dose of clonazepam is excreted unchanged in the urine.

Mode of action
Clonazepam is believed to exert its antiepileptic effects by a similar mechanism to that reported for clobazam.

Interactions
The interactions with clonazepam are of a similar nature to those with clobazam. Once again, all enzyme inducers will accelerate its breakdown.[41]

Children
Children absorb and eliminate clonazepam rapidly with a higher relative clearance value than adults.[42] They show an almost linear relationship between serum plasma levels and oral dose, although the mean ratio tends to be lower than in adults.

Ethosuximide

Pharmacokinetics
Ethosuximide is a substituted succinimide, which is absorbed rapidly on oral administration. Maximal concentrations are reached 3-7 hours after dosing.[43] The bioavailability is 90-95%, and this is not altered by the use of capsules or syrup. Ethosuximide is only minimally bound to plasma proteins. It is evenly distributed throughout the body, producing a volume of distribution of 0.6-0.9 L/kg. The elimination half-life of ethosuximide is 20-60 hours, with faster clearance rates occurring in children than in adults. Due to its slow clearance, once daily dosing may be sufficient to achieve stable serum concentrations,[44] although dividing the dose may be necessary if side effects occur.[45] Ethosuximide is subject to hepatic metabolism, although 25% of a dose is excreted unchanged by the kidneys.[46]

Mode of action
The clinical expression of generalised absence seizures is believed to rely on unique low threshold activated calcium-mediated burst firing in thalamocortical relay neurones.[47] Voltage-clamp recordings from isolated thalamic neurones suggest that these cells express both L- and T-type voltage-sensitive calcium channels[48] and that they are unusual within the

central nervous system in having a relatively large T-type current.[21] Ethosuximide selectively blocks voltage-sensitive calcium influx through the T-type calcium channel in these cells.[49] The block is voltage-dependent and is markedly reduced at depolarised potentials.

Interactions
Ethosuximide neither inhibits nor induces drug metabolism, but circulating levels are decreased by enzyme inducers such as phenytoin and carbamazepine and increased by sodium valproate.[50]

Children
Browne et al.[51] demonstrated a statistically significant relationship between dosage and plasma ethosuximide concentrations in children with absence seizures. In children under 10, the rise in plasma concentration with increasing dosage was less than in older children.

Phenobarbital

Pharmacokinetics
Phenobarbital is a 5,5-substituted barbituric acid. It has relatively low lipid solubility and is absorbed slowly following oral administration. Peak plasma levels occur at an average of two hours after ingestion, although this may be delayed for up to 12 hours in some circumstances.[52] The majority of oral phenobarbital is absorbed in the small intestine due to its large surface area and slow transit time. The bioavailability of phenobarbital is high at 95-100%. The volume of distribution is 0.51-0.57 L/kg and it is 48-54% protein bound. Its half life is 72-144 hours, elimination occurring by a combination of hepatic metabolism and renal excretion, with delays in the presence of impaired hepatic or renal function. In the liver, phenobarbital is subject to parahydroxylation, then conjugation with glucuronic acid. The parahydroxylated metabolite accounts for up to 20% of excreted drug. 25% of a phenobarbital dose is excreted unchanged by the kidneys. Urinary pH and flow affect phenobarbital reabsorption in the distal nephron, with alkalinisation and increased urinary flow raising its excretion.

Mode of action
Traditionally, the barbiturates were believed to exert their pharmacological effects via an action at the $GABA_A$-receptor,[21] and this has been elegantly confirmed from single channel patch clamp studies in cultured mouse spinal cord neurones.[38] The barbiturates shift the receptor to a state in which the affinity for GABA is increased and the rate at which GABA dissociates is decreased.[53] This leads to an increased probability of opening the chloride ion channel and a prolonged duration of opening, without affecting the single channel conductance.[38] Additional mechanisms have been proposed to confer the relative antiepileptic potency

of this drug over other more sedative barbiturate compounds. These include an inhibitory action on calcium influx through voltage-sensitive N- and L-type calcium channels,[54] a blockade of non-N-methyl-D-aspartate (NMDA) type glutamate receptors,[55] and an enhancement of voltage-dependent potassium currents.[56]

Interactions
Phenobarbital is a powerful enzyme inducer that can accelerate the metabolism of many lipid soluble drugs.[57] Phenobarbital levels are themselves increased by sodium valproate[24] and by dextropropoxyphene.[58]

Children
In newborns, absorption of oral phenobarbital is delayed and incomplete.[59] In infants older than 6 weeks and in children, the extent and rates of absorption are similar. Relative hypoalbuminaemia is thought to be the cause of lower phenobarbital protein binding in this patient population.[60] The volume of distribution in newborns is larger, while in older infants and children, the values are similar to those in adults. The elimination half life is longest in newborns and shortest in infants aged 6 weeks to 12 months.

Phenytoin

Pharmacokinetics
Phenytoin is the generic name for 5,5-diphenylhydantoin. It is a weak organic acid that is poorly soluble in water. Maximum absorption occurs in the duodenum following oral administration due to increased solubility at alkaline pH and the presence of bile salts.[61] Absorption is slow as a consequence of low solubility in gastrointestinal fluids, with peak plasma levels occurring 4-8 hours after a single dose, but with wide interindividual variation on long-term treatment. Phenytoin is 90-93% protein bound and has a bioavailability of 85-95%. In its unbound form, it undergoes rapid diffusion with distribution to cerebrospinal fluid, saliva, tears and other body fluids, giving a distribution volume of 0.5-0.7 L/kg.

Phenytoin is metabolised in the liver to its main metabolite, 5-(*p*-hydroxyphenyl)-5-phenylhydantoin, which is conjugated with glucuronic acid and as such excreted in the urine. This conversion accounts for 50-70% of a given phenytoin dose, with 5% of the parent drug being excreted unchanged by the kidney. The elimination half-life of phenytoin is 9-40 hours. Phenytoin is subject to zero order kinetics due to saturable hepatic metabolism within the therapeutic range of serum levels. A small increase or decrease in dose may, therefore, result in a large change in serum concentration (figure 2).

The metabolism of phenytoin is not affected by mild hepatic impairment, but dosage reduction is necessary in patients with advanced liver disease. Phenytoin binding is affected by renal disease. In patients with impaired renal function, the total phenytoin concentration can fall to half the anticipated level,[62] although standard doses can still be used with the aim of keeping the concentration of unbound drug in the range 1-2 g/ml.[3] Febrile illness may increase phenytoin metabolism.[63]

Mode of action
Like carbamazepine, phenytoin has been shown to stabilise the inactive form of the sodium channel in a voltage-dependent manner, to slow the rate of recovery from sodium channel inactivation, and to shift the steady-state sodium inactivation curve to more negative voltages.[20] This is the only known action of phenytoin that can adequately explain its ability to suppress seizures without causing generalised depression of the central nervous system.[21] Of interest is the finding that phenytoin has a stronger slowing effect than carbamazepine,[20] which may result in slightly dissimilar actions under different conditions of repetitive firing and may help to explain the subtle clinical difference between these drugs. Other proposed mechanisms of phenytoin's action include an inhibitory effect on T-type[64] and non-T-type calcium channels[22] and a potentiation of GABA-mediated synaptic inhibition.[65]

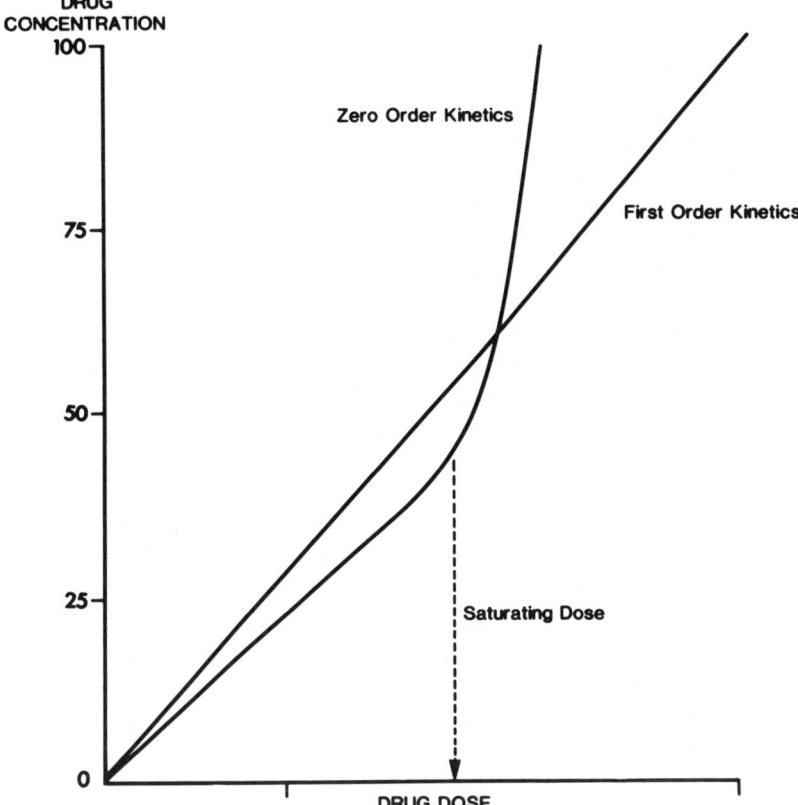

Figure 2. Comparison of first and zero order kinetics.

Interactions
Phenytoin is a powerful enzyme inducer, which interacts with many other drugs. It can reduce the concentration and, therefore, the efficacy of other anticonvulsants, corticosteroids, anticoagulants, oral contraceptives, cyclosporin and theophylline.[66] As its own metabolism is saturable, it is a prime target for enzyme inhibitors such as allopurinol, amiodarone, chloramphenicol, cimetidine, imipramine, isoniazid, metronidazole, phenothiazines and sulphonamides.[50] The absorption of phenytoin is affected by agents that affect gastric pH, gastric emptying and gastrointestinal motility such as antacids, cytotoxics, sucralfate and some foods.[24]

Children
In newborns and infants younger than 3 months, phenytoin absorption is slow and incomplete.[67] Hepatic enzyme activity at this time is low and so intravenous phenytoin supplementation may be necessary. Hepatic function improves rapidly in the first few weeks of life and so the elimination half-life falls. Infants and children have faster metabolic rates than adults leading to the requirement for relatively higher doses given at shorter dosage intervals.[68]

Primidone

Pharmacokinetics
Primidone is a weakly acidic cyclic amide. It is absorbed rapidly on oral administration and has a bioavailability of 90-100%.[69] Primidone is 20-30% protein bound and distributes relatively slowly to nonvascular tissues, with a volume of 0.4-0.6 L/kg. Its elimination half-life varies between 4-12 hours. It is metabolised in the liver with the production of two active metabolites,[70] phenobarbital, formed by the oxidation of the methylene group, and phenylethylmalonamide (PEMA), produced by the cleavage of the pyrimidine ring. Primidone and its metabolites are mainly eliminated renally, with 40% of a dose excreted unchanged by the kidneys. Clearance is influenced by age, duration of treatment, and concomitant drug therapy.

Mode of action
The mechanism of action of primidone is less well documented than that of other barbiturate compounds.[69] Nevertheless, a direct action at the $GABA_A$ receptor, similar to that proposed for phenobarbital, is likely to play a major role in its antiepileptic properties. In the absence of a clear understanding of the cellular effects of the other active metabolite, PEMA, the mechanism of primidone's action has often been assumed to reflect that of phenobarbital.[21] Whether such an assumption is well founded remains to be determined.

Interactions
As phenobarbital is a principle metabolite of primidone, their drug interactions are similar. However, when primidone metabolism is stimulated by, for example other antiepileptic drugs, its effect is increased due to the production of more phenobarbital which has a longer half-life and therefore accumulates.[69]

Children
The metabolism of primidone in children has not been studied to the same extent as in adults. Absorption is known to be faster in infants,[71] and so peak levels tend to occur earlier. In children over age 7, the elimination half-life of primidone is similar to that of adults.

Valproic acid

Pharmacokinetics
Valproic acid is a branched fatty acid, which is slightly soluble in water.[72] The drug is available in several forms, including the parent compound, its sodium salt, its amide derivative, and a combination of valproic acid and sodium valproate (divalproex sodium). Following oral administration, absorption rates are dependent upon the dosage and the formulation. Valproate is absorbed efficiently from syrups, capsules and tablets with peak plasma levels rapidly attained. Enteric coated preparations have variable absorption rates with peak levels occurring after 3-6 hours and with controlled-release tablets 10-12 hours after dosing. Valproate has a bioavailability of 100% and is highly bound to circulating albumin (88-92%), the binding being concentration-dependent. In patients with low albumin levels, the protein bound fraction is less. The distribution volume of valproate is 0.09-0.17 L/kg.

Valproate's elimination half life is relatively short at 7-17 hours, with 96% of a dose being eliminated by hepatic metabolism. It undergoes metabolism by a variety of oxidative and conjugation processes, with the formation of a number of active and inactive metabolites.[73] Some of these metabolites are believed to contribute significantly to both the anticonvulsant properties and hepatotoxic and teratogenic side effects of the drug.[74] Renal excretion forms a minor route of unchanged valproate excretion, with the mean fraction of a dose excreted in this way ranging from 1.8-3.2%. Valproate use should be avoided if possible in the presence of liver disease due to its potential hepatotoxic effects.

Mode of action
Sodium valproate has such a wide spectrum of antiepileptic activity that it has been attractive to speculate that it may exert its pharmacological effects by a combination of mechanisms. At present several different cellular effects have been reported, although the extent to which any one

of these might contribute to the antiepileptic action of the drug remains to be determined. Valproate has been shown to increase brain GABA levels in experimental animals[75] and CSF GABA levels in man[76] by an inhibitory action on the enzymes of the GABAergic pathway.[77] The drug has also been proposed to increase the synthesis of GABA via a stimulatory action on glutamic acid decarboxylase activity.[78] Like phenytoin and carbamazepine, valproate has been shown to exert a use- and voltage-dependent limitation of sustained repetitive firing of sodium-dependent action potentials at therapeutically relevant concentrations.[79] However, detailed biochemical studies have failed to link these observations to a direct action at the neuronal sodium channel.[21]

The observation that valproate is one of the most effective drugs against generalised absence seizures has suggested that it may interact with the voltage-sensitive T-type calcium channel in a manner analogous to ethosuximide. Initial studies failed to demonstrate any such effect in isolated thalamic neurones,[80] although a subsequent investigation revealed a modest reduction of T-type calcium current in rat primary afferent neurones, albeit at supra-therapeutic concentrations.[81] Acute doses of valproate have been found to decrease brain levels of the excitatory amino acid aspartate, without affecting those of glutamate or GABA.[82] This decrease in aspartate concentration has been shown to correspond with the period of anticonvulsant action of the drug.[83]

Interactions
Valproate is the only first-line anticonvulsant that is not an enzyme inducer. It inhibits oxidative metabolism to a minor extent, increasing the circulating concentrations of other antiepileptic drugs such as carbamazepine epoxide, phenytoin, phenobarbital, and ethosuximide.[50] Its withdrawal can lead to a fall in the concentration of the target drug, with possible attenuation of its pharmacological effect. Enzyme induction by other antiepileptic drugs may result in higher dose requirements of valproate. Valproate is displaced from its plasma protein binding sites by aspirin, which also inhibits its metabolism.[84] Valproate does not interfere metabolically with the hormonal components of the oral contraceptive pill.

Children
The elimination half-life of valproate is related to age, being longest in neonates and shortest in children.[3] Relatively higher doses may be necessary at shorter intervals in children due to the combination of lower volume of distribution and more efficient clearance. Tablet, syrup and sprinkle formulations of valproate are available for children. Sprinkle formulations have also been suggested to have benefits in terms of ease of administration, twice daily dosing, reduction in fluctuations in serum concentration, and enhanced compliance due to improved palatability.[85] A controlled-release preparation has been reported to provide more stable pharmacokinetic features in children with intractable epilepsy than

conventional formulations.[10]

NEW ANTIEPILEPTIC DRUGS

The pharmacokinetic properties of a range of new antiepileptic drugs are summarised in *Table 2*. Each will be discussed individually in alphabetical order.

Table 2. Pharmacokinetics of new antiepileptic drugs

Drug	Absorption bio-availability	Distribution volume (l/kg)	Protein binding (%)	Half-life (hours)	Route(s) of elimination
FELBAMATE	Slow absorption (95-100%)	0.7 - 0.9	22 - 26	13 - 23	Hepatic metabolism Renal excretion
GABAPENTIN	Rapid absorption (60%)	0.7 - 0.9	0	5 - 7	Renal excretion
LAMOTRIGINE	Rapid absorption (95-100%)	0.8 - 1.2	55	22 - 36	Metabolised to glucuronide conjugate
OXCARBAZEPINE	Rapid absorption (95-100%)	0.7 - 0.8*	40*	8 - 10*	Hepatic conversion to active moiety
VIGABATRIN	Rapid absorption (60-80%)	0.6 - 1.0	0	5 - 7	Renal excretion

* parameters for active metabolite 10, 11-dihydro-10-hydroxycarbamazepine

Felbamate

Felbamate is a dicarbamate, which is only slightly soluble in water.[86] On oral administration it is slowly absorbed, has a bioavailability of 95-100%, and a volume of distribution of 0.7-0.9 L/kg. Peak concentrations occur around 4 hours after dosing. The elimination half-life varies between 13 and 23 hours. Metabolism by hydroxylation occurs in the liver with the production of a number of inactive metabolites. These are excreted renally along with a substantial proportion of unchanged felbamate.

The mechanism by which felbamate exerts its antiepileptic effects remains unclear. Although the drug has failed to demonstrate any effect on ligand binding to the $GABA_A$ receptor,[20] its potentiation of diazepam's effects against experimentally-induced seizures[87] and its facilitation of GABA responses in cultured hippocampal neurones under voltage-clamp conditions[88] has suggested some indirect action at this site. The drug has also been shown to attenuate NMDA responses in cultured hippocampal neurones by a channel blocking mechanism[88] and to inhibit binding of 5,7-dichlorokynurenic acid to the strychnine-sensitive glycine recognition site on the NMDA subtype of glutamate receptor.[89]

Phenytoin, sodium valproate and phenobarbital levels rise by 20-30% when felbamate is added into the treatment regimen, whereas serum carbamazepine levels fall by a similar amount.[90] The latter effect is offset by greater production of the active metabolite of carbamazepine, the 10,11 epoxide.[91] The rate of clearance of felbamate is increased by enzyme inducing antiepileptic drugs.[92] The effect of valproate on felbamate levels, however, is less well defined.[93]

Gabapentin

Gabapentin, an analogue of γ-aminobutyric acid, is hydrophilic and can cross the blood-brain barrier.[94] It is rapidly absorbed on oral administration, has a bioavailability of around 60% at therapeutic dosage, and a volume of distribution of 0.7-0.9 L/kg. The lack of proportionality between doses exceeding 1200 mg daily and plasma levels are due to saturation of the amino acid transport mechanism responsible for its absorption.[95] Gabapentin is not protein bound. It has an elimination half-life of 5-7 hours and is excreted unchanged in the urine. Renal elimination of the drug mirrors that of creatinine. Gabapentin lacks any important drug interactions, although antacids and cimetidine reduce its bioavailability and clearance respectively by around 10%.[24]

At present, gabapentin's precise mechanism of action remains unclear. Despite its structural similarity to GABA, gabapentin has often failed to influence GABA pharmacology.[96] It has, however, been shown to increase GABA turnover in certain regions of the brain.[97] There is accumulating evidence to suggest that gabapentin binds with high affinity to a specific site within the brain, proposed to be the plasma membrane L-amino acid transporter, where it is potently displaced by the anticonvulsant compound 3-isobutyl GABA.[98] Investigations are ongoing to characterise this interaction more fully and to determine its relevance to the antiepileptic effects of the drug. Gabapentin also limits high frequency action potential firing by central neurones in cell culture.[99]

Lamotrigine

Lamotrigine, a phenyltriazine derivative, is absorbed rapidly and completely following oral administration.[100] Its bioavailability is 95-100% with peak concentrations occurring around 3 hours after dosing. Lamotrigine is approximately 55% protein bound and has a distribution volume of 0.8-1.2 L/kg. It is subject to first order kinetics both in healthy volunteers[101] and in treated epileptic patients.[102] It has an elimination half-life of 22-36 hours and is biodegraded mainly by hepatic metabolism to the glucuronide conjugate.

Lamotrigine is believed to stabilise presynaptic neuronal membranes by blocking voltage-dependent sodium channels, thereby preventing the release of excitatory neurotransmitters, particularly glutamate and aspartate (figure 3).

This was demonstrated in rat cerebral cortical slices incubated with veratrine, a depolarising agent that promotes neurotransmitter release by opening sodium channels. Lamotrigine inhibited the release of glutamate and aspartate stimulated by veratrine, but did not influence potassium-mediated amino acid secretion.[103] In addition, the neurotoxic effect of intrastriatal kainic acid in rats mediated by glutamate could be prevented by prior administration of lamotrigine, but not that of ibotenate, a neurotoxin which acts independently of glutamate.[104]

In vitro intracellular recording has also supported an important action of lamotrigine on sodium channels. Sustained repetitive firing of sodium-dependent action potentials in mouse spinal cord cultured neurones was blocked by treatment with lamotrigine.[105] Burst firing from neurones in primary neuroglial cell culture was antagonised by the effect of lamotrigine predominantly on sodium and to a lesser extent on calcium currents[106] As lamotrigine is effective in the primary generalised epilepsies including absences and myoclonic jerks, this is unlikely to be the whole story.[107]

The metabolism of lamotrigine is induced by enzyme inducing anticonvulsants, such as carbamazepine and phenytoin to about double that in naive patients dropping the elimination half-life to around 15 hours.[108] Valproate, however, inhibits its conjugation leading to a half-life around 60 hours.[109] A combination of valproate and an inducer brings it back to near 30 hours. In children aged 5-11 years with treated epilepsy, these wide variations in the half-life of lamotrigine have been confirmed.[110] Those on concurrent sodium valproate had a lamotrigine half-life of 54-94 hours, compared to 11 hours on enzyme inducing drugs. Lamotrigine itself neither induces nor inhibits hepatic enzymes and so does not affect the metabolism of lipid soluble drugs, including warfarin or the hormonal components of the oral contraceptive.[100]

Lamotrigine does not alter the concentrations of other antiepileptic drugs. There have been, however, reports of symptoms of neurotoxicity (headache, nausea, dizziness, diplopia, ataxia) in patients taking carbamazepine in whom lamotrigine has been introduced.[111] These disappeared when the dose of either drug was reduced.

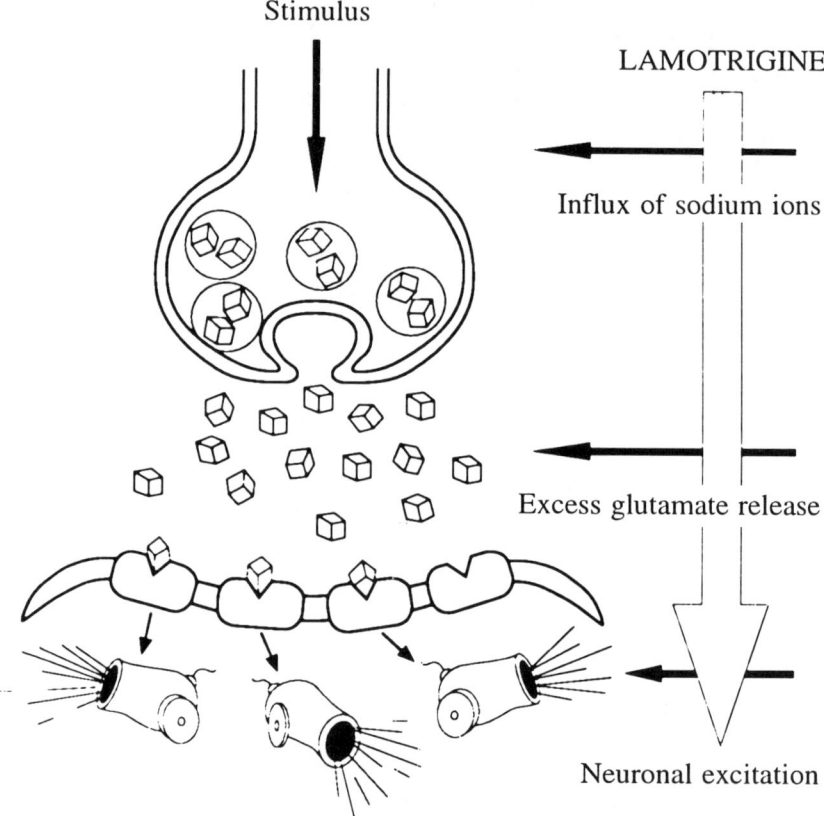

Figure 3. Putative mechanism of action of lamotrigine

The suggestion from an open study that carbamazepine epoxide levels were elevated by lamotrigine[112] has not been supported by results from a placebo-controlled, dose-ranging study of additional lamotrigine in patients with refractory epilepsy.[113] This important interaction would appear, therefore, to have a pharmacodynamic basis.

Oxcarbazepine

Oxcarbazepine, the 10-keto analogue of carbamazepine, is functionally a pro-drug.[114] Following oral administration it is rapidly reduced to the active metabolite 10,11-dihydro-10-hydroxy-carbamazepine (figure

4), which is itself conjugated with glucuronic acid.[115]

Serum concentrations and urinary excretion are best measured in terms of the active moiety rather than the parent oxcarbazepine. The 10-hydroxy metabolite has a bioavailability of 95-100% and maximal serum concentrations occur within 4-6 hours of a single oral dose. It is 40% protein bound, has a volume of distribution of 0.7-0.8 L/kg and an elimination half-life of 8-10 hours. Oxcarbazepine, unlike carbamazepine, does not undergo autoinduction of metabolism,[116] and its administration has no effect on the metabolism of other anticonvulsants.[117] However, in patients taking carbamazepine, phenytoin and phenobarbital, the area under the concentration-time curve of the active metabolite is lower than in controls. This suggests a small, probably clinically irrelevant, induction effect of these drugs on its elimination.[117-118]

Although little data exist on the mechanism of oxcarbazepine's action, similar experimental and clinical anticonvulsant profiles to carbamazepine would suggest that it also exerts it effects by blockade of voltage-sensitive sodium channels. Its ability to limit sustained repetitive firing in cultured mouse central neurones would support this hypothesis.[119] The drug has also demonstrated 4-aminopyridine-sensitive effects on penicillin-induced epileptiform discharges in the hippocampal slice suggesting an additional action on neuronal potassium channels.[119]

Volunteer studies have suggested that oxcarbazepine does not interfere with warfarin metabolism.[120] However, in some patients receiving oxcarbazepine, decreased bioavailability of the oral contraceptive pill has been noted.[121] Oxcarbazepine reduced the area under the concentration-time curve of the calcium antagonist felodipine by 28%.[122] It appears to induce selectively a single isoform of cytochrome p450, namely IIIA.[123]

Vigabatrin

Vigabatrin (γ-vinyl GABA) is a synthetic derivative of the inhibitory neurotransmitter GABA.[124] It has two enantiomers one of which, the S-enantiomer, is pharmacologically active. Vigabatrin is water soluble. It is well absorbed after oral administration with a bioavailability of 60-80%. Peak concentrations occur around 3 hours after dosing. It is excreted unchanged in the urine and has an elimination half-life of around 6 hours. Vigabatrin is not significantly protein bound, and does not induce or inhibit hepatic metabolism.[24]

Vigabatrin exerts its antiepileptic effect by an enzyme-activated suicide inhibition of GABA-transaminase (figure 5), the enzyme directly responsible for the metabolic degradation of the inhibitory neurotransmitter GABA.[126] As such, it has been shown to elevate GABA levels in the plasma, CSF, and brain of experimental animals,[127] and the plasma and CSF in man.[128] An inhibitory action on cellular GABA uptake has also been reported with the drug.[129]

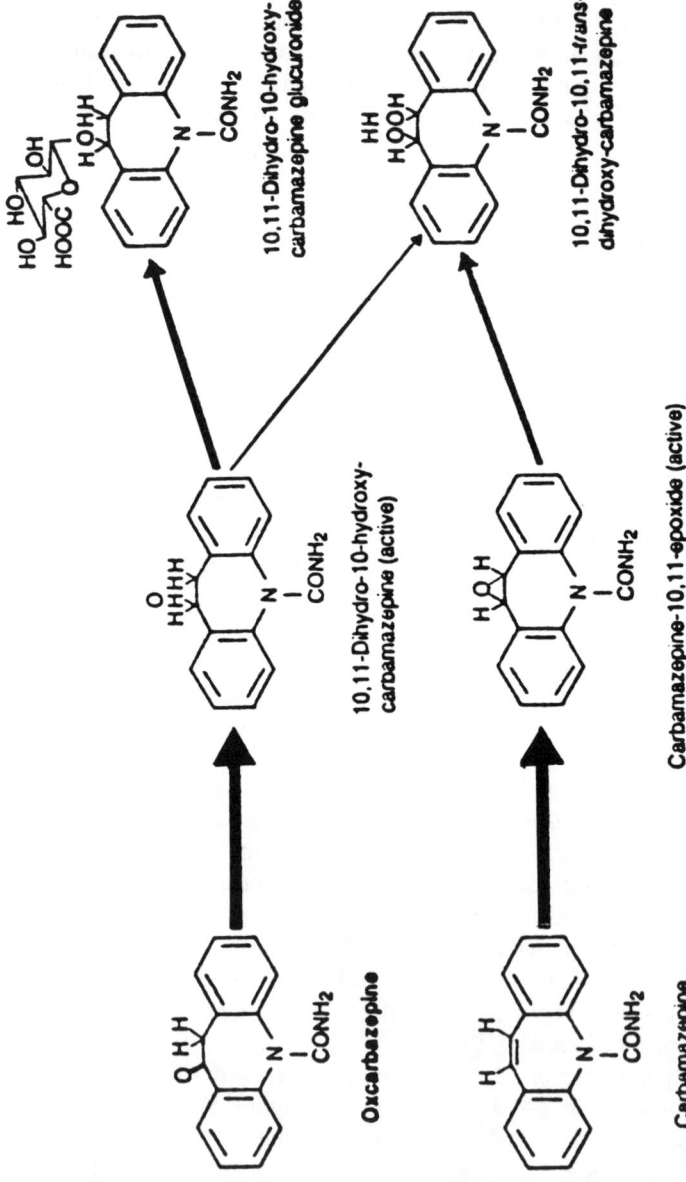

Figure 4. Metabolism of oxcarbazepine and carbamazepine

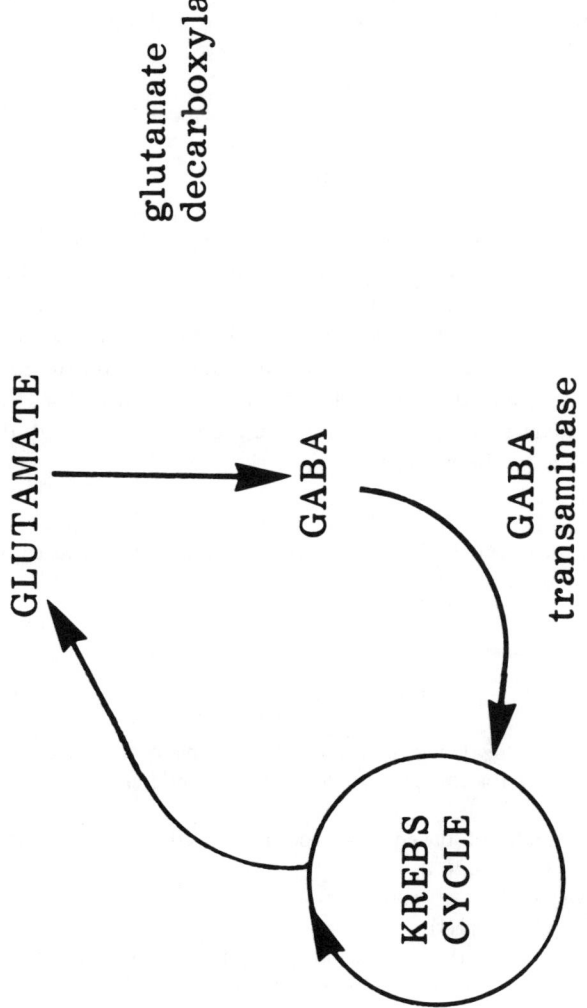

Figure 5. Synthesis and breakdown of GABA. Vigabatrin binds irreversibly to GABA transaminase

Vigabatrin's renal excretion is unchanged by concomitant medication. However, vigabatrin decreases phenytoin levels by around 20% by an unknown mechanism.[130] A 7% and 11% decrease respectively in phenobarbital and primidone levels has been reported in patients receiving additional vigabatrin,[131] although there were no associated alterations in the patients' clinical status. It has not been found to affect the concentrations of carbamazepine or sodium valproate.[132] Likewise, no change has been noted in clonazepam, clobazam, ethosuximide or oxcarbazepine levels.

CONCLUSION

Knowledge of differences in drug handling and response in children with epilepsy is limited. Many of the recommendations for treating children are based on studies in adults. In general, however, more efficient clearance rates in children of most antiepileptic drugs results in a higher ratio of the steady-state drug level to dose than in adults. Information on pharmacokinetic profiles in children remains an area for further investigation and research in order to maximise the potential use of already available and novel antiepileptic drugs.

Acknowledgement
We are grateful to Mrs. Moya Dewar for expert secretarial assistance in the production of this manuscript.

REFERENCES

1. Commission on Antiepileptic Drugs of the International League Against Epilepsy. Guidelines for antiepileptic drug trials in children. Epilepsia, 1994; 35: 94-100.
2. Rylance GW. Treatment of epilepsy and febrile convulsions in children. Lancet, 1990; 336: 488-491.
3. Leppik IE. Metabolism of antiepileptic medication: newborn to elderly. Epilepsia, 1992; 33 (suppl 4): 32-40.
4. Thomson AH, Brodie MJ. Pharmacokinetic optimisation of anticonvulsant therapy. Clin. Pharmacokinet., 1992; 23: 216-230.
5. Gilman JT, Duchowny M. Childhood epilepsy: current therapeutic recommendations. CNS Drugs, 1994; 1: 180-192.
6. Blain PG, Mucklow JC, Bacon CJ, et al. Pharmacokinetics of phenytoin in children. Br. J. Clin. Pharmac., 1981; 12: 659-661.
7. Eeg-Olofsson O, Nilsson HL, Tonnby B, et al. Diurnal variation in carbamazepine and carbamazepine 10,11-epoxide in plasma and saliva in children with epilepsy: a comparison between conventional and slow-release formulations. J. Child Neurol., 1990; 5: 159-165.
8. Ryan SW, Forsythe I, Hartley R, et al. Slow release carbamazepine in

treatment of poorly controlled seizures. Arch. Dis. Child, 1990; 65: 930-935.
9. Brouwer OF, Pieters MS, Edelbroek PM, et al. Conventional and controlled release valproate in children with epilepsy: a cross-over study comparing plasma levels and cognitive performances. Epilepsy Res., 1992; 13: 245-253.
10. Imaizumi T, Izumi T, Fukuyama Y. A comparative clinical and pharmacokinetic study of a new slow-release versus conventional preparations of valproic acid in children with epilepsy. Brain Dev., 1992; 14: 304-308.
11. Gilman JT. Intractable childhood epilepsy: issues in pharmacotherapy. J. Epilepsy, 1990; 3 (suppl): 21-24.
12. Morselli PL. Carbamazepine absorption, distribution and excretion. In: Levy RH, Dreifuss FE, Mattson RH, et al., editors. Antiepileptic Drugs (third edition), New York: Raven Press, 1989: 473-490.
13. Macphee GJA, Butler E, Brodie MJ. Intradose and circadian variation in circulating carbamazepine and its epoxide in epileptic patients: a consequence of auto-induction of metabolism. Epilepsia, 1987; 28: 286-294.
14. Belluci G, Berti G, Chiappe C, et al. The metabolism of carbamazepine in humans: steric course of the enzymatic hydrolysis of the 10,11-epoxide. J. Med. Chem., 1987; 30: 768-773.
15. Larkin JG, McLellan A, Munday A, et al. A double-blind comparison of conventional and controlled release carbamazepine in healthy subjects. Br. J. Clin. Pharmac., 1989; 27: 313-322.
16. McKee PJW, Blacklaw J, Carswell A, et al. Double-dummy compariso between once and twice daily dosing with modified-release carbamazepine in epileptic patients. Br. J. Clin. Pharmac., 1993; 36: 257-261.
17. McKee PJW, Blacklaw J, Butler E, et al. Monotherapy with conventional and controlled-release carbamazepine: a double-blind, double dummy comparison in epileptic patients. Br. J. Clin. Pharmac., 1991; 32: 99-104.
18. Johnson FN, Brodie MJ. Pharmacokinetics of carbamazepine: implications for clinical use. Rev. Contemp. Pharmacother., 1990; 1: 85-98.
19. Macdonald RL. Antiepileptic drug actions. Epilepsia, 1989; 30 (suppl 1): S19-S28.
20. Macdonald RL, Kelly KM. Antiepileptic drug mechanisms of action. Epilepsia, 1993; 35 (suppl 5): S1-S8.
21. Rogawski MA, Porter RJ. Antiepileptic drugs: pharmacological mechanisms and clinical efficacy with consideration of promising developmental stage compounds. Pharmacol. Rev., 1990; 42: 223-286.
22. Crowder JM, Bradford JF. Common anticonvulsants inhibit Ca^{2+} uptake and amino acid neurotransmitter release in vitro. Epilepsia, 1987; 28: 378-382.
23. Marangos PJ, Patel J, Smith KD, et al. Adenosine antagonist properties of carbamazepine . Epilepsia, 1987; 28: 387-394.
24. McKee PJW, Brodie MJ. Pharmacokinetic interactions with antiepileptic drugs. In: Trimble MR, editor. New anticonvulsants: advances in the treatment of epilepsy. Chichester, England: John Wiley, 1994: 1-33.
25. Zielinski JJ, Haidukewych D. Dual effect of carbamazepine-phenytoin interaction. Ther. Drug Monit., 1987; 9: 21-23.
26. Browne TR, Szabo GK, Evans JK, et al. Carbamazepine increases serum concentration and reduces phenytoin clearance. Neurology, 1988; 38: 1146-1150.
27. Gillham RA, Williams W, Weidmann K, et al. Concentration-effect relationships with carbamazepine and its epoxide on psychmotor and

cognitive function in epileptic patients. J. Neurol. Neurosurg. Psychiatry, 1988; 51: 929-933.
28. Brodie MJ, Forest G, Rapeport WG. Carbamazepine 10,11 epoxide concentrations in epileptics on carbamazepine alone and in combination with other anticonvulsants. Br. J. Clin. Pharmac., 1983; 16: 747-750.
29. Macphee GJA, Mitchell J, Wiseman L, et al. Effect of sodium valproate on carbamazepine disposition and psychomotor profile in man. Br. J. Clin. Pharmac., 1988; 25: 59-66.
30. McKee PJW, Blacklaw J, Butler E, et al. Variability and clinical relevance of the interaction between sodium valproate and carbamazepine in epileptic patients. Epilepsy Res., 1992; 11: 193-198.
31. Battino D, Bossi L, Croci D, et al. Carbamazepine plasma levels in children and adult influence of age and associated therapy. Ther. Drug Monit., 1980; 2: 315-322.
32. Pieters MS, Jennekens-Schinkel A, Stijen T, et al. Carbamazepine controlled release compared with conventional carbamazepine: a controlled study of attention and vigilance in children with epilepsy. Epilepsia, 1992; 33: 1137-1144.
33. Shorvon SD. Benzodiazepines. Clobazam. In: Levy RH, Mattson RH, Meldrum BS, et al. editors. Antiepileptic Drugs (third edition), New York: Raven Press, 1989: 821-840.
34. Marcia Divoll BS, Greenblatt D, Ciraulo D, et al. Clobazam kinetics: intrasubject variability and effect of food absorption. J. Clin. Pharmacol., 1982; 22: 69-73.
35. Haigh JRM, Pullar T, Gent JP, et al. N-desmethylclobazam: a possible alternative to clobazam in the treatment of refractory epilepsy? Br. J. Clin. Pharmac., 1987; 23: 213-218.
36. Ehlert FJ. Inverse agonists, cooperativity and drug action at benzodiazepine receptors Trends Pharmacol. Sci., 1986; 7: 28-32.
37. Macdonald RL, Barker JL. Benzodiazepines specifically modulate GABA-mediated postsynaptic inhibition in cultured mammalian neurones. Nature, 1978; 271: 563-564.
38. Twyman RE, Rogers CJ, Macdonald RL. Differential regulation of gamma-aminobutyric acid receptor channels by diazepam and phenobarbital. Ann. Neurol., 1989; 25: 213-220.
39. Haefely W. Benzodiazepines: mechanisms of actionIn: Levy RH, Mattson RH, Meldrum BS, et al. editors. Antiepileptic Drugs (third edition), New York: Raven Press, 1989: 721-734.
40. Taerber K, Badian M, Brettel H, et al. Kinetic and dynamic interaction of clobazam and alcohol. Br. J. Clin. Pharmac., 1979; 7 (suppl 1): 91S-97S.
41. Pisani F, Perucca E, Di Perri R. Clinically relevant antiepileptic drug interactions. J. Int. Med. Res., 1990; 18: 1-15.
42. Sato S. Benzodiazepines Clonazepam In: Levy RH, Mattson RH, Meldrum BS et al. editors Antiepileptic Drugs (third edition), New York: Raven Press, 1989: 765-784.
43. Buchanan RA, Fernandez L, Kinkel AW. Absorption and elimination of ethosuximide in children. J. Clin. Pharmacol., 1969; 9: 393-398.
44. Buchanan RA, Kinkel AW, Turner JL, et al. Ethosuximide dosage regimens. Clin. Pharmacol. Ther., 1976; 19: 143-147.
45. Dooley JM, Camfield PR, Camfield CS, et al. Once daily ethosuximide in the treatment of absence epilepsy. Pediatr. Neurol., 1990; 6: 38-39.

46. Glazko AJ. Antiepileptic drugs: biotransformation, metabolism and serum half-life. Epilepsia, 1975; 16: 367-391.
47. Vergnes M, Marescaux C, Depaulis A, et al. Spontaneous spike and wave discharges in thalamus and cortex in a rat model of genetic petit mal-like seizures, Exp Neurol. 1987; 96: 127-136.
48. Hernandez-Cruz A, Pape HC. Identification of two calcium currents in acutely dissociated neurons from the rat lateral geniculate nucleus. J. Neurophysio.,l 1989; 61: 1270-1283.
49. Coulter DA, Huguenard JR, Prince DA. Characterization of ethosuximide reduction of low-threshold calcium current in thalamic neurons. Ann. Neurol., 1989; 25: 582-593.
50. Brodie MJ. Drug interactions and epilepsy. Epilepsia, 1992: 33 (Suppl 1): 13S-22S.
51. Browne TR, Dreifuss FE, Dyken PR, et al. Ethosuximide in the treatment of absence (petit mal) seizures. Neurology, 1975; 25: 515-524.
52. Nelson E, Powell JR, Conrad K, et al. Phenobarbital pharmacokinetics and bioavailability in adults. J. Clin. Pharmacol., 1982; 22: 141-148.
53. Yang JSJ, Olsen RW. Gamma-aminobutyric acid receptor binding in fresh mouse brain membranes at 22°C: ligand-induced changes in affinity. Mol. Pharmacol., 1987; 32: 266-277.
54. Gross RA, Macdonald RL. Barbiturates and nifedipine have different and selective effects on calcium currents of mouse DRG neurones in culture: a possible basis for differing clinical actions. Neurol., 1988; 38: 443-451.
55. Miljkovic Z, Macdonald RL. Voltage-dependent block of excitatory amino acid currents by pentobarbital. Brain Res., 1986; 376: 396-399.
56. Huguenard JR, Wilson WA. Barbiturate-induced alterations in the kinetic parameters of slow outward current in aplysia giant neurones. J. Pharmacol. Exp. Ther., 1985; 234: 821-829.
57. Patsalos PN, Duncan JS. Antiepileptic drugs. A review of clincally significant drug interactions. Drug Safety, 1993; 9: 156-184.
58. Hansen BS, Dam M, Brandt J, et al. Influence of dextropropoxyphene on steady-state serum levels and protein big of three antiepileptic drugs. Acta Neurol. Scand., 1980; 61: 357-367.
59. Jalling B. Plasma concentrations of phenobarbital in the treatment of seizures in the newborns. Acta Paediatr. Snd., 1975; 64: 514-524.
60. Rust RS, Dodson WE. Phenobarbital absorption, distribution, and excretion In: Levy RH, Mattson RH, Meldrum BS, et al. editors Antiepileptic Drugs (third edition), New York: Raven Press, 1989: 293-304.
61. Woodbury DM. Phenytoin. Absorption, distribution, and excretionIn: Levy RH, Mattson RH, Meldrum BS, et al. editors Antiepileptic Drugs (third edition), New York: Raven Press, 1989: 177-195.
62. Reidenberg MM, Odar-Cederlof I, von Bahr C, et al. Protein binding of diphenylhydantoin and desmethylimipramine in patients with poor renal function. N. Engl. J. Med., 1971; 285: 264-267.
63. Leppik IE, Fisher J, Kriel R, et al. Altered phenytoin clearance with febrile illness. Neurology, 1986; 36: 1367-1370.
64. Twombly DA, Yoshii M, Narahashi T. Mechanisms of calcium channel block by phenytoin. J. Pharmacol. Exp. Ther., 1988; 246: 189-195.
65. Raabe W, Ayala GF. Diphenylhydantoin increases cortical postsynaptic inhibition. Brain Res., 1976; 105: 597-601.
66. Kutt H. Phenytoin interactions with other drugs In: Levy RH, Mattson RH,

Meldrum BS, et al. editors Antiepileptic Drugs (third edition), New York: Raven Press, 1989: 215-232.
67. Jalling B, Boreus LO, Rane A, et al. Plasma concentrations of diphenylhydantoin in young infants. Clin. Pharmacol., 1970; 2: 200-202.
68. Dodson WE. Antiepileptic drug utilization in pediatric patients. Epilepsia, 1984; 25 (suppl 2): 132S-139S.
69. Cloyd JC, Leppik IE. Primidone absorption, distribution, excretion. In: Levy R, Mattson R, Meldrum B, et al. editors. Antiepileptic Drugs (third edition),. New York: Raven Press, 1989: 391-400.
70. Butler TC, Waddell WJ. Metabolic conversion of primidone (mysoline) to phenobarbital. Proc. Soc. Exp. Biol. Med., 1956; 93: 544-546.
71. Smith DB. Primidone. Clinical use. In: Levy RH, Mattson RH, Meldrum BS, et al. editors Antiepileptic Drugs (third edition), New York: Raven Press, 1989: 423-438.
72. Davis R, Peters DH, McTavish D. Valproic acidA reappraisal of its pharmacological properties and clinical efficacy in epilepsy. Drugs, 1994; 47: 332-372.
73. Levy RH, Shin DD. Valproate. Absorption, distribution, and excretion In: Levy RH, Mattson RH, Meldrum BS, et al. editors. Antiepileptic Drugs (third edition), New York: Raven Press, 1989: 583-599.
74. Nau H, Löscher W. Valproic acid and metabolites: pharmacological and toxicological studies. Epilepsia, 1984; 25 (suppl.1): 14S-22S .
75. Loscher W. Valproate-induced changes in GABA metabolism at the subcellular level. Biochem, Pharmacol., 1981; 30: 1364-1366.
76. Loscher W, Siemes H. Valproic acid increases gamma-aminobutyric acid in CSF of epileptic children. Lancet, 1984; ii: 225.
77. Loscher W. Comparative study of the inhibition of GABA aminotransferase by different anticonvulsant drugs. Arch. Int. Pharmacodyn. Ther., 1980; 243: 48-55.
78. Phillips NI, Fowler LJ. The effects of sodium valproate on gamma-aminobutyrate metabolism and behaviour in naive and ethanolamine-O-sulphate pretreated rats and mice. Biochem. Pharmacol., 1982; 31: 2257-2261.
79. McLean MJ, Macdonald RL. Sodium valproate, but not ethosuximide, produces use- and voltage-dependent limitation of high frequency repetitive firing of action potentials of mouse central neurones in cell cultureJ Pharmacol. Exp. Ther., 1986; 237: 1001-1011.
80. Coulter DA, Huguenard JR, Prince DA. Differential effects of petit mal anticonvulsants and convulsants on thalamic neurones: calcium current reduction. Brit. J. Pharmac., 1990; 100: 800-806.
81. Kelly KM, Gross RA, Macdonald RL. Valproic acid selectively reduces the low threshold (T) calcium current in rat nodose neurones. Neurosci. Lett., 1990; 116: 233-238.
82. Schechter PJ, Tranier Y, Grove J. Effect of n-dipropylacetate on amino acid concentrations in mouse brain: correlation with anticonvulsant activity. J. Neurol., 1978; 31: 1325-1327.
83. Chapman AG, Meldrum BS, Mendes E. Acute anticonvulsant activity of structural analogues of valproic acid and changes in brain GABA and aspartate content. Life Sci., 1983; 2023-2031.
84. Goulden KJ, Dooley JM, Camfield PR, et al. Clinical valproate toxicity induced by acetylsalicylic acid. Neurology, 1987; 37: 1392-1394.

85. Cloyd JC, Kriel RL, Jones-Saete CM, et al. Comparison of sprinkle versus syrup formulations of valproate for bioavailability, tolerance and preference. J. Pediatr., 1992; 120: 634-638.
86. Dodson WE. Felbamate. In: Trimble MR editorNew anticonvulsants. Advances in the treatment of epilepsy, Chichester, England: John Wiley, 1994: 61-73.
87. Gordon R, Gels M, Diamantis W, et al. Interaction of felbamate and diazepam against maximal electroshock seizures and chemoconvulsants in mice. Pharmacol. Biochem. Behav., 1991; 40: 109-113.
88. Rho JM, Donevan SD, Rogawski MA. Mechanism of action of the anticonvulsant felbamate: opposing effects on N-methyl-D-aspartate and gamma-aminobutyric acid$_A$ receptors. Ann. Neurol., 1994; 35: 229-234.
89. McCabe RT, Wasterlain CG, Kucharczyk N, et al. Evidence for anticonvulsant and neuroprotectant action of felbamate mediated by strychnine-insensitive glycine receptors. J. Pharmacol. Exp. Ther., 1993; 264: 1248-1252.
90. Brodie MJ. Felbamate: a new antiepileptic drug. Lancet, 1993; 341: 1445-1446.
91. Albani F, Theodore WH, Washington P, et al. Effect of felbamate on plasma levels of carbamazepine and its metabolites. Epilepsia, 1991; 32: 130-132.
92. Wagner ML, Graves NM, Marienau K, et al. Discontinuation of phenytoin and carbamazepine in patients receiving felbamate. Epilepsia, 1991; 32: 398-406.
93. Wagner ML, Graves NM, Leppik IE, et al. The effect of felbamate on valproate disposition. Epilepsia, 1991; 32: 15.
94. Chadwick D. Gabapentin. Lancet, 1994; 343: 89-91.
95. Stewart BH, Kugler AR, Thompson PR, et al. A saturable transport mechanism in the intestinal absorption of gabapentin is the underlying cause of lack of proportionality between increased dose and levels in plasma. Pharmacol. Res., 1993; 102: 276-281.
96. Rock DM, Kelly KM, Macdonald RL. Gabapentin actions on ligand- and voltage-gated responses in cultured rodent neurones Epilepsy Res., 1993; 16: 89-98.
97. Loscher W, Honack D, Taylor CP. Gabapentin increases aminooxyacetic acid-induced GABA accumulation in several regions of rat brain. Neurosci Lett., 1991; 128: 150-154.
98. Taylor CP, Vartanian MG, Yuen PW, et al. Potent and stereospecific anticonvulsant activity of 3-isobutyl GABA relates to in vitro binding. Epilepsy Res., 1993; 14: 11-15.
99. Wamil AW, McLean MJ. Limitation by gabapentin of high frequency action potential firing by mouse central neurones in cell culture. Epilepsy Res., 1994; 17: 1-11.
100. Brodie MJ. Lamotrigine. Lancet, 1992; 339: 1397-1400.
101. Cohen AF, Land GS, Breimer DD, et al. Lamotrigine, a new anticonvulsant: pharmacokinetics in normal humans. Clin. Pharmacol. Ther., 1987; 42: 535-541.
102. Ramsay RE, Pellock JM, Garnett WR, et al. Pharmacokinetics and safety of lamotrigine (Lamictal) in patients with epilepsy. Epilepsy Res., 1991; 10: 191-200.
103. Leach MJ, Marden CM, Miller AA. Pharmacological studies on lamotrigine, a novel potential antiepileptic drug: 2 neurochemical studies on the

mechanism of action. Epilepsia, 1986; 27: 490-497.
104. McGeer EG, Zhu SG. Lamotrigine protects against kainate but not ibotenate lesions in rat striatum. Neurosci. Lett., 1990; 112: 348-357.
105. Cheung H, Kamp D, Harris E. An in vitro investigation of the action of lamotrigine on neuronal voltage-activated sodium channels. Epilepsy Res., 1992; 13: 107-112.
106. Lees G, Leach MJ. Studies on the mechanism of action of the novel anticonvulsant lamotrigine (Lamictal) using primary neuroglial cultures from rat cortex. Brain Res., 1993; 612: 190-199.
107. Brodie MJ. Lamotrigine versus other antiepileptic drugs: a star rating system is born. Epilepsia, 1994; 35 (suppl 5): S41-S46.
108. Jawad S, Yeun WC, Peck AW, et al. Lamotrigine: single dose pharmacokinetics and initial one week exposure in refractory seizures. Epilepsy Res., 1987; 1: 194-201.
109. Binnie CD, Debets RM, Engelsman M, et al. Double-blind, crossover trial of lamotrigine as add-on therapy in intractable epilepsy. Epilepsy Res., 1989; 4: 222-229.
110. Wallace SJ. Add-on trial of lamotrigine in resistant childhood seizures. Brain Dev., 1990; 12: 734.
111. Wolf P. Lamotrigine: preliminary clinical observations on pharmacokinetics and interactions with traditional antiepileptic drugs. J. Epilepsy, 1992; 5: 73-79.
112. Warner T, Patsalos PN, Prevett M, et al. Lamotrigine-induced carbamazepine toxicity: an interaction with carbamazepine-10,11 epoxide. Epilepsy Res., 1992; 11: 147-150.
113. Stolarek I, Blacklaw J, Thompson GG, et al. Vigabatrin and lamotrigine in refractory epilepsy. J. Neurol. Neurosurg. Psychiatry, 1994; 57: 921-924.
114. Editorial Oxcarbazepine. Lancet, 1989; ii: 196-198.
115. Bialer M. Comparative pharmacokinetics of the newer antiepileptic drugs. Clin. Pharmacokinet., 1993; 24: 441-452.
116. Larkin JG, McKee PJW, Forrest G, et al. Lack of enzyme induction with oxcarbazepine (600mg daily) in healthy subjects. Br. J. Clin. Pharmacol., 1991; 31: 65-71.
117. McKee PJW, Blacklaw J, Forrest G, et al. A double-blind, placebo-controlled interaction study between oxcarbazepine and carbamazepine, sodium valproate and phenytoin in epileptic patients. Br. J. Clin. Pharmac., 1994; 37: 27-32.
118. Tartara A, Galimberti CA, Manni R, et al. The pharmacokinetics of oxcarbazepine and its active metabolite 10-hydroxy-carbamazepine in healthy subjects and in epileptic patients treated with phenobarbitone or valproic acid. Br. J. Clin. Pharmac., 1993; 36: 366-368.
119. McLean MJ, Schmutz M, Wamil AW, et al. Oxcarbazepine: mechanisms of action. Epilepsia, 1994; 35 (suppl 3): S5-S9.
120. Kramer G, Tettenborn B, Klosterkov-Jensen B, et al. Oxcarbazepine does not affect the anticoagulant activity of warfarin. Epilepsia, 1992; 33: 1145-1148.
121. Klosterkov Jensen P, Saano V, Haring P, et al. Possible interaction between oxcarbazepine and an oral contraceptive. Epilepsia, 1992; 33: 1149-1152.
122. Zaccara C, Gangemi PF, Bendoni L, et al. Influence of single and repeated doses of oxcarbazepine on the pharmacokinetic profile of felidipine. Ther. Drug Monit., 1993; 15: 39-42.
123. Grant SM, Faulds D. Oxcarbazepine. A review of its pharmacology and

therapeutic potential in epilepsy, trigeminal neuralgia and affective disorders. Drugs, 1992; 42: 873-888.
124. Editorial Vigabatrin. Lancet, 1989; 1: 532-533.
125. Schechter PJ. Clinical pharmacokinetics of vigabatrin. Br. J. Clin. Pharmacol., 1989; 27 (suppl 1): S19-S22.
126. Lippert B, Metcalf BW, Jung MJ, et al. 4-Amino-hex-5-enoic acid, a selective catalytic inhibitor of 4-aminobutyric acid aminotransferase in mammalian brain. Eur. J. Biochem., 1977; 74: 441-445.
127. Bohlen P, Huot S, Palfreyman MG. The relationship between GABA concentrations in brain and cerebrospinal fluid. Brain Res., 1979; 167: 297-305.
128. Ben-Menachem E, Persson LI, Schechter PJ, et al. The effect of different vigabatrin treatment regimens on CSF biochemistry and seizure control in epileptic patients. Brit. J. Clin. Pharmac., 1989; 27: S79-S85.
129. Sills GJ, Thompson GG, Carswell A, et al. Anticonvulsant blockade of GABA uptake into cortical astrocytes in primary culture. Epilepsia, 1993; 34: 92.
130. Rimmer EM, Richens A. Interaction between vigabatrin and phenytoin. Br. J. Clin. Pharmacol., 1989; 37: (suppl 1): S27-S33.
131. Browne TR, Mattson RH, Penry JK, et al. Vigabatrin for refractory complex partial seizures: multicentre single-blind study with long-term follow-up. Neurology, 1987; 37: 184-189.
132. Grant SM, Heel RC. Vigabatrin. Drugs, 1991; 41: 889-926.

8 Why and when to start treatment, when to stop treatment

BERNARD ECHENNE

Service de Neuropédiatrie, Centre Gui de Chauliac, Montpellier, France

Basic principles of the treatment of epilepsy are discussed: diagnosis, differential diagnosis between epileptic seizures and non-epileptic events, the role of investigations. The risk factors for recurrence after a first seizure are examined, as well as some unusual situations with unfrequent seizures. Finally we will focus on the possibilities to withdraw therapy and the re-seizure risks after remission.

INTRODUCTION

During the last decade, a number of changes developed in our methods to treat children with epilepsy.[1-4] Several findings have contributed to this change:

- the observation that many children, who experienced a single, unprovoked seizure will not experience a second seizure. Many clinicians will therefore start treatment only after a second, unprovoked seizure
- the emphasis in treatment has evolved from seizure control to a more comprehensive approach in which the prevention of side-effects is given due attention
- the reduction of the risks of long-term exposure to withdrawal of the antiepileptic drugs (AED's) after seizure remission is generally started earlier.

BASIC PRINCIPLES OF TREATMENT

A number of studies have shown that an early start of AED-treatment is positively correlated with a better prognosis of seizure control.[5-7] The shorter the interval before treatment is started, the higher the chance of achieving long-term remission.[5,8,9] Moreover, epileptic seizures are a potential risk for the patient: in generalized seizures with loss of consciousness, a high incidence of head injury is reported; prolonged seizures or status epilepticus may be life threatening; chronic epilepsy entails the development of cognitive impairment or even intellectual deterioration; seizures may have an impact on family function and academic achievement. These are, thus, a number of good reasons to start the treatment of epileptic seizures. Such treatment can only be started after a thorough diagnosis of the epilepsy, following a number of basic principles:[10]

- The diagnosis of epilepsy has to be based on clinical evidence. If there is any doubt, treatment should be withheld until clear and convincing descriptions of the seizure are available. Overdiagnosis of epilepsy is a frequently occurring problem and may have serious consequences for the development of the child, caused by the adverse effects of the AED's and the psychological consequences of a diagnosis.[3,10]
- The diagnosis of epilepsy must be based on detailed description of seizure-related events by the patients, or by an eye-witness. Especially in infants the diagnosis of epilepsy is difficult. The symptoms must therefore be analyzed and weighted with caution. Sudden loss of consciousness without provoking factors, blank stare or ocular rotation, loss of tone or slight clonic or myoclonic localized or generalized jerks, postictal vomiting and confusion of prolonged duration, headache and muscular aches and pain: all these symptoms are of great value as they may, in some combination signify the occurrence of an epileptic event. In cases of partial seizures, direct questioning, if possible, is important, particularly to elicit the characteristic features associated with complex partial seizures: fixed motionless stare, automatisms that may include fidgeting, repetitive movements of hands, chewing or swallowing movements of mouth and face.
- A number of related questions have to be answered before a definite diagnosis can be made, such as the relationship with acute encephalopathy, with extracerebral factors or with acute brain insults as these may pertain to occasional seizures.
- The differential diagnosis between epileptic seizures and non-epileptic events is difficult in young infants in whom faints or reflex syncopes are frequent. Paroxysmal episodes are often labeled convulsions, but may also be episodes of apnea and/or bradycardia of various origin. The apneas may be accompanied by motor phenomena such as hypertonia or hypotonia, at times by acute dystonic posturing and by changes in colour that may be suggestive of epileptic seizures.[10] Such accidents may be related to transient upper airway obstruction, to gastro-oesophageal reflux or to oesophageal spasm. Other episodes may be mild forms of near-miss sudden infant death syndrome, and may easily be mistaken for epileptic events as they can be followed by status epilepticus, induced by the anoxia.[10,12,13] In these situations, digestive system radiological and manometric studies, cardiorespiratory monitoring during sleep and prolonged video-EEG monitoring is imperative. In older children reflex psychogenic syncopes and psychogenic attacks are frequently misdiagnosed as seizures. In adolescents, pseudoseizures represent a significant problem: up to 30% of the

patients admitted in epilepsy centres ultimately appear to have such pseudoseizures sometimes in combination with epileptic seizures (see Table 1).[14]
- When the seizures are clinically accepted as being a symptom of an epilepsy, adequate classification of the seizures and the epilepsy syndrome must be attempted, as this may have important prognostic, therapeutic and aetiologic implications.

Table 1. Differential diagnosis of epilepsy.

▸	Syncopes:	
	in young infants	gastro-esophageal reflux esophageal spasm apneic-bradycardic episodes upper airway obstruction near-miss sudden death syndrome anoxic seizures
	in older children	reflex syncopes and faints anoxic seizures cyanotic breath-holding attacks sobbing spasms pallid breath-holding attacks febrile reflex anoxic seizures
	of cardiac origin	sick sinus syndrome prolonged Q.T syndrome syncopes in cyanotic cardiac patients cardiomyopathies
▸	Paroxysmal vertigo	
▸	Psychogenic attacks	acute psychiatric manifestations hyperventilation syndrome pseudo-epileptic seizures
▸	Migraine and periodic syndrome	
▸	Transient ischaemic attacks	
▸	Tics and other abnormal movements	
▸	Paroxysmal disturbances occuring during sleep	

A basic principle concerns the EEG. EEG recordings are required to confirm the clinical diagnosis and to establish clinical EEG correlates that may help to classify the epileptic syndrome. It must be emphasized, however, that epilepsy is a condition characterized by recurrent clinical events. Therefore, the mere presence of epileptic interictal EEG discharges is not sufficient to assert the diagnosis of epilepsy. In 10 to 15% of the general population, EEG-abnormalities are found, especially during sleep.[15-17]

Even precise definitions of such abnormalities (i.e. focal or generalized spikes or polyspike and wave abnormalities) may lead to a false positive diagnosis in approximately 5% of the general population. On the other hand, a single routine EEG is likely to show interictal discharges only in 50% of the epileptic patients,[17] although the diagnostic value of EEG recordings can be improved significantly, using e.g. long-term monitoring, cassette EEG or sleep-deprivation. On the other hand, clinical information is often sufficient to allow a tentative diagnosis: a history of occasional nocturnal focal motor seizures involving face and upper limb in children below the age of 12 strongly suggests benign Rolandic epilepsy; the onset, in adolescence, of myoclonic jerks on awakening, with occasional tonic-clonic seizures point to juvenile myoclonic epilepsy.

The aetiology must be questioned in every child. In addition to complete historoy-taking and physical examination, imaging procedures are essential in many cases.

WHEN TO START TREATMENT

The probability of recurrence after a single seizure is still in debate. Estimates range from a 25% recurrence rate within 2 years[1,18] up to 71% recurrence within 2½ years.[1,2,19-21] The majority of the studies recommend that treatment does not have to be started after a single seizure, but probably should be started when two or more unprovoked seizures have occurred within a short interval (i.e. 6 months to one year[3,14]). Other criteria that have to be weighted before a treatment is started are:
- The type of seizure and the type of epilepsy syndrome is related to the risk of seizure recurrence. Atonic seizures and absence seizure for example have a high recurrence risk and treatment may be started in these children without further delay.[10,20] On the other hand, the risk of recurrence is low for patients with a generalized tonic-clonic seizure (without neurological and EEG abnormalities).[10,20,21]
- Even when the risk of recurrence is high, treatment may be delayed in some benign conditions such as for patients with rolandic seizures, if the seizures occur only during the night, if sleep quality is not affected, if there are no repercussions for behaviour or family function and if seizure frequency is low.[22,23]

Several studies have identified factors that increase the risk of seizure recurrence:[10,24]
- a history of prior neurological insult
- a positive family history of seizures
- mental or neurological abnormalities on clinical examination
- the presence of interictal paroxysmal activities in the EEG

Other criteria that may be used in the decision making process about treatment are:
- parental anxiety about the risk of recurrence
- concomitant learning or behavioral disorders
- the possible consequences of a seizure during school or sport activities
- the expectations about compliance. Some patients are reluctant to comply with any long-term therapy, e.g., adolescents.

In some conditions, antiepileptic treatment may be started although the seizures may be rare:
- the occurrence of epileptiform EEG abnormalities interfering with cognitive function[25-30]
- major paroxysmal EEG abnormalities with unfrequent and apparanetly minor seizures but with severe behavioural disturbances, frequently seen in hypsarrhythmia, in the Lennox-Gastaut syndrome, or in other severe epileptic encephalopathies
- seizures associated with EEG background activity disorganisation and frequent interictal discharges in children with severe retardation such as mental retardation or autism[30]
- seizures associated with symptoms of the Landau-Kleffner syndrome, or continuous spike-wave activity during sleep, or in certain forms of developmental dysphasias[30]

In these cases the condition may improve by antiepileptic treatment.

WHEN TO STOP TREATMENT

Withdrawal of AED's after a seizure-free period of 2 to 5 years leads to a remission rate of 50 to 88%. The follow-up period varies however from 4 to 20 years.[31-36] Approximately 70% of the children who are seizure-free for a period longer than 2 or 4 years, while still on AED's, will remain seizure-free after withdrawal of the medication.[4,36-39] In a study of 264 children, Shinnar et al.[39] have recently shown that only 36% had a seizure recurrence when AED's were discontinued after a mean seizure-free interval of 2.9 years. The mean time to recurrence was 9.5 months with a median of 4.3 months. The risk of recurrence is highest in the first few months after medication withdrawal. This study is particularly important, because the authors have classified the patients according to epileptic syndromes. The factors differentiating children with good prognosis from those in whom epilepsy relapsed are:
- *Etiology.* Patients with remote symptomatic seizures (i.e: children with a static encephalopathy prior to the seizure, or with a prior neurological insult such a stroke) have a higher recurrence risk than those with idiopathic seizures (47% versus 29%[39]).

- *Age.* Literature shows conflicting results. Some authors found no significant correlation between age at seizure onset and risk of recurrence.[34,36] Most studies, however, showed an elevated risk of relapse in patients with adolescent-onset epilepsy compared with childhood onset.[32,35,37-39] The risk for seizure relapse is substantially increased (up to 73%) when onset of epilepsy occurs after the age of 12 years.[39] The risk percentage is 45% in children with seizure onset before the age of 2 and a mere 26% in those with seizure onset between the age of 2 and 12. It is important to emphasize that the increased recurrence risk in the group with onset after the age of 12 is the same for both patients with idiopathic and remote symptomatic epilepsies.
- *A family history of seizures.* A family history of unprovoked seizures in a first degree relative is associated with an increased risk of recurrence in children with idiopathic epilepsy.[39]
- *Seizure type.* The risk of recurrence is increased if a subject experiences a high number of seizures before seizure control,[36] and when multiple seizure types or generalized tonic-clonic seizures associated with other seizure types occur.[35,36,40] In remote symptomatic epilepsies, the risk of recurrence is increased when atypical febrile seizures (63%), akinetic seizures (54%) multiple tonic-clonic and partial seizures occur (up to 51%). The risk is not increased in cases of status epilepticus (25%) or atypical absences (25%).[39] In idiopathic epilepsies, the risk of recurrences is high in cases of status epilepticus (63%), atypical febrile seizures (50%), moderate in cases of multiple seizure types, generalized tonic-clonic seizures, atypical absences, or myoclonic seizures (32 to 44%).[39]
- *Epileptic syndromes.* A high remission rate for generalized tonic-clonic seizures and a low rate for remission with complex partial seizures have been reported by many authors.[41,42] Benign rolandic epilepsy and juvenile myoclonic epilepsy are associated with, respectively, particularly favourable and unfavourable outcome.
- *Intelligence and motor abnormalities.* Severity of mental retardation is directly correlated with the risk of seizure recurrence.[35] The presence of a motor deficit has a statistically significant association with outcome as it is a marker for remote symptomatic seizures, which are associated with an increased recurrence risk.
- *EEG-abnormalities.* The predictive value of epileptiform EEG-abnormalities is controversial. The majority of the studies showed an increased risk of recurrence in children with EEG-abnormalities, prior to drug withdrawal,[32,33,35,37,41,43,44] although other reports deny this relationship.[20,21,31,45] Our conclusion is that, although the risk of recurrence seems increased for patients with interictal EEG-abnormalities, the absence of such abnormalities is not a prerequisite for discontinuation of treatment.[10] An example is the benign rolandic epilepsy, in which EEG-abnormalities may persist until adult life,

despite good prognosis.
- *Seizure-free interval.* The seizure-free interval, after which discontinuation of treatment can be considered is still under debate. Most studies propose a seizure-free period of 2 to 3 years. The decision to continue or to stop treatment must, however, be individualized. The following factors must be taken into account:
  - The statistical probability of seizure recurrence following the different prognostic factors as previously described
  - The consequences of a seizure relapse: risk of injury, psychosocial impact, risk of status epilepticus[46]
  - The risk that previously well-controlled seizures will become refractory after seizure relapse, a very rare but possible event
  - The fact that even, if the patient is seizure-free for many years, and his/her medication is continued, the risk of seizure relapse is not completely absent[37,41]
  - The risks of adverse effects of AED's, particularly cognitive and behavioral side-effects.[47-49,50]

Table 2 gives a summary of the relationship between the risk of seizure relapse and the number of risk factors.

Table 2. Risk of seizure recurrence 2 years after withdrawal of AEDs in children as a function of number of risk factors[39]

Nr of risk factors	% of recurrences	2 years recurrence risk
Idiopathic seizures		
0	14	12%
1	50	46%
2	86	71%
Remote symptomatic seizures		
0	11	11%
1	34	35%
2	62	51%
3	89	78%

Risk factors are those from multivariate analysis and are different for idiopathic and remote symptomatic groups (idiopathic group: age at onset >12 years, family history of seizures, slowing on EEG, atypical febrile seizures; remote symptomatic seizures: age at onset >12 years, moderate severe mental retardation, absence seizures, atypical febrile seizures)

REFERENCES

1. Shinnar S, Berg AT, Moshe S. Risk of seizure recurrence following unprovoked seizure in childhood: a prospective study. Pediatrics, 1990; 85: 1076-1085.
2. Berga T, Shinnar S. The risk of seizure recurrence following a first unprovoked seizure: a meta-analysis. Neurology, 1991; 41: 965-972.
3. Chadwick D. Epilepsy. J. Neurol. Neurosurg. Psy., 1994; 57: 264-277.
4. Holmes GL. Stopping anti-epileptic drugs in children: when and why. Ann. Neurol., 1994; 35: 509-510.
5. Elwes RDC, Reynolds EH. The early prognosis of epilepsy. In: Dam M, Gram L, editors. Comprehensive Epileptology. New-York: Raven Press, 1981: 715-727.
6. Reynolds EH, Elwes RDC, Shorvon SD. Why does epilepsy become intractable. Lancet, 1983; 2: 952-954.
7. Shorvon SD. The temporal aspects of prognosis in epilepsy. J. Neurol. Neurosurg. Psychiatr., 1984; 47: 1157-1165.
8. Rodin EA. The prognosisof patients with epilepsy. CC. Thomas, Springfield, Ill, 1968.
9. Gram L. How do you treat epilepsy? In: DAM M. editor 'A practical approach to Epilepsy' - Pergamon Press, 1991; 7: 105-135.
10. Aicardi J. Epilepsy in children. Second edition International Review of Child Neurology Series - New York, Raven Press, 1994.
11. Echenne B, Humbertclaude V, Cheminal R. Traitement de l'epilepsie de l'enfant - Pediatrie, 1993; 48: 883-887.
12. Fontan JP, Helot GP, Heyman MB, et al. Esophageal spasm associated with apnea and bradycardia in an infant. Pediatrics, 1984; 73: 52-55.
13. Aubourg P, Dulac O, Prouin P, et al. Infantile status epilepticus as a complication of 'nearmiss' sudden infant death. Develop. Med. Child Neurol., 1985; 27: 40-48.
14. Chadwick D. How do you diagnose epilepsy? In: A practical approach to epilepsy. DAM M. editor- Pergamon Press, 1991; 4: 61-93.
15. Cavazzutti GB, Cappella L, Nalin A. Longitudinal study of epileptiform EEG patterns in normal children. Epilepsia, 1980; 21: 43-55.
16. Eeg-Olofson 0, Petersen I, Selden U. The development of the electro-encephalogram in normal children from the age of 1 through 15 years paroxysmal activity. Neuropediatrics, 1971; 2: 375-404.
17. Billiard M, Echenne B, Besset A, et al. Intérêt de l'enregistrement polygraphique de sommeil de nuit chez l'enfant suspect de crises epileptiques Iorsqueles EEG de routine et apres privation de sommeil demeurent normaux. Rev. Electro-encephalogr. Clin. Neurophysiol., 1981; 11: 450-456.
18. Hauser WA, Rich SS, Annegers JF, et al. Seizure recurrence after a first unprovoked seizure: an extended follow-up. Neurology, 1990; 40: 1163-1170.
19. Elwes RDC, Chesterman P, Reynolds EH. Prognosis after a first untreated tonic-clonic seizure. Lancet, 1985; 2: 752-753.
20. Camfield PR, Campfield CS, Dooley JM, et al. Epilepsy after a first unprovoked seizure in childhood. Neurology, 1985; 35: 1657-1660.
21. Boulloche J, Leloup P, Mallet E, et al. Risk of recurrence after a single

unprovoked generalized tonic clonic seizure. Dev. Med. Child Neurol., 1989; 31: 620-632.
22. Beaussart M. Benign epilepsy of children with rolandic (centro-temporal) paroxysmal loci: a clinical entity-study of 221 cases. Epilepsia, 1992; 3: 795-811.
23. Ambrosetto G, Giovanardi Rossi P, Tassinari CA. Predictive factors of seizure frequency and duration of anti-epileptic treatment in rolandic epilepsy: a retrospective study. Brain Dev., 1987; 9: 300-304.
24. Robinson R. When to start and stop anticonvulsivants. In: Recent advances in Pediatrics. Meadow R. editor, Edinburgh Churchill Livingstone, 1984: 155-174.
25. Aarts JP, Binnie CD, Smit AM, et al. Selective cognitive impairment during focal any generalized epileptiform EEG activity. Brain, 1984; 107: 293-308.
26. Aldenkamp AP, Gutter T, Beun AM. The effects of seizure activity and paroxysmal electro-encephalographic discharges or cognition. Acta Neurol. Scand., 1992; 86 (suppl. 140): 111-121.
27. Kastelein-Nolst Trenite DGA, Bakker DJ, Binnie CD, et al. Psychological effects of subclinical EEG discharges in children. I: Scholastic skills. Epilepsy Res., 1988; 2: 111-116.
28. Kastelein-Nolst Trenite DGA, Siebelink BM, Berends SGC, et al. Lateralized effects of subclinical epileptiform EEG discharges on scholastic perromances in school children. Epilepsia, 1990; 31: 740-746.
29. Echenne B. Dysphasies et epilepsie. Approche neuropsychologique des apprentissages chez l'enfant. ANAE, 1990; 3: 138-143.
30. Echenne B, Cheminal R, Rivier F, et al. Epileptic electro-encephalographic abnormalities and developmental dysphasias. A study of 32 patients. Brain Dev., 1992; 14: 216-225.
31. Juul Jensen P. Frequency of recurrence after discontinuance of anticonvulsivant therapy in patients with epileptic seizures: a new followup study after 5 years. Epilepsia, 1968; 9: 11.
32. Shinnar S, Vining EPG, Millits ED. et al. Discontinuing antiepileptic medication in children with epilepsy after two years with out seizures: a prospective study. N. Engl. J. Med., 1985; 313: 976-980.
33. Bouma PAD, Peters ACB, Arts RJHM, et al.: Discontinuation of anti-epileptic therapy: a prospective trial in children. J. Neurol. Neurosurg. Psychiatry, 1987; 50: 1579-1583.
34. Arts WFM, Visser WFM, Visser LH, et al. Follow-up of 146 children with epilepsy after withdrawal of anti-epileptic therapy. Epilepsia, 1988; 29: 244-250.
35. Matricardi M, Brinciott M, Benedetti P. Outcome after discontinuation of anti-epileptic drug therapy in children with epilepsy. Epilepsia, 1989; 30: 582-589.
36. Gherpelli JLD, Kok F, Dalforno S, et al. Discontinuing medication in epileptic children: a study of risk factors related to recurrence. Epilepsia, 1992; 33: 681-686.
37. Medical Research Council Anti-epileptic Drug Withdrawal study Group. Randomised study of anti-epileptic drug withdrawal in patients with remission. Lancet, 1991; 337: 1175-1180.
38. Gross-Tsur V, Shinnar S. Discontinuing anti-epileptic drug therapy in patients with epilepsy. In: Wyllie E. editor The treatment of epilepsy: principles and practice. Lea and Febiger, Philadelphia,]993: 858-866.

39. Shinnar S, Berg AT, Moshe S.L. et al. Discontinuing anti-epileptic drugs in children with epilepsy: a prospective study. Ann. Neurol., 1994; 35: 534-45.
40. Callaghan N, Garrett A, Coggin T. Withdrawal of anti-convulsivant drugs in patients free of seizures for two years. N. Engl. J. Med.,]988; 318: 942-946.
41. Annegers JF, Hauser WA, Alveback LR. Remission of seizures and relapse in patients with epilepsy. Epilepsia, 1979; 20: 729-737.
42. Sofijanov NG. Clinical evolution and prognosis of childhood epilepsies. Epilepsia, 1982; 23: 61-69.
43. Todt H. The late prognosis of epilepsy in childhood: results of a prospective follow-up study. Epilepsia, 1984: 137-144.
44. Emerson R, d'Souza BJ, Vining EP, et al. Stopping medication in children with epilepsy: predictors of outcome. N. Engl. J. Med., 1981; 304: 1125-1129.
45. Holowach-Thurston JH, Thurston DL, Hixon BB, et al. Prognosis in childhood epilepsy: additional follow-up of 148 children 15 to 23 years after withdrawal of anticonvulsivant therapy. N. Engl. J. Med., 1982; 306: 831-836.
46. Gross Tsur V, Shinnar S. Convulsive status epilepticus in children. Epilepsia, 1993; 34 (suppl. 1): 512-520.
47. Vining EPG, Mellits D, Dorsen et al. Psychological and behavioral effects of anti-epileptic drugs in children: a double blind comparison between phenobarbital and valproic acid. Pediatrics, 1987; 80: 165-174.
48. Aldenkamp AP, Alpherts WCJ, Blennow G, et al. Withdrawal of anti-epileptic medication in children - effects on cognitive function: the multicenter HOLMFRID study. Neurology, 1991; 41: 141-143.
49. Meador KJ, Loring DW, Huh K, et al. Comparative cognitive effects of anticonvulsivants. Neurology, 1990; 40: 391-394.
50. Dreifuss FE. Cognitive function - Victim of disease or hostage to treatment. Epilepsia, 1992(S1): 7-12.

9 Clinical aspects of antiepileptic treatment

STEPHEN W. BROWN

The David Lewis Centre, Cheshire, and Manchester Royal Infirmary, Manchester, UK

The aims of treating epilepsy in childhood are considered, together with the importance of involving children and their families in decisions about the treatment process. Basic clinical aspects of antiepileptic treatment are described, with emphasis both on choosing medication according to syndrome diagnosis, and also on consideration of the effect of long-term treatment on the individual. While monotherapy remains the ideal, seizures are sometimes refractory to first-line treatment, and in such cases there may be a role for adjunctive treatments and synergistic polytherapy. The use of rectal preparations of diazepam and paraldehyde minimises the need for excessive medical intervention, but there may be difficulties in administration by non-medical staff. Some other non-surgical treatments are briefly mentioned.

> "the chief factor in the cure of epilepsy in the young is change, especially that due to growing up, but seasonal change of climate, or change of place or mode of life are also important."

Hippocrates [1]

> " ...I know of a child that had not been attacked at all for 8 months since wearing the root (of the peony). When however, somehow the amulet slid off, he was immediately seized..."

Galen [2]

WHY TREAT EPILEPSY?

Aim of treatment

The aim of treatment is to stop seizures, and to neutralise any cognitive or psychosocial penalty associated with epilepsy or its therapy. Ideally the process by which this is achieved should be interactive and based on a balanced relationship between the child, the family and the professionals involved. It should not be appropriate to consider someone's epilepsy to be 'satisfactorily controlled' if seizures continue to occur or if treatment-related adverse effects are interfering with school progress or the activities of everyday life.

Consumer attitudes

In one postal survey where 896 British schoolchildren with epilepsy were asked about their attitudes to their seizures and treatment,[3] comments from the subjects included statements such as:

> 'The drug copes with the epilepsy but I feel lousy'

> 'I ... feel sick a lot and mum says this is my tablets - is this right?'

> 'I feel very tired at school. I feel I should be able to concentrate better. This makes me want to stop taking the tablets.'

In the same study more than a third of the respondents claimed that their doctor had never explained anything to them about epilepsy, while another third reported that they had received an explanation that they did not understand.

> 'When I was first diagnosed ... nothing was explained to me or my mum. The doctor ... told us to get a book out of the library and that was all he said.'

Furthermore nearly half of the children felt that their drug treatment was ineffective, and nearly one-fifth admitted not taking their medication regularly. Such findings may not of course be representative of all children receiving treatment for epilepsy. However, they underline the importance of good two-way communication between physicians and consumers. Proper explanation will increase compliance, while obtaining feedback will enable appropriate modification to the treatment plan.

First medical contact

The initial presentation is likely to be to a family physician, a specialist paediatrician or neurologist, depending on the availability of local services. The functions of the initial medical assessment are to make a provisional diagnosis, where possible uncover an aetiology, initiate treatment and monitor progress, and provide information and counselling to patients and their families.[4] In order to draw up an initial epilepsy treatment plan, it is

necessary to:
- make a diagnosis at the epileptic syndrome level (even if only provisional), especially distinguishing localisation-related from generalised epilepsy. This will guide initial treatment. For example, carbamazepine is unlikely to be effective in Juvenile Myoclonic Epilepsy and may make the myoclonus worse.
- consider how the child's future development might be affected by treatment. For example valproate and phenytoin may have adverse cosmetic effects which could become serious issues around puberty. Potential future teratogenicity, especially of valproate, phenytoin and carbamazepine, may also be important.
- consider in each case how the family's attitude to epilepsy and ability to cope with the diagnosis might affect compliance or future attitudes towards changes of treatment. Some families may need to be seen more frequently than others, and consideration given to convenience of dosing schedules, use of controlled release preparations, and use of explanatory literature. Others may resent being patronised. In either case, a therapeutic alliance has to be established.

BASIC CLINICAL ASPECTS OF ANTIEPILEPTIC MEDICATION

The use of medication is part of an overall treatment package that attends to biological, psychological and social aspects of epilepsy. It is generally accepted that treatment with a single antiepileptic drug (AED), properly used, will achieve remission in about 70% of cases overall. The best results seem to occur in the idiopathic epilepsies of age-related onset, both generalised and partial, such as juvenile absence epilepsy or benign partial epilepsy with centrotemporal spikes. More difficult to control are symptomatic localisation-related syndromes such as temporal lobe epilepsy, although the majority of patients should still obtain remission with appropriate treatment. The most problematic are the symptomatic generalised epilepsies such as the Lennox-Gastaut Syndrome, where complete seizure control may not be possible.

Indications

It has been suggested that sodium valproate, as a broad-spectrum, generally well tolerated agent, could be used as first choice in all newly diagnosed epilepsies, and further investigation to delineate the precise syndrome is only necessary where seizure control is not achieved in this way.[5] There are potential pitfalls in taking this approach with children. Valproate can cause weight gain and hair loss, which disturbs teenagers who are sensitive about their appearance. It is potentially teratogenic.

Valproate induced hepatotoxicity, although rare overall in children taking monotherapy, is still more likely to occur in those under 10 years old, and especially in those under 2. Most patients will be taking medication for several years. There is an argument therefore for starting in childhood with a drug which will still be suitable in adulthood. In some cases therefore (e.g. girls with localisation-related epilepsy) it is appropriate to commence treatment with carbamazepine which seems to be marginally less teratogenic,[6] and does not affect physical appearance. Valproate remains to date the undisputed first choice for juvenile myoclonic epilepsy. It is also a first choice in other idiopathic generalised epilepsies, such as Juvenile Absence Epilepsy, where there is an advantage over ethosuximide in that it is effective against tonic-clonic as well as absence seizures, and is less likely to cause psychosis. This latter syndrome often remits permanently before childbearing becomes an issue.

There are broadly speaking three generations of AEDs in use today. The first generation, all in use before 1960, include phenobarbitone, phenytoin, primidone and ethosuximide. They are all relatively cheap, and are effective as AEDs, but have been claimed to show problems due to sedative and cognitive side-effects, adverse cosmetic effects, and teratogenicity. The second generation (1960-1989) is represented by carbamazepine and valproate. Oxcarbazepine is included here with this group because of its chemical and therapeutic similarity to carbamazepine. These drugs are mostly more expensive than the first generation, and are regarded as having less adverse sedative, cognitive and cosmetic effects. They still have problems with teratogenicity. The third generation (1989-present) consists of vigabatrin, lamotrigine, gabapentin, topiramate and tiagabine (a sixth, felbamate, has been withdrawn due to reports of aplastic anaemia). They are all expensive. Possible adverse or beneficial effects on cognitive functioning and mood are still under evaluation, although claims are made of neutral or even beneficial effects in these parameters.[7,8] No adverse cosmetic effects have yet been described, and despite a shortage of human data, it has been suggested that these agents are much less teratogenic than previous therapies.[9] It is interesting that the emergence of newer AEDs with claims of fewer psychological side-effects has seen a re-evaluation of those effects in the older drugs, which may modify the previous consensus.[10-13]

The third generation all carry licence indications as 'add-on' treatment, which might give the impression that they are only effective as adjunctive therapies. This is in fact a consequence of the way in which drugs are developed and licensed rather than a picture of their true potential use. In practice, after adding on has had a beneficial effect on seizures, it is possible in many cases to reduce successfully to monotherapy with these agents. Furthermore reports are now emerging of efficacy as monotherapy in newly diagnosed patients.[14] In due course it is likely that licence indications and data sheets will change to reflect these uses.

Steroids and ACTH have been used in the management of certain epileptic syndromes, most notably in the acute treatment of West and Landau-Kleffner syndromes, and to obtain short term remission in some cases of Lennox-Gastaut syndrome.[15-17] It seems likely that the anti-seizure effect is a consequence of steroid induced remission of some underlying pathological process, rather than a direct effect on the epilepsy itself.

Other AEDs include the 1,4 benzodiazepines (diazepam, clonazepam, nitrazepam), the 1,5 benzodiazepine clobazam, and the carbonic anhydrase inhibitor acetazolamide. All show the phenomenon of pharmacological tolerance to a greater or lesser degree. Their main uses therefore are as adjunctive or intermittent treatments, although clonazepam and nitrazepam are occasionally used in monotherapy, especially in cryptogenic generalised epilepsies. Strategies for their use are discussed further below. The choice of one drug over another as initial antiepileptic therapy will depend on:

- known efficacy in the relevant epileptic syndrome
- known long-term safety
- known adverse effects on mood, cognitive functioning, cosmetic appearance
- cost

For these reasons first-line treatment will tend to be with valproate, carbamazepine or oxcarbazepine. If these fail, either because of lack of efficacy or other adverse effects, second-line treatment could either be with the older agents (phenobarbitone, phenytoin, primidone, ethosuximide) or the very new (vigabatrin, lamotrigine, topiramate, tiagabine, gabapentin). The choice will still depend on syndrome diagnosis, known side-effects and cost, but will also be influenced by availability of the drugs and personal preference and experience of the physician.

Table 1. First and second line treatments

	First Line	Second Line
GENERALISED		
idiopathic	valproate	ethosuximide (if absence seizures only) lamotrigine, clonazepam,
cryptogenic (symptomatic)		
West syndrome:	ACTH (or vigabatrin, especially if tuberose sclerosis)	prednisolone, prednisone
Others: (e.g. Lennox-Gastaut syndrome)	valproate	lamotrigine, clonazepam, intermittent prednisone
LOCALISATION-RELATED		
idiopathic	carbamazepine or oxcarbazepine	
cryptogenic	carbamazepine, oxcarbazepine or valproate	gabapentin, lamotrigine, vigabatrin (expensive), phenytoin (cheap)

A view of first and second line drugs for different syndromes is given in Table 1, with prescribing recommendation for the major antiepileptic drugs in table 2. Some mention is made of relevant chronic side-effects in this table. The acute toxic effects of AEDs (cerebellar signs, confusion), due to the serum level being too high for that individual, are discussed elsewhere in this volume.

Table 2 a: Features of antiepileptic drugs

	Indication	starting dose (mg/kg/day)	maintenance dose mg/kg/day
First Line drugs			
valproate*	broad spectrum	10	10-40 →
carbamazepine*	localisation-related	5	10-30 →
oxcarbazepine	localisation-related	10	15-45 →
Newer drugs			
vigabatrin	localisation-related; also symptomatic West Syndrome	25	25-80 →
lamotrigine	broad spectrum	0.2 (with valproate) 2 (without valproate)	1-5 → 5-15
gabapentin	localisation-related	10	20-40 →
Older drugs			
phenobarbitone	localisation-related	4	4-10 →
primidon	localisation-related	10	15-30 →
phenytoin	localisation-related; ?also JME?	5	5-15 →
ethosuximide	absence seizures only	10	10-15 →

*controlled-release preparation available

Table 2 b: Features of antiepileptic drugs

cognitive side-effects	behavioural side-effects	cosmetic side-effects	teratogenicity	other
not usually significant	rare	weight gain, hair loss, hair becoming finer	1-2% risk of neural tube defect	tremor, gastrointestinal, hepatoxicity, hyperammonaemia
	not significant	none	probably <1% risk neural tube defect	neutropenia, oedema, hyponatremia, skin reactions (5%)
not usually significant	not significant	none	insufficient data	hyponatremia, less likely to cause skin reaction than carbamazepine
none reported	behaviour change in up to 5%, may be frank psychosis	possible weight gain	none reported: safe in animal studies	initial sedation
none reported	not significant	none reported	none reported: safe in animal studies	skin reactions (3-10%), lessened by slow dose escalation
none reported	not significant	none reported	none reported: safe in animal studies	initial sedation: generally well tolerated
yes	insomnia, hyperkinesis	none	yes, probably less than phenytoin	hepatotoxicity, skin reactions
yes	as PHB	as PHB	as PHB	as PHB
yes	psychomotor slowing	hirsutes, coarsening of facies, acne	yes	hepatotoxicity, rickets
probably not significant	behaviour changes, psychosis	none reported	insufficient evidence	skin reactions

General principles of prescribing

The received view since the early 1980s has been that good practice in epilepsy management is to use one AED at a time (monotherapy).[18-20] The first drug will be chosen according to seizure type and syndrome and is most likely to be either valproate or carbamazepine. Oxcarbazepine is an alternative to carbamazepine in countries where it is licensed.[21] A general principle is to introduce at a low dose and increase gradually until either seizure control is obtained or toxic effects appear. If the first choice drug is ineffective or not tolerated, therapy should be changed to one of the second line drugs. In localisation-related epilepsy, either valproate or carbamazepine is likely to be effective,[22,23] and failure with one might be followed by success with the other before the more expensive or the more side-effect laden drugs are tried. Serial monotherapy should therefore follow the sequence suggested in Table 1.

Generally speaking a new AED should be introduced alongside the existing therapy and the dose built up to establish efficacy or any idiosyncratic adverse reactions. Efficacy should not be judged before steady state has been reached, which will take at least 5 half-lives.[24] If seizure frequency improves, the previous drug can then be slowly withdrawn. If seizures continue as before, it is reasonable to assume that the new drug is ineffective, and there is the option of tailing this off instead, which may be easier to withdraw than the previous agent because of the short exposure time. The alternative strategy, building up a new drug while simultaneously tailing off the old, can lead to management problems. The changeover period may include a point where neither AED is at a therapeutic level, and there will be no protection against seizures, which may be further exacerbated by a rebound or withdrawal effect due to the first drug. This approach recommended here requires the patient to undergo a period of dual therapy, while the recently added AED is being evaluated and before the first AED is withdrawn. When the time comes to tail off one of these, the change can usually be accomplished fairly rapidly. In particular, valproate and phenytoin seem to be generally safe in withdrawal under supervision, (over 2-4 weeks) while carbamazepine may take longer.[25]

Examples

Initiation with first line drugs
Carbamazepine should be introduced at a low dose because initial clearance will be low until autoinduction occurs, so central nervous system side-effects may occur. Consequently the dose may need to be increased during the first few weeks of therapy. Valproate however can be started at a relatively higher daily dose, although an escalation up to the maintenance

level is still advisable because of potential initial gastrointestinal side-effects. Full clinical effectiveness may take several weeks, so dose adjustments should probably be less frequent at first than with carbamazepine.[26] Oxcarbazepine, despite its chemical similarity to carbamazepine, appears not to be involved in autoinduction,[27] and so dose adjustment in the early stages of treatment need not be so frequent as with the latter drug.

Adding newer drugs
Because at the time of writing these are used as second line agents, interactions with concomitant AEDs have to be taken into account. We here discuss lamotrigine as an example for two reasons. Firstly, its serum level is greatly enhanced by the presence of valproate,[8] necessitating a much lower dose regime for lamotrigine where the two are used together, or else clinical toxicity may be seen. Furthermore, withdrawal of concomitant valproate will typically require the lamotrigine to be increased to maintain seizure control. Experience suggests that this is often the case when the dose of valproate has been reduced to a low level, near the end of the reduction period. Secondly, adding on to carbamazepine may produce clinical features of carbamazepine toxicity. This may be due to a kinetic effect on the active metabolite carbamazepine-10,11-epoxide, although a pharmacodynamic interaction may also play a part.[28] Enzyme-inducing AEDs such as carbamazepine and phenytoin shorten the half life of lamotrigine, so that when these are withdrawn the serum lamotrigine level will rise. In clinical practice this rarely produces any problems.

Vigabatrin may produce a 20-30% decrease in the serum level of co-administered phenytoin,[29] but this is not usually of clinical significance, and gabapentin seems to have no significant effect on the serum levels of concomitant AEDs.[30] Both these drugs may be associated with initial sedation, and best results are obtained by adding on a low dose, and building up slowly. This may be especially relevant in patients with learning disability or structural brain abnormality. If seizure control is achieved, and concomitant AEDs are being tailed off, it is not usually necessary to alter the maintenance dose of vigabatrin or gabapentin. In countries where gabapentin is licensed, the data sheet indication does not yet cover children, but does include adolescents. Clinical trials in younger children are currently nearing completion.

Adding older drugs
Phenytoin takes about 2-4 weeks to achieve steady state concentrations, and it can reduce the efficacy of co-administered carbamazepine and valproate by enzyme induction. These facts, combined with its saturation kinetics and variability in rate of metabolism between subjects make out a case for monitoring serum levels during early stages of treatment. The change in serum level after withdrawal of concomitant AEDs is not usually clinically significant. Phenobarbitone is associated with behavioral

and cognitive changes[17,31] to the extent that it cannot be routinely recommended for children. These effects are shared by primidone.

Dose timing & formulations

Compliance is best when drugs only have to be taken once or twice a day, and preferably when the same dose is given each time. It is also preferable to minimise the actual number and type of tablets, capsules, or spoonfuls of liquid that have to be taken. Most AEDs can be taken twice daily, and a regime in which the doses are evenly divided can be devised. However, used in this way carbamazepine occasionally gives rise to double vision, or other acute side-effects. Until recently the adverse effects could often be lessened by taking the same total daily amount in smaller individual doses, up to four times daily. This could of course lead to compliance problems. The problem has largely been resolved by the appearance of a controlled-release preparation of carbamazepine, which is associated with less side-effects. This can be given twice daily. It has always been possible to give both valproate and phenytoin as once daily regimes, so the recent appearance of sustained-release valproate merits an explanation. In fact, this once daily regime is associated with fewer side-effects[32] and improved compliance.[33]

Controlled-release carbamazepine and sustained-release valproate serve another useful function in epilepsy management, guaranteed continuity of formulation. Generic substitution is widely practised in many countries, and evidence is accumulating that loss of seizure control or toxicity may result, especially with carbamazepine,[34-37] and phenytoin,[38] which have lead to calls for closer monitoring of these phenomena.[39] The problem may not only lie in bioavailability. A different presentation of medication may alter compliance, either due to misunderstanding, mistrust or unpalatability, for example:

"I don't mind taking Epilim, but I hate ordinary sodium valproate because the tablets disintegrate in my mouth, it tastes horrible, and gives me an upset stomach." (13 year old girl with idiopathic generalised epilepsy)

There is therefore a case for attempting to continue with the same formulation of an AED throughout treatment.

Monitoring treatment

General management
Treatment will not be effective unless the consumers understand the reasons for treatment and are in agreement with it. Parents and children should be given the opportunity to ask questions about treatment and prognosis and given an explanation that they understand. Clinical consultations are often anxious situations for them, and details of the initial

interview are often not easily recollected later. Information may therefore need to be repeated at follow-up appointments. Patient information materials are available from voluntary organisations, and it should be routine to put children newly diagnosed with epilepsy and their families in touch with the relevant group. Where more than one agency is involved with the child, such as education authorities or social services, appropriate liaison will ensure that the physician obtains accurate feedback about seizure frequency and behavioral and cognitive problems to assess treatment progress. Simple, specially printed diaries to record seizures can be provided from the clinic. Also, 'co-operation cards' containing diagnostic, investigative and prescribing summaries can be supplied to families. These are useful when the child comes into contact with other medical agencies.

Laboratory monitoring
The initial investigation of the aetiology of the seizure disorder will have already resulted in baseline of blood tests such as full blood count and liver function tests being carried out. Although AEDs can affect these parameters, usually benignly but sometimes dangerously, routine monitoring in the absence of clinical symptoms is nevertheless highly unlikely to be of benefit.[40] Carbamazepine often causes neutropenia and changes in hepatic enzymes which are clinically irrelevant, while serious problems such as valproate-induced liver failure cannot reliably be predicted in this way. There is a case for identifying children at higher risk of developing valproate hepatotoxicity (e.g. aged under 3 years, with metabolic disorder, progressive neurological disease, history of previous drug reactions) and monitoring these more closely.[41] An alternative would be to avoid using valproate in this group.

Plasma level monitoring is available for most AEDs, and in children salivary levels may be more convenient to obtain.[42] Their role in management should not be over-estimated. 'Therapeutic ranges' quoted by laboratories are at best statistical concepts which may not have much bearing on the situation of an individual, and can therefore be misleading. The zero-order kinetics of phenytoin mean that a knowledge of plasma level is helpful in guiding the rate of dose escalation while avoiding toxic effects,[43] and it may be difficult to manage otherwise. Plasma level monitoring is somewhat less useful with carbamazepine, while with valproate, efficacy and toxicity do not clearly relate to blood levels. The reasons for measuring plasma AED levels are:
- to predict how much to increase or decrease phenytoin (and to a lesser extent carbamazepine)
- to decide which drug is the culprit if toxic symptoms occur on combination therapy (but this can be confusing with carbamazepine unless the epoxide is measured as well)
- to assess compliance

An exception to this general principle of benign indifference to routine laboratory monitoring may be made in the child with learning disability or speech and language problems. Here the patient may not be able to communicate symptoms to carers or the physician, and changes in behaviour or general health may be incorrectly attributed to the co-existing psychiatric or neurological condition. The rate of speech and language problems in children with complex epilepsy such as this may be much higher than previously realised.[44]

Monitoring and managing adverse events
Central nervous system related toxic symptoms, such as ataxia, nystagmus, tremor and confusion are typically dose-related and respond to dose adjustment. The chronic adverse effects of phenytoin, including those on gums, skin, hair and bones are discussed elsewhere. Other idiosyncratic clinically relevant effects of AEDs which may occur early in treatment include a) allergic skin rashes and b) tremor.

a) Phenytoin, carbamazepine, phenobarbitone, lamotrigine and ethosuximide are all known to cause idiosyncratic skin reactions. Oxcarbazepine is much less likely to cause this than carbamazepine. The skin rashes tend to occur early on in treatment (especially within the first month) and although most are mild and may even be transient, some have more sinister implications such as the development of Stevens-Johnson syndrome. Although drug data sheets may suggest withdrawal of the AED, clinical practice is often to continue unless the reaction is severe, and the rash may or may not disappear. For example, even in clinical trials of lamotrigine, only one-third of patients experiencing a rash were actually withdrawn.[45] The rash rate seems to be influenced by initial dosing schedule, so that lower initial concentrations are associated with lower skin problems. This is true for carbamazepine and phenytoin[46] as well as lamotrigine. In at least some cases it seems possible to successfully re-challenge with lamotrigine after initial rash if the patient's epilepsy appeared to benefit.[3, 47]

b) Valproate can cause a symptom identical to benign essential tremor. This may present as a trembling of the hands but the complaint may be instead that handwriting skills have deteriorated or failed to develop properly. The phenomenon is often dose related.[48]

POLYTHERAPY - 'RATIONAL' AND IRRATIONAL

Despite monotherapy being the ideal, many patients whose seizures continue after trying their first AED accumulate 2 or more therapeutic agents. This polytherapy does not usually develop as part of a carefully planned program of therapy. Since 1989 five major new AEDs have been licensed around the world; vigabatrin, oxcarbazepine, lamotrigine, felbamate (now withdrawn because of a risk of aplastic anaemia), and gabapentin. This means that the number of effective drug treatments available has practically doubled. New medications are often licensed as 'add-on' therapies in the first instance. Consequently we may be observing polytherapy making a comeback.

How irrational polytherapy develops

Polytherapy may be patient-led; if seizure control improves on adding a second drug, there may be patient and family resistance to withdrawing the first. Even when this is attempted, seizures may occur causing the withdrawal to be halted. It is difficult to be sure in such a case whether the increase in seizures is due to a rebound effect, which would settle with time, or whether both drugs are needed to maintain optimum control. In this case the resulting polytherapy may be led by either patient or physician. If this situation occurs when the second AED was only partially successful in controlling seizures, the possibility is there of triple therapy arising by the same mechanism. Much polytherapy probably occurs by default, where treatment plans are not initiated properly or carried through due to lack of communication between the specialist, patient, relatives, and family physician.

Is there rational polytherapy?

Sometimes using more than one AED is a deliberate part of a treatment plan. This has usually been in the context of 'adjunctive' treatment, where one AED is regarded as the mainstay of treatment, but an additional agent is used to enhance seizure control, usually over a limited time period. Recently there has been greater understanding of the mechanism of action of AEDs, and of the way seizures develop and spread at a cellular level. This provides a theoretical opportunity for combining AEDs to act synergistically according to their different actions. These two aspects are considered in turn.

Adjunctive therapy
The two main adjunctive agents in antiepileptic treatment are acetazolamide and clobazam. Acetazolamide may be helpful in treating complex partial seizures alongside carbamazepine.[49] Clobazam may be especially useful in catamenial epilepsy, where 20 mg/day added to background medication for the 10 most vulnerable days each cycle may be highly effective.[50] One alternative to clobazam is clonazepam, but the former is preferred because it has less effect on psychomotor function. All three drugs have antiepileptic properties, and all may display the phenomenon of tolerance, so that beneficial effects are often short-lived, although the duration of effect may vary from days to months. Withdrawal

of the adjunctive treatment typically results in reversal of tolerance, so that seizures respond to re-introduction. In some cases, a prolonged remission may occur with continuous treatment.[51,52]

Synergistic therapy
Some authorities have suggested that a combination of 2 drugs may be helpful in patients who have failed to respond to either in monotherapy.[53] Proposed combinations have included carbamazepine with phenytoin[25] in adults and oxcarbazepine with valproate in children.[27] It has yet to be demonstrated whether this approach can be recommended as standard practice in children, where effects of polytherapy on learning cause such concern. One recognised exception is the use of combined ethosuximide and valproate in resistant childhood absence epilepsy if neither drug on its own should prove effective.[54]
Two rationales for combined therapy are sometimes advanced:
- Drugs acting on the *same* receptor site or biochemical system may act synergistically. Vigabatrin has been noted to produce better seizure control in combination with valproate than with other AEDs in children; this effect being ascribed to a greater degree of GABA-T inhibition.[55]
- Drugs acting on *different* receptor sites or biochemical systems may act synergistically. In particular, combining glutamate inhibition with GABA enhancement could be a powerful antiepileptic combination. This has been proposed for lamotrigine and vigabatrin.[56]

Future research needs to address this issue more closely. In the meantime the physician still has a duty to protect children with epilepsy from the hazards of over-zealous treatment.

OTHER ASPECTS OF DRUG MANAGEMENT

Rectal diazepam & paraldehyde

Intravenous diazepam has long been the treatment of choice for aborting serial seizures or early status epilepticus. Rectal administration is potentially as effective, quickly reaching an equivalent blood level to that obtained by intravenous injection, and proprietary rectal preparations are available. The advantages of rectal diazepam are 1) it does not have to be given by someone trained in intravenous technique, and so can be given by parents or responsible non-medical professionals at home or in the school setting, and so can therefore be given without delay, and 2) superficial veins are not subject to repeated puncture; the procedure is therefore less traumatic for the child, and venous access is preserved for future use. Although there is no obvious disadvantage, an issue arises where educational staff or social carers are reluctant to administer rectal

diazepam citing 1) that they lack the specialist training, and /or 2) that the person administering the drug could be vulnerable to an accusation of child abuse. These problems are potentially remediable if the following policy is adopted:
- care staff are instructed in use by trained nursing or medical staff
- the physician provides a written statement setting out the particular circumstances in which the rectal preparation is to be used for each patient
- parents, and where appropriate the patient, give care staff written permission to administer rectal diazepam according the physician's instructions.

Some children may show idiosyncratic reactions to diazepam, or it may be ineffective. In such cases, intermittent rectal paraldehyde may be used for the same indications. Because this is an irritant to mucosal surfaces, dilution with a suitably innocuous solvent such as oil of arachis is necessary. Paraldehyde leaves an unpleasant after-taste, and its use has to be carefully weighed up against the risks of not using it.

OTHER NON-SURGICAL TREATMENTS AVAILABLE

Ketogenic diet

There is an ancient observation that starvation can reduce seizure frequency,[57] and this seems to be related to the production of ketone bodies.[58,59] Diets based on this principle are sometimes effective in reducing seizure frequency in severe, refractory epilepsies in childhood. However, the remission is often only temporary. Also, children typically find the diet unpalatable, and therefore their parents may find it difficult to enforce compliance.

Vagal nerve stimulation

This is a recently developed procedure which may assist management of partial seizures. Although results of long-term use in large patient numbers are not yet available, some early reports are encouraging. The treatment does not apparently cause significant cardiovascular or gastrointestinal adverse effects. This may or may not yet turn out to be a significant therapeutic development.[60,61]

Psychological treatments

Some children spontaneously learn ways of starting and stopping seizures. This is presumably by a global change in arousal, or by redirecting attention and thus altering the level of incidental traffic in neurones which constitute the focus.[62] This observation carries an implication that epilepsy may be treated by the techniques of behaviour therapy. There is now a growing literature on this subject,[63,64] including a therapists' manual.[65]

REFERENCES

1. Hippocrates of Kos, Aphorisms, Section II no 45. In Lloyd GER, editor. Hippocratic Writings. Harmondsworth: Penguin Books, 1983.
2. Galen, De Simpl. Med. Temp. Ac facult. VI c3 col 11 p 859. Quoted in: Temkin O. The Falling Sickness. 2nd rev. ed. Baltimore: Johns Hopkins Press, 1971.
3. Brown SW. Quality of Life: a view from the playground. Seizure, 1994; 3 (Suppl. A): 11-15.
4. Brown S, Betts T, Chadwick D, et al. An epilepsy needs document. Seizure, 1993; 2: 91-103.
5. Richens A, Perucca E. Clinical pharmacology and medical treatment. In: Laidlaw J, Richens A, Chadwick D, editors. A textbook of epilepsy. 4th ed. Edinburgh: Churchill Livingstone, 1994: 495-559.
6. Lindhout D, Omtzigt JG. Pregnancy and the risk of teratogenicity. Epilepsia, 1992; 33: (Suppl. 4), S41-S48.
7. Besag FMC. Difficult-to-treat childhood epilepsies. In: Reynolds EH, editor. Lamotrigine - a new advance in the treatment of epilepsy, London, Royal Society of Medicine, 1994: 53-62.
8. Gillham RA, McKee PJW, Blacklaw J, et al. The effect of vigabatrin on cognitive function: a double blind placebo controlled crossover study. Seizure, 1992; 1: Suppl. A P7/20.
9. Richens A. Pharmacokinetics of lamotrigine. In: Richens A, editor. Clinical update on lamotrigine: a novel antiepileptic agent. Royal Tunbridge Wells: Wells Medical, 1992: 21-30 (in discussion, page 29).
10. Dodrill CB, Troupin AS. Neuropsychological effects of carbamazepine and phenytoin: a reanalysis. Neurology, 1991; 41: 141-143.
11. Forsyth I, Butler R, Berg I, et al. Cognitive impairment in new cases of epilepsy randomly assigned to carbamazepine, phenytoin and sodium valproate. Dev. Med. Child Neurol., 1991; 33: 524-534.
12. Berg I, Butler A, Ellis M, et al. Psychiatric aspects of epilepsy treated with carbamazepine, phenytoin or sodium valproate: a random trial. Dev. Med. Child Neurol., 1993; 35: 149-157.
13. Meador KJ, Loring DW, Huh K et al. Comparable cognitive effects of anticonvulsants. Neurology, 1990; 40: 391-394.
14. Goa KL, Ross SR, Chrisp P. Lamotrigine. A review of its pharmacological properties and clinical efficacy in epilepsy. Drugs, 1993; 46: 152-176.

15. Lerman P, Kivity S. The efficacy of corticotrophin in primary infantile spasms. J. Pediatr., 1982; 101: 294-296.
16. Lombroso CT. A prospective study of infantile spasms: Clinical and therapeutic correlations. Epilepsia, 1983; 24: 135-158.
17. Snead OC, Benton JW, Myers GJ. ACTH and prednisone in childhood seizure disorders. Neurology, 1983; 33: 966-970.
18. Heller AJ, Chesterman P, Elwes RDC, et al. Monotherapy for newly diagnosed epilepsy: a comparative trial and prognostic evaluation. Epilepsia, 1989; 30: 648-651.
19. McGowan MEL, Neville BGR, Reynolds EH. Comparative monotherapy trial in children with epilepsy. Brit. J. Clin. Pract., 1983; Suppl. 27: 115-119.
20. Da Silva M, McArdle B, McGowan M, et al. Monotherapy for newly diagnosed childhood epilepsy: a comparative trial and prognostic evaluation. Epilepsia, 1989; 30: 662A.
21. Dam M, Ekberg R, Loyning Y, et al. A double-blind study comparing oxcarbazepine and carbamazepine in patients with newly diagnosed, previously untreated epilepsy. Epilepsy Res., 1989; 3: 70-76.
22. Da Silva M, McArdle B, McGowan M, et al. A prospective randomised comparative monotherapy clinical trial in childhood epilepsy. In: Chadwick D, editor. Fourth international symposium on sodium valproate and epilepsy. London, Royal Society of Medicine, 1989: 81-86.
23. Verity CM. Comparative multicentre trial of sodium valproate and carbamazepine in newly diagnosed childhood epilepsy (paediatric EPITEG trial). In: Brodie M, editor. Valproate: New Milestones, Medical Meeting Series No. 1. London, Franklin Scientific, 1993: 12-15.
24. Kriel RL, Cloyd JC. Antiepileptic drug therapy in children: Pharmacokinetics, adverse effects and monitoring. In: Swaiman KF, editor. Pediatric neurology: principles and practice, 2nd ed. St Louis: Mosby, 1994: 583-609.
25. Duncan JS. Strategies of antiepileptic drug treatment in patients with chronic epilepsy. In: Trimble MR, editor. Chronic epilepsy, its prognosis and management. Chichester, John Wiley, 1989: 143-149.
26. Thomson AF, Brodie MJ. Pharmacokinetic optimisation of anticonvulsant therapy. Clin. Pharmacokinet., 1992; 23: 216-230.
27. Johannessen SI, Loyning Y, Munthe-Kaas AW. Rationale for antiepileptic drug therapy: clinical and pharmacokinetic aspects. In Sillanpaa M, Johannessen SI, Blennow G, Dam M, editors. Paediatric epilepsy. Petersfield, Wrightson Biomedical Publishing, 1990: 243-268.
28. Warner T, Patsalos PN, Prevett M, et al. Lamotrigine-induced carbamazepine toxicity: An interaction with carbamazepine-10,11-epoxide. Epilepsy Res., 1992; 11: 147-150.
29. Herranz JL, Arteaga R, Farr IN, et al. Dose-response study of vigabatrin in children with refractory epilepsy. J. Child Neurol., 1991; 6: Suppl. 2: 2S45-2S51.
30. Richens A. Clinical pharmacokinetics of gabapentin. In: Chadwick D, editor, New trends in epilepsy management: The role of gabapentin. London, Royal Society of Medicine, 1993: 41-46.
31. Wolf M, Forsyth A. Behaviour disturbance, phenobarbitone and febrile seizures. Pediatrics, 1979; 61: 728-731.
32. Rentmeester Th, Hulsman J. Sustained release valproate versus conventional formulation valproate. A study of the tolerance and efficacy of LA 40220.

In: Chadwick D, editor. Fourth international symposium on sodium valproate and epilepsy. London, Royal Society of Medicine, 1989: 185-191.
33. Brouwer OF, Pieters MSM, Edelbroek PM, et al. Conventional and controlled release valproate in children with epilepsy: a cross over study comparing plasma levels and cognitive performance. Epilepsy Research, 1992; 13: 245-253.
34. Sachdeo RC, Belendiuk G. Generic versus branded carbamazepine. Lancet, 1987; 1: 1432.
35. Koch G, Allen JP. Untoward effects of generic carbamazepine therapy. Arch. Neurol., 1987; 44: 578-579.
36. Welty TE, Pickering PR, Hale BC, et al. Loss of seizure control associated with generic substitution of carbamazepine. Ann. Pharmacotherapy, 1992; 26: 775-777.
37. Gilman JT, Alvarez LA, Duchowny M. Carbamazepine toxicity resulting from generic substitution. Neurology, 1993; 43: 2696-2697.
38. Tyrer JH, Eadie MJ, Sutherland JM, et al. Outbreak of anticonvulsant intoxication in an Australian city. Brit. Med. J., 1970; 4: 271-273.
39. Nuwer MR, Browne TR, Dodson WE, et al. Generic substitutions for antiepileptic drugs. Neurology, 1990; 40: 1647-1651.
40. Camfield C, Camfield P, Smith E, et al. Asymptomatic children with epilepsy: little benefit from screening for anticonvulsant-induced liver, blood or renal damage. Neurology, 1986; 36: 838-841.
41. Willmore LJ, Triggs WJ, Pellock JM. Valproate toxicity: risk screening strategies. Child Neurol., 1991; 6: 3-6.
42. Knott C. Measurement of saliva drug concentrations in the control of antiepileptic medication. In: Pedley TA, Meldrum BS, editors. Recent advances in epilepsy. Edinburgh, Churchill Livingstone, 1983: 57-73.
43. Brodie MJ. Established anticonvulsants and treatment of refractory epilepsy. Lancet, 1990; 336: 350-354.
44. Parkinson GM, Whitfield SP. Undiagnosed language impairment in children with complex epilepsy. Epilepsia, 1994; 35 (S 7): 48.
45. Yuen A. Safety issues. In: Richens A, editor. Clinical update on lamotrigine: a novel antiepileptic agent. Royal Tunbridge Wells: Wells Medical, 1992: 69-75
46. Chadwick D, Shaw M, Foy P, et al. Serum anticonvulsant concentrations and the risk of drug induced skin eruptions. J. Neurol. Neurosurg. Psychiatry, 1984; 47: 642-644.
47. Tavernor SJ, Newton ER, Brown SW. Rechallenge with lamotrigine after initial rash. Epilepsia, 1994; 37 (S 7): 72.
48. Karas BJ, Wilder BJ, Hammond EJ, et al. Valproate tremors. Neurology, 1982; 32: 428-432.
49. Oles KS, Penry JK, Cole DLW, et al. Use of acetazolamide as an adjunct to carbamazepine in refractory partial seizures. Epilepsia, 1989; 30: 74-78.
50. Feely M, Gibson J. Intermittent clobazam for catamenial epilepsy: tolerance avoided. J. Neurol. Neurosurg. Psychiatry, 1984; 47: 1279-1282.
51. Allen JW, Oxley J, Robertson MM, et al. Clobazam adjunctive treatment in refractory epilepsy. Brit. Med. J., 1983; 286: 1246-1247.
52. Callaghan N, Goggin T. Adjunctive treatment in resistant epilepsy. In: Trimble MR, editor. Chronic epilepsy, its prognosis and management. Chichester, John Wiley, 1989: 165-176.
53. Schmidt D. Two antiepileptic drugs for intractable epilepsy with complex

partial seizures. J. Neurol. Neurosurg. Psychiatry, 1982; 45: 1119-1124.
54. Rowan AJ, Meijer JWA, De-Beer Pawlikowski N, et al. Valproate-ethosuximide combination therapy for refractory absence seizures. Arch. Neurol., 1983; 40: 797-802.
55. Armijo JA, Arteaga R, Valdizan EM, et al. Coadministration of vigabatrin and valproate in children with refractory epilepsy. Clinical Neuropharmacology, 1992; 15: 459-469.
56. Stewart J, Hughes E, Kirker S, et al. Combined vigabatrin and lamotrigine for very intractable epilepsy. Seizure 1992; 1: Suppl. A P13/39.
57. Scott DF. The history of epileptic therapy. Carnforth: Parthenon Publishing Group, 1993.
58. Wilder RM. The effect of ketonemia on the course of epilepsy. Mayo Clinic Bull., 1921; 2: 307.
59. Kinsman SL, Vining EPG, Quaskey SA, et al. Efficacy of the Ketogenic Diet for Intractable Seizure Disorders: Review of 58 Cases. Epilepsia, 1992; 33: 1132-1136.
60. Ben-Menachem E, Manon-Espaillat R, Ristanovic R, et al. Vagus nerve stimulation for treatment of partial seizures: 1. A controlled study of effect on seizures. Epilepsia, 1994; 35: 616-626.
61. Ramsay RE, Uthman BM, Augustinsson LE, et al. Vagus nerve stimulation for treatment of partial seizures: 2. Safety, side effects, and tolerability. Epilepsia, 1994; 35: 627-636.
62. Fenwick PBC, Brown SW. Evoked and Psychogenic Seizures: I Precipitation. Acta Neurologica Scandinavica, 1989; 80: 535-540.
63. Dahl JC. A behavioural approach to the treatment of epilepsy. In: Sillanpaa M, Johannessen SI, Blennow G, Dam M, editors. Paediatric epilepsy. Petersfield, Wrightson Biomedical Publishing, 1990: 285-294.
64. Gillham RA: Refractory Epilepsy - an Evaluation of Psychological Methods in Outpatient Management Epilepsia, 1990, 31: 427-432.
65. Dahl JC. Epilepsy: A behavior medicine approach to assessment and treatment in children. Seattle, Hogrefe & Huber, 1992.

10 Cognitive side-effects of antiepileptic drugs

ALBERT P. ALDENKAMP
Department of Neuropsychology "Meer en Bosch", Epilepsy Center, Heemstede, The Netherlands

Antiepileptic treatment may be accompanied by the development of cognitive side-effects. Such cognitive impairment represents the long-term outcome of the chronic toxicity of antiepileptic drugs (AED's). This may contribute to the impact on daily life functioning, especially in refractory epilepsies, as the effects may increase with prolonged therapy. Using the DIMDI computer-database, 358 studies were evaluated. These studies showed most impairment for phenobarbital and phenytoin and -to a lesser extent- valproate. Carbamazepine and its related compound oxcarbazepine have a more favourable cognitive profile.

INTRODUCTION

Antiepileptic drug treatment usually requires several years or even decades. Such chronic treatment may be accompanied by a variety of side-effects. A number of these side-effects are potentially fatal such as some of the idiosyncratic reactions. Other effects appear immediately after the start of drug exposure but are relatively benign because they show habituation, or are reversible when they are dose dependent. In this chapter we will concentrate on the cognitive effects of AED's, i.e. the adverse effects of drugs on higher-order cerebral functions, such as attention, reaction speed or memory. At first glance this type of side-effect seems less dramatic than e.g. some of the idiosyncratic reactions to drugs. Nonetheless, a number of studies have claimed that drug-induced cognitive impairment may have a much greater impact on daily life function, especially in children, than had hitherto been suspected.[1-9] The cognitive side-effects represent the long-term outcome of chronic toxicity of the AED's. This may contribute to the impact on daily life functioning, especially in refractory epilepsies, as the effects may increase with prolonged therapy.[10]

As a consequence, cognitive side-effects now represent a major issue in clinical assessment and are an outcome measure in many trials with new antiepileptic drugs. Fortunate as this is, the literature still shows serious controversies about type and severity of drug-induced cognitive impairment, even in recent and well controlled studies. These controversies are due, among others to differences between the studies with respect to the type of subjects investigated and the design used to detect the drug effects.

In this chapter, we will attempt to review the existing literature to evaluate the findings, especially in the light of their clinical relevance. We will mainly concentrate on the most commonly used (and studied) AED's:

phenobarbital (PHB), phenytoin (PHT), valproate (VPA) and carbamazepine (CBZ). Necessarily, this review can not be limited to children. Most of the evidence for cognitive side-effects come from studies in adults. We will use these findings in an attempt to interpret the results for the treatment in children.

One of the most authoritative reviews of the recent years[2] gave the following table (Table 1), summarising the results of a large number of studies evaluating the effects of AED's on general behaviour and on cognitive function. The suggestion from this summary, but also from other reviews[1-14] is that PHT and CZP are harmful for cognitive function whereas CBZ has a more favourable profile.

Table 1. Summary of the effects of anticonvulsants on cognitive function and behaviour

	Behaviour	Cognitive function
Carbamazepine	Minimal	Minimal
Clonazepam	Impairs	Impairs
Ethosuximide	?	?
Phenobarbitone	Impairs	Minimal
Phenytoin	Minimal	Impairs
Sodium Valproate	Minimal	?

From Trimble, M.R. and Reynolds, E.H., Eds. *Epilepsy, Behaviour and Cognitive Function*, Copyright 1987 John Wiley & Sons. Reprinted by permission of John Wiley & Sons, Ltd.

These conclusions, however, came back into debate when some studies [15-19] failed to reproduce the cognitive effects of, e.g., PHT when serum concentrations were sufficiently controlled. These studies suggest that some of the reported differences between for example, CBZ and PHT actually may have been due to an artifact, i.e. differences in drug concentration. Moreover, some studies suggested that subject selection bias may have influenced the results, as there is evidence that, at least in some countries, PHT may be given to other types of patients than CBZ. PHT is cheaper and it is considered a 'simple drug', because it can be given once a day.[14,20-23] Observed cognitive differences between groups after exposure to PHT and CBZ may thus simply reflect differences in patient characteristics rather than drug effects. Finally it is suggested that the adverse effects of PHT on cognitive functions may actually have been due to a contaminating factor: motor speed. In one study[19] motor slowing appeared to be the only factor that discriminated between PHT and CBZ. As almost all cognitive tests use motor output, 'peripheral' motor effects may erroneously have been interpreted as 'central', cognitive effects. A critical review of the existing literature is therefore useful, both for future research and for the proper interpretation of existing data, a highly relevant issue to clinical practice.

IDENTIFICATION OF RELEVANT STUDIES

As a first step in identifying relevant studies, the DIMDI computer-database was used. Next, reviews[1-14,20-23] were checked for additional references. The resulting 1256 papers were screened for information about cognitive side-effects of AED's. This resulted in 358 potentially relevant papers. Then the inclusion criteria given in Table 2 were applied.

Table 2. Literature selection: inclusion and exclusion criteria

1. **Type of article**: report of original research, in English, in peer reviewed journals or proceedings, published after 1970.
   Excluded: reviews, abstracts, brief communications, internal journals of the pharmaceutical industry

2. **Subjects**: epilepsy or healthy volunteers.
   Excluded: psychiatric patients, delinquents, mentally handicapped, children with behavioural or mood problems, animal studies

3. **Treatments**: current AED's.
   Excluded: studies on experimental drugs that have failed to prove efficacy and will not be introduced in clinical practice, such as flunarizine or loreclezole

4. **Outcome measures**: psychometrically assessed cognitive functions.
   Excluded: clinical observations, studies employing complaint indices

Included studies should be readily accessible, and allow the readers to judge the validity of the results; abstracts, for example, usually do not provide sufficient detail. The target population is the 'regular' epilepsy patient, receiving long term AED treatment. AED treatment in other groups is associated with complicating interactions between condition and therapy. An exception was made for studies in healthy volunteers, on the assumption that such studies might suggest hypotheses worth further exploration in epilepsy. The constraint on publication date was set fairly arbitrarily at 25 years ago. Studies after that date were all done in a time when serum level monitoring and modern cognitive tests had come in widespread use. After application of these criteria, a database of **73** articles remained.

In weighing the evidence from these studies, the approach taken here is first to disregard studies that contain certain basic deficiencies that render them uninformative with regard to cognitive AED effects, because they fail to rule out too many plausible alternative explanations. The problems examined here are not minor annoyances: their impact on the validity of the conclusions drawn is such that a useful AED, not harmful

to cognitive functioning might look bad, while genuine adverse cognitive effects might be missed. Although a large number of methodological comments are in order, we will first discuss three types of studies that are arguably not sufficient for permitting reasonably valid inferences regarding the cognitive AED-effects:
- Polytherapy studies
- Post-test only studies
- Studies that provide insufficient information

With only high quality data to go on, the last section of this review attempts to draw conclusions in which some confidence can be placed.

DESCRIPTION OF STUDIES THAT DO NOT ALLOW VALID INFERENCES

Polytherapy studies

Although polytherapy is the most common treatment in refractory epilepsies, it introduces complications in identifying the exact cause of observed cognitive changes. Interactions between antiepileptic drugs became evident soon after routine measurement of serum levels came into practice. Such interactions can alter therapeutic efficacy and thus, conceivably, cognitive functioning. Moreover polytherapy is typically given to patients suffering from refractory epilepsy, and the threat of a 'seizure confound' is thus always serious. 'Seizure confound' refers to a major validity concern: the failure to separate seizure effects from 'genuine' AED-effects.

Typically, cognitive AED effects are studied in add-on studies, where a new drug is introduced into a polytherapy regime (e.g.[24]). Nonetheless, in this type of study the seizure confound is even stronger and the results on cognitive tests are a potpourri of positive and negative seizure effects, AED-effects and drug-drug interactions that can never be disentangled. It is therefore impossible to use this type of study for inferences about cognitive side-effects of AED's.

Posttest-only studies

In this type of study, patients are selected on clinical grounds and on a non-randomized basis. The patients are then divided in groups (i.e. those on PHT and those on CBZ) and their cognitive test results are compared. Although this 'one-shot' design is not uncommon in clinical research, it holds a major threat to the validity of the results. Due to the lack of pretest observations in this design, there is no way of knowing whether the treatment as such is related to any kind of cognitive change. Posttest differences can be attributed either to a treatment effect or to various selection differences between PHT and CBZ groups. In other words, any observed difference between the groups may already have been present before the treatment started. The plausibility of such selection differences renders this design uninterpretable.

A concrete example of a validity threat in this design is found in the study by Andrewes et al.[25] The authors conclude that PHT shows an overall trend towards poorer performance on several cognitive tasks when compared to CBZ. This study is frequently cited as evidence for the cognitive side-effects of PHT. A complication in the study is the difference in drug exposure between CBZ (3.6 years) and PHT (5.8 years). It is conceivable that longer drug exposure in itself, regardless of type of drug, leads to the differences between the investigated groups. The study used, however, a posttest-only design that does not allow the evaluation of such alternative explanations. Thus, results of studies with a post-test only design do not allow valid inferences about cognitive side-effects of AED's.

Studies that provide insufficient information

Some papers do not provide the information to enable the reader to evaluate the appropriateness of the methods and the reliability and validity of the results. For example, certain articles fail to give the number of subjects studied, do not report the results of the statistical analysis to support the conclusions, or mention only statistical results for cognitive variables yielding significant effects. One is thus forced into the unfortunate position of either accepting such results uncritically, or specifying certain minima with respect to the methodological and statistical information provided that must be met before evaluation is possible. In this review 'numbers of patients' and 'number of cognitive variables' are used as presentation minima. These data are essential to evaluate the statistical power of the study and, consequently, the validity of the reported results, especially in the case of 'no-effect' reports.

One statistical comment may illustrate the relevance of such presentation minima. Authors may wish to emphasize *'no effect'* findings, particularly if these are in line with the research hypothesis that a certain AED has no adverse cognitive effects. In contrast to studies in other areas, non-significant results are seen as good news here, because they represent the absence of cognitive side-effects. This conclusion lacks meaning, however, unless the a priori probability of obtaining a significant result (*'the power'*) is sufficiently high (80% being a conventional value). In other words, the study should have been able to detect genuine impairment and non-significance should not be due to insufficient power or to other failures in the study. The power decreases especially if small samples are used, or if the effect size, i.e. the magnitude of the cognitive effects under study, is small. In order to achieve an 80% chance of detecting small, medium and large differences between two independent means(.20, .50 and .80 standard-deviation respectively in line with the convention proposed by e.g. Cohen[26,27]), the necessary number of patients per group, at a 5 % significance level, is 393, 64 and 26 respectively. Most reviews claim that, although their long-term impact on daily life function may be considerable, the cognitive effects of AED's, as assessed by cognitive tests, are generally small. Nonetheless, limited sample sizes (20 patients or less) are used in the majority of studies that claim 'no effect'. Consequently, these studies could only have detected cognitive effects of such magnitude that they would presumably be obvious to the clinician, even without any psychological tests. With "no-effect" claims it is therefore worth checking whether they might not simply reflect inadequate statistical power. Studies that fail to present the necessary data for such evaluation were thus considered elusive to evaluation due to insufficient information.

An example of such a study is the study by Smith et al.,[17] investigating no less than 618 patients on AED's. The patients were, however, distributed among different drug groups and the study fails to give data on: number of patients in each drug group; data on the separate tests; data for the separate assessments; inferential statistics (in fact not one results of the statistical analysis is presented); essential data on patients characteristics (e.g. the 'untreated baseline' is characterized 'relatively drug-free'). It is clear that the validity of the results of such a study - although its design and the number of patients are impressive - can not be evaluated.

NORMAL-VOLUNTEER STUDIES

The remaining studies are potentially useful to evaluate the cognitive effects of AED's. These studies can be divided into normal-volunteer studies and monotherapy studies in patients with epilepsy.

Table 3. Normal volunteer-studies

AEDs	Subjects N	Period on drug	Type of impairment
PHT versus placebo			
Ideström et al, 1972	15	2 hours	n.i.
Houghton et al, 1973	6	1½-7 hours	n.i.
Smith & Lowrey, 1975	10	3 weeks	impairment of mental concentration
Thompson et al, 1980	8	2 weeks	global memory impairment
Thompson et al, 1981	8	2 weeks	Impairment of memory, concentration, mental and motor speed
CBZ versus placebo			
Thompson et al, 1980	8	2 weeks	no impairment
Other AED's versus placebo			
Houghton et al, 1973 (PHB)	6	1-7 hours	n.i.
Thompson & Trimble. 1981 (VPA)	10	2 weeks	impairment of decision making speed
Cohen et al, 1985 (LTG)	12	1 day	n.i.
Saletu et al, 1986 (VGB)	10	1-8 hours	n.i.
Curran & Java, 1993 (OXC)	12	2 weeks	no impairment; improvement of focused attention and speed
PHT versus CBZ			
Meador et al, 1991	21	1 month	impairment of choice reaction time and attention for CBZ and PHT
Meador et al, 1993	15	1 month	impairment of memory for CBZ and PHT
Other types of comparisons			
* PHT low dose versus PHT high dose:			
Stevens et al, 1974	107	2 weeks	no difference
* CBZ versus CBZ controlled release:			
Tedeschi et al., 1989	6	1 day	n.i.

All studies use a randomized double blind cross-over design. **CBZ**=Carbamazepine; **VGB**=Vigabatrin; **LTG**=Lamotrigine; **OXC**=Oxcarbazepine; **PHT**=Phenytoin; **PHB**=Phenobarbital; **VPA**=Valproate; n.i. = results not interpretable

Table 3 summarises the normal-volunteer studies. All these studies use double-blind randomized cross-over designs. Three types of comparisons were found: AED versus placebo, AED versus AED, and dose-effect studies. Most of the information is available for the comparison of PHT versus placebo. The last column in table 3 gives the reported findings of the studies. Before interpreting these claims, we must make a few comments. Firstly, the use of a normal-volunteer study has some clear advantages. The problems of the intervening contribution of seizure related variables on cognitive function is absent and manipulations of drug and dose are not limited by clinical considerations. A clear disadvantage is, however, that the period of drug exposure is mostly restricted. There is evidence, that in most AED's 'early' cognitive side-effects may develop, only during a short period, i.e. during the first few days or weeks of drug exposure. After this period normalization occurs, possibly due to the development of so-called positive tolerance or habituation.[28] Although little is known about how tolerance to the cognitive effects of AED's develops, a failure to take this factor into account may lead to overestimation of the negative effects of drugs on cognition.[20,21] This important point has to be taken into account in studies with healthy volunteers who are typically given AED's during a few weeks at most, long term studies not being feasible.

Our examination of the 13 normal-volunteer studies[15,16,29-40] that were carried out to assess the cognitive effects of AED's revealed that the majority of these studies use fairly short periods of drug exposure, often no longer that 1 day. Given the half-lives of most of the AED's it was decided to ignore all claims from studies that used periods of drug exposure of one day or less. In the table this is indicated as n.i. (not interpretable).

For the comparison of PHT with placebo, three studies[31-33] remain, reporting -without exception- the impairment of memory, concentration, mental and motor speed. For the comparison of other AED's (including CBZ) only one study per AED is available.[32] These studies suggest that valproate (VPA) impairs decision making speeds, whereas CBZ and its related compound oxcarbazepine do not impair cognitive function. The studies by Meador et al.[16,38] did not yield differences between PHT and CBZ, although in both studies significant differences were found (despite the small number of subjects) between the AED's and the non-drug baseline. Especially memory is mentioned as an area that may be similarly affected by both drugs.[38] Finally, two dose-relation studies are available.[39,40] Only the study by Stevens et al.[39] allows evaluation as it uses a 2 weeks period of drug exposure. This study did not find dose-dependent differences for PHT.

The overall conclusion from the normal-volunteer studies is that PHT impairs cognitive function. Some controversies remain as to the cognitive effects of CBZ, whereas reliable and reconfirmed information for other AED's is lacking.

MONOTHERAPY STUDIES IN PATIENTS WITH EPILEPSY

Table 4. Monotherapy studies - single drug comparisons

AEDs	comparisons	Subjects on AED	Control subjects	Type of impairment
Single drug comparisons against a non-drug condition (untreated control or Placebo)				
Phenobarbital				
MacLeod et al, 1978	PHB↓-PHB↑-UTC	19	20	impairment of short-term memory
Camfield et al, 1979	PHB-PLA	24/18	--	n.i.
Wolf et al, 1981	PHB-UTC	25	25	n.i.
Phenytoin				
Callassi et al, 1987	PHT-UTC	10	10	n.i.
Carbamazepine				
O'Dougherty et al, 1987	CBZ↓-CBZ↑-UTC	11	11	n.i.
Aldenkamp et al, 1987	CBZ-CBZ^{cr}-UTC	11	11	increase of variability in CBZ compared to CBZ^{ct}
Aldenkamp et al, 1993	CBZ-UTC	56/17/10	83	no impairment for CBZ
Valproate				
Callassi et al, 1990	VPA-UTC	20	20	n.i.
Single drug comparisons without a non-drug condition				
Phenytoin				
Dodrill&Temkin, 1989	PHT↑ with PHT↓	36/34	--	n.i.
Carbamazepine				
Macphee et al, 1986	CBZ↑ with CBZ↓	8	--	n.i.
Amman et al, 1990	CBZ↑ with CBZ↓	50	--	improvement at peak serum levels
Valproate				
Amman et al, 1987	VPA↑ with VPA↓	46	--	no effect of VPA serum level

CBZ=Carbamazepine; **CBZ^{cr}**=Carbamazepine; controlled release;
VGB=Vigabatrin; **LTG**=Lamotrigine; **OXC**=Oxcarbazepine; **PHT**=Phenytoin;
PHB=Phenobarbital; **VPA**=Valproate; **UTC**= untreated controls;
PLA=Placebo. ↑=high dose; ↓=low dose;
n.i. = results not interpretable

Table 4 evaluates 12 studies[41-52] in which a single drug is investigated in patients with epilepsy with or without a non-drug condition. Although these studies *potentially* allow valid inferences about the cognitive effects of AED's, some comments are in order. MacLeod et al.[41] studied low and high doses of phenobarbital (PHB) in 19 patients. The results show impairment of short-term memory with a high dose of PHB. Camfield et al.[42] used intelligence tests in very young children (average age is 15 months). Intelligence testing is, however, not reliable before the age of 4 years. Therefore the results of this study are difficult to interpret. Wolf et al.[43] also used intelligence testing in somewhat older children (ranging from 45 to 71 months). In this study children with febrile seizures were included, using phenobarbital as prophylaxis. This study did not find any negative effects on cognitive tests, despite the long period of treatment. However, compliance seems low (50% had serum level < 15 µg/ml) and generalization to patients with chronic epilepsy is difficult. The studies by the Italian group of Gallassi et al.[44,48] require some detailed comments. All studies from this group used the same design: patients were tested on monotherapy and retested after removal of the AED. The assumption of this 'withdrawal design' is that reversible side-effects are demonstrated by an *improvement* after a drug is stopped. A point to be made here is that other factors can also cause such improvement, one of the most obvious being the practice effect due to retesting. In other words, if we retest a patient at short intervals, he may 'learn' or recognize the tests which in itself causes improved test scores. Thus, the validity of the results from this type of study largely depends on the extent to which retesting effects are controlled. A minimal requirement is a non-medication control group that is retested with exactly the same intervals as the drug groups.[27] Moreover, retesting effects are dependent on the type of tests and the number of retests. Gallassi and coworkers used four retests and at least three of the tests are known to be extremely sensitive to such retesting effects (e.g. 'Raven Progressive Matrices', a Visuo-Motor Test and a Verbal Learning Task). Improvement produced by retesting may thus erroneously be interpreted as evidence for the elimination of cognitive side-effects in these studies. Nonetheless the control subjects were only tested once. Therefore, although these studies used a potential powerful design, the results can not used to evaluate cognitive effects of AED's. In the study by O'Dougherty et al.[45] patients were investigated before and during drug treatment and with high or low serum levels of CBZ. However, 3 of the 11 children were reported to be already on medication at the 'untreated pretest'. This may well have caused the observed variability at pretest. The study by Aldenkamp et al.[46] used a repeated measurement design, comparing CBZ with CBZ controlled release (CR) and observing more variability in ordinary CBZ (relative to CBZ-CR), especially in the episodes with peak serum levels. The Swedish 'Holmfrid study' by Aldenkamp et al.[47] also used a withdrawal design: assessing patients on monotherapy PHT, VPA or CBZ, followed by reassessment after complete withdrawal of the medication. The results have specific value for this review as they concern children. Practice effects were controlled by the use of a non-medication control group that followed the same (re)testing schedule as used for the drug groups. Although the study compared several AED's, the sample size was too limited for VPA and PHT to allow valid inferences. Therefore only the comparison between CBZ (n=56) and the non-drug condition (n=83) is evaluated here. This comparison does not yield evidence for CBZ-induced impairment. The study by Dodrill[49] evaluates patients on high doses of PHT, relative to low doses PHT. The original study reported impairment on several cognitive tasks for higher doses of PHT. A reanalysis 14 years later[49] suggests that the effects are restricted to motor factors that are defined as 'peripheral' and not cognitive. This study, however, shows only relative effects of PHT (high serum level versus low serum level) as an untreated control group or untreated baseline was lacking. The group on higher doses is difficult to evaluate in the light of our current knowledge about the 'therapeutical window' of PHT. The average serum level of 42.1 µg/ml is within the toxic range. Given the standard deviation, some patients may have had serum levels up to 60 µg/ml. Nonetheless it is remarkably that even with these high levels of medication the number of cognitive effects, measured with a comprehensive test battery were neglectable. This may have been caused by the fact that the assignment to the groups was not randomized. Therefore a serious seizure confound may occur: patients probably received high doses on clinical grounds, that may also interfere with cognitive function. Indeed, more seizures were found in the period preceding the study in the high doses group. Thus, cognitive functions may already have been impaired by the seizures, limiting the potential negative effects of the AED's. The study by MacPhee et al.,[50] used a 1 day evaluation of increasing doses of CBZ after the administration of a single 400 mg dose. This period is apparently too short to evaluate chronic effects of the medication. Furthermore it is doubtful whether scaling effects, caused by frequent retesting (5 times during one day) were sufficiently controlled. In both studies by Amman et al.[51,52] serum levels of CBZ or VPA were experimentally changed by given the daily dose before or after the testing procedure. They report no effect of time of medication for VPA, whereas children performed better at peak serum levels of CBZ on measures of attention and motor function.

Table 5 summarises the 6 studies[15,18,53-57] that use multiple drug comparisons, with or without a non-drug condition. The methodological problems associated with the withdrawal studies by Gallassi et al. have been amply reviewed in the previous paragraph. Due to the lack of a retested control group the 'non-drug - drug comparisons' do not allow to distinguish withdrawal effects from retesting effects. The multiple drug study[53], mentioned in table 5 may nevertheless give valid information, i.e. on drug-drug differences. The withdrawal of the drugs was investigated in this study for each separate drug with the same retesting procedures.

Table 5. Monotherapy studies - multiple drug comparisons

AEDs	AEDs	Subjects on AED	Control subjects	Type of impairment
Multiple comparisons relative to other AEDs and to untreated controls or Placebo				
PHT/CBZ				
Calandre et al, 1990	PHB-VPA	26/23	60 UTC	n.i.
Multiple comparisons relative to other AEDs (without controls of placebo)				
PHT/CBZ				
Dodrill & Troupin 1991	CBZ-PHT	40	--	n.i.
PHT/OXC				
Aikiae et al, 1992	OXC-PHT	14/15	--	n.i.
PHB/VPA				
Vining et al, 1987	PHB-VPA	21	--	VPA favourable than PHB
PHB/PHT/CBZ				
Meador et al, 1990	PHB-PHT-CBZ	15	--	no extreme differences between PHB-PHT-CBZ
PHB/PHT/CBZ/VPA				
Gallassi et al, 1992	PHB-CBZ-PHT-VPA	27/18/-16/29	--	lower scores, relative to the other AED's for PHB and VPA on tests of visuomotor performance and memory lower scores, relative to the other AED's for PHT on intelligence tests not conclusive on the reversibility of the side-effects
ZSM/CBZ				
Wilensky et al, 1985	ZSM-CBZ	4	--	n.i.

CBZ=Carbamazepine; CBZ^{cr}=Carbamazepine; controlled release; VGB=Vigabatrin; LTG=Lamotrigine; OXC=Oxcarbazepine; PHT=Phenytoin; PHB=Phenobarbital; VPA=Valproate; ZSM= Zonisamide; UTC= untreated controls; PLA=Placebo. ↑=high dose; ↓=low dose; n.i. = results not interpretable

Relative differences (e.g the differences between CBZ and PHT) may therefore be interpreted. Table 5 shows only one[53] of the three multiple drug studies by Gallassi et al. that we found with our literature search. All these studies had the same number of controls and used increasing number of patients. Presumably, they are all integral parts of the same investigation, presented as different studies. We therefore used only the most recent study that also included most patients.[53] The study reports no evidence of impairment for CBZ, lower scores compared to the other drugs on tests of visuomotor performance and memory for PHB and VPA and lower scores on intelligence tests for PHT. No differences between the four drugs remained one year after complete withdrawal. The study by Calandre et al.[54] used an intelligence tests that does not allow an interpretation of underlying cognitive impairment. The study by Dodrill & Troupin[18] is one of the pioneer studies in this field. The study compared CBZ with PHT in a randomized cross-over design. Patients were assessed at the end of a four month treatment period. The study gives data on relative effects of the drugs, i.e. differences between PHT and CBZ. The patients have not been tested pre-treatment or compared to an untreated control group. In their original publication the authors reported impairment of attention and problem solving in PHT (relative to CBZ). However, 14 years later the results were reanalysed[18] and the authors argued that no differences between PHT and CBZ remained, after removing patients with high PHT serum levels. The rational behind this procedure was that, in the light of our current knowledge, some patients on PHT may have been treated in the 'toxic range' during the study. One comment should be made about this procedure. The procedure of removing patients resulted in smaller sample sizes, i.e. when removing patients with serum levels > 30 µg/ml, the sample size decreased from 40 to 15 patients. The original statistical differences were lost with these smaller sample sizes but the meaning of the 'no-difference' conclusion of the authors (true absence of differences, or an effect of decreased power of the study due to the lowered sample size) can not be established. The study by Aikia et al.[55] used a randomized double blind parallel group design. Patients were assessed before treatment and after either PHT or oxcarbazepine (OXC). The study reports both the absence of PHT-OXC differences and drug-nondrug differences, suggesting a complete absence of cognitive drug effects. An essential problem with this study is the limited sample size (14 versus 15), combined with a 20% drop-out rate. An illustration of the effect of the small sample size is that even a difference between the groups of approximately 1 standard-deviation that was found at pretest baseline (an effect size that is generally considered as extremely large[26]) does not yield statistical significance. Thus the absence of effects in this study probably reflects the limited power, causing a failure to detect genuine impairment. The study by Vining et al.[56] is a randomized double blind cross-over study comparing PHB with VPA in 21 children. The study reports lower scores on several intelligence tests for PHB compared to VPA. Due to the lack of a non-drug condition no data on absolute effects could be reported. Finally, the study by Meador et al.[15] used a randomized double-blind triple cross-over design comparing CBZ with PHT and PHB in 15 patients. The study lacks a no-treatment condition: the patients were not tested during a pretreatment phase and the study did not use an untreated control group. The overall conclusion of the study, i.e. 'no differences between the drugs' can only be given limited credit. The study uses a small sample (15 patients; 7 cognitive variables) allowing only extreme cognitive changes to be detected.

RESULTS OF THE INVESTIGATED STUDIES; INFERENCES ABOUT THE COGNITIVE SIDE-EFFECTS OF ANTIEPILEPTIC DRUGS

The tables 6a to 6c summarise the results of the studies that allow valid interpretation of the results. Table 6a summarises those studies that allow the evaluation of absolute effects, i.e. the difference between a specific drug and a non-drug condition (untreated baseline, untreated control or placebo). Table 6b summarises relative effects: the difference between two or more drugs. These latter differences must be interpreted with caution of course. The absence of a difference between two drugs does not exclude the possibility that *both* impair cognitive function to the same extent. Table 6c summarises the studies that investigated dose-effect relationships. Tables 6a-6c will be discussed for separate AED's.

Phenobarbital

Only one study[41] allows the evaluation of absolute effects (differences between PHB and a non-drug condition). Short-term memory impairment was found in 19 patients with epilepsy. Differences with other AED's are available from three studies,[15,53,56] all with patients with epilepsy. One of these shows more impairment for PHB than for PHT or CBZ on visuomotor and memory tests[53] and an other study shows lower intelligence scores than VPA.[56] Only the study by Meador et al.[15] does not show differences between PHB and PHT or CBZ. We must therefore conclude that the majority of these studies show evidence for PHB-induced cognitive impairment.

Phenytoin

Five studies[16,31-33,38] report PHT-induced cognitive impairment compared with a non-drug condition. Regrettably, these are all normal-volunteer studies. Until a study in patients with epilepsy reconfirms this finding, we must accept the possibility that these effects represent short-term outcomes of the drug. The comparison of PHT with other AED's shows a controversy between the studies by Meador et al.[15,16,38] and the Italian study.[53] Contrary to the Italian study by Gallassi and coworkers,[53] Meador et al.[15,16,38] do not find differences with CBZ or even with PHB. This latter finding suggests a 'centre bias' and there is clearly a need for a well-controlled monotherapy study in patients with epilepsy. No relations between dose and type of cognitive impairment was found.[39]

Table 6a. Summary of demonstrated cognitive side-effects

	Type of impairment	Number of subjects on AED Type of patients (e=epilepsy; nv=normal-volunteers)
A. Absolute Effects (comparison with a non-drug condition)		
PHB		
MacLeod et al, 1978	impairment of short-term memory	19(e)
PHT		
Smith & Lowrey, 1975	impairment of mental concentration	10(nv)
Thompson et al, 1980	global memory impairment	8(nv)
Thompson et al, 1981	impairment of memory, concentration, mental and motor speed	8(nv)
Meador et al, 1991	impairment of choice reaction time and attention	21(nv)
Meador et al, 1993	impairment of memory	15(nv)
CBZ		
Thompson et al, 1980	no impairment	8(nv)
Meador et al, 1991	impairment of choice reaction time and attention	21(nv)
Meador et al, 1993	impairment of memory	15(nv)
Aldenkamp et al, 1993	no impairment for CBZ	56(e)
OXC		
Curran & Java, 1993	no impairment; improvement of focussed attention and speed	12(nv)
VPA		
Thompson & Trimble, 1981	impairment of decision making speed	10(nv)

CBZ=Carbamazepine; **OXC**=Oxcarbazepine; **PHT**=Phenytoin; **PHB**=Phenobarbital; **VPA**=Valproate;

Table 6b. Summary of demonstrated cognitive side-effects

	Type of difference	Number of patients (e=epilepsy; nv=normal-volunteers)
B. Relative Effects (comparison with other AEDs)		
PHT with CBZ		
Meador et al, 1991	no difference, both induce impairment of choice reaction time and attention	21(nv)
Meador et al, 1993	no difference; both induce impairment of memory for CBZ and PHT	15(nv)
PHT with CBZ with PHB		
Meador et al, 1990	no extreme differences between PHB-PHT-CBZ	15(e)
PHT with CBZ with PHB with VPA		
Gallassi et al, 1992	lower scores for PHB and VPA on tests of visuomotor performance and memory lower scores for PHT on intelligence tests	27/18/16/29(e)
PHB with VPA		
Vining et al, 1987	PHB lower scores on intelligence tests	21(e)

Table 6c. Summary of demonstrated cognitive side-effects

	Type of difference	Number of patients (e=epilepsy; nv=normal-volunteers)
C. Dose-relation studies		
PHT low dose vs PHT high dose		
Stephens et al, 1974	no difference	107(nv)
CBZ versus CBZ^{ct}		
Aldenkamp et al, 1987	increase of variability in CBZ compared to CBZ^{ct}	11(e)
CBZ↑ with CBZ↓		
Amman et al, 1990	improvement at peak serum levels	50(e)
VPA↑ with VPA↓		
Amman et al, 1987	no effect of VPA serum level	46(e)

CBZ=Carbamazepine; **CBZ**^{cr}=Carbamazepine; controlled release; **PHT**=Phenytoin; **PHB**=Phenobarbital; **VPA**=Valproate; ↑=high dose; ↓=low dose;

Carbamazepine

The summarising tables 6A and 6B show the same controversy between Meador et al.[15,16,38] and other studies as illustrated for PHB and PHT. All studies[32,47,53] except for Meador et al. report 'no impairment' in both normal-volunteer and epilepsy studies compared to a non-drug condition and a favourable profile compared to PHT, PHB and VPA. There is no agreement about the effect of dose increase: both cognitive improvement and increased variability are reported at higher serum levels. The only study that used the related compound oxcarbazepine[37] reported cognitive improvement.

Valproate

One study[34] allows the interpretation of absolute effects and shows impairment of mental speed. The comparison with other drugs shows lower performance of memory and visuomotor function compared to CBZ[53] and a favourable profile compared to PHB.[56] The cognitive impairment is dose-independent.[52]

No interpretable data are available for other types of AED's which were mostly evaluated in add-on polytherapy designs or other designs that were excluded form this review.

CONCLUSIONS

A disappointing number of studies pass criteria of design, methodology and statistical analysis that are in line with common scientific conventions.[27] From a database of 358 papers on cognitive effects of AED's, a total of 16 studies remained that potentially allow valid inferences about the cognitive effects of AED's. This necessarily leads us to our first conclusion: our current knowledge about this type of side-effect is far from complete, even for the most commonly prescribed antiepileptic drugs. The main reason is that most studies fail to meet the criteria for design, methodology and type of analysis that are indispensable for the rather complex studies on cognitive effects of AED's. An illustration of the complexity is the clear controversy between study groups, that is described above. With our current knowledge it is therefore only possible to draw conclusions about the cognitive side-effects of AED's with caution. The often cited claim for PHT and CBZ that ..."*both drugs have an impact on cognitive function, PHT to a larger degree than CBZ*"[2] or in contrast ..."*drug-induced cognitive effects of these AED's on cognitive function are probably mild or even negligible*"[14,20,21] are simply not supported by valid 'high quality' data.

Most evidence for cognitive impairment is obtained for phenobarbital that appears to have a significant impact on memory and on intelligence. Its cognitive profile is inferior compared to PHT, CBZ and VPA. Phenytoin may also cause cognitive impairment, although it is still under debate whether this claim also holds for chronic treatment in patients with epilepsy. Especially mental speed and memory are mentioned as the areas with most impairment. Its cognitive profile is reported to be favourable compared to PHB but inferior to that of CBZ. Carbamazepine does not seem to induce cognitive side-effects, and the same conclusion seems valid for its related compound oxcarbazepine. For valproate the data are still inconclusive, but as yet most of the studies report impairment of mental speed and memory. Except for carbamazepine most studies report the observed side-effects to be dose-independent.

REFERENCES

1. Trimble MR. Anticonvulsant drugs and psychosocial development: phenobarbitone, sodium valproate, and benzodiazepines. In: Morselli PL, Pippenger CE, Penry JK, editors. Antiepileptic Drug Therapy in Pediatrics. New York: Raven Press, 1983: 201-217.
2. Trimble MR. Anticonvulsant drugs: mood and cognitive function. In: Trimble MR, Reynolds EH, editors. Epilepsy, Behaviour and Cognitive function. Chichester: John Wiley & Sons, 1987: 135-145.
3. Trimble MR. Anticonvulsant drugs and cognitive function: a review of the literature. Epilepsia, 1987; 28(S3): 37-45.
4. Trimble MR, Thompson PJ. Memory, anticonvulsant drugs and seizures. Acta Neur. Scand., 1981; 64: 31-41.
5. Trimble MR, Thompson PJ. Anticonvulsant drugs, cognitive function and behaviour. Epilepsia, 1983; 24(S1): 55-63.
6. Trimble MR, Thompson PJ. Sodium Valproate and cognitive function. Epilepsia, 1984; 25(S1): 60-64.
7. Trimble MR, Thompson PJ. Anticonvulsant drugs, cognitive function and behaviour. In: Ross E, Reynolds E, editors. Paediatric perspectives on Epilepsy. Chicester: John Wiley & Sons, 1985: 141-148.
8. Trimble MR, Cull C. Children of school age: the influence of antiepileptic drugs on behavior and intellect. Epilepsia, 1988; 29(S3): 15-19.
9. Trimble MR, Thompson PJ, Huppert F. Anticonvulsant drugs and cognitive abilities. In: Canger R, Angeleri F, Penry JK, editors. Advances in Epileptology. New York: Raven Press, 1980: 199-204.
10. Committee on Drugs. Behavioral and cognitive effects of anticonvulsant therapy. Pediatrics, 1985; 76: 644-647.
11. Evans RW, Gualtieri CT. Carbamazepine: a neuropsychological and psychiatric profile. Clinical Neuropharmacology, 1985; 8(3): 221-241.
12. Parnas J, Gram L, Flachs H. Psychopharmacological aspects of antiepileptic treatment. Prog. Neurobiol., 1980; 15: 119-138.
13. Parnas J, Flachs H, Gram L. Psychotropic effect of antiepileptic drugs. Acta Neurol. Scand., 1979; 60: 329-343.
14. Smith DB. Cognitive effects of antiepileptic drugs. Advances in Neurology, 1991; 55: 197-212.

15. Meador KJM, Loring DW, Huh K, et al. Comparative cognitive effects of anticonvulsants. Neurology, 1990; 40: 391-394.
16. Meador KJM, Loring DW, Allen, ME, et al. Comparative cognitive effects of carbamazepine and phenytoin in healthy adults. Neurology, 1991; 41: 1537-1540.
17. Smith DB, Mattson RH, Cramer JA, et al. Results of a nationwide Veterans Administration Cooperative Study comparing the efficacy and toxicity of carbamazepine, phenobarbital, phenytoin, and primidone. Epilepsia, 1987; 28(S3): 50-58.
18. Dodrill CB, Troupin AS. Neuropsychological effects of carbamazepine and phenytoin; a reanalysis. Neurology, 1991; 41: 141-143.
19. Dodrill CB, Temkin NR. Motor Speed is a contaminating Factor in evaluating the "cognitive" effects of Phenytoin. Epilepsia, 1989; 30: 453-457.
20. Dodrill CB. Behavioral effects of antiepileptic drugs. Advances in Neurology, 1991; 55: 213-224.
21. Dodrill CB. Problems in the assessment of cognitive effects of antiepileptic drugs. Epilepsia, 1992; 33(S6): 29-32.
22. Wilder BJ, Rangel RJ. Phenytoin, clinical use. In: Dreifuss FE, Mattson RH, Meldrum BS, et al., editors. Antiepileptic drugs - Third Edition, New York: Raven Press, 1989: 233-241.
23. Novelly RA, Schwartz MM, Mattson RH, et al. Behavioral toxicity associated with antiepileptic drugs: concepts and methods of assessment. Epilepsia, 1986; 27(4): 331-340.
24. McKee PJW, Blacklaw J, Forrest G, et al. A double-blind placebo-controlled interaction study between oxcarbazepine and carbamazepine, sodium valproate and phenytoin in epileptic patients. Br. J. Clin. Pharmac., 1994; 37: 27-32.
25. Andrewes DG, Bullen JG, Tomlinson L, et al. A comparative study of the cognitive effects of phenytoin and carbamazepine in new referrals with epilepsy. Epilepsia, 1986; 27: 128-134.
26. Cohen J. Statistical Power analysis for the behavioral sciences. New York: Academic Press, 1977.
27. Cook D, Campbell, DT. Quasi-Experimentation; Design & Analysis Issues for Field Settings. Boston: Houghton Mifflin Company, 1979.
28. Kulig B, Meinardi H. Effects of antiepileptic drugs on motor activity and learned behavior in the rat. In: Meinardi H, Rowan AJ, editors. Advances in Epileptology. Amsterdam: Swets & Zeitlinger, 1977: 98-104.
29. Idestrom, CM, Schalling, D., Carlquist, U., et al. Behavioral and psychological studies: accute effects of diphenylhydantoin in relation to plasma levels. Psych. Med., 1972; 2: 111-120.
30. Houghton GW, Latham AN, Richens, A. Differences in the central actions of phenytoin and phenobarbitone in man, measured by critical flicker fusion treshold. Eur. Journ Clin Pharmacology., 1973; 6: 57-60.
31. Smith WL, Lowrey JB. Effects of diphenylhydantoin on mental abilities in the elderly. J. Am. Geriatr. Soc. 1975; 23: 207-211.
32. Thompson PJ, Huppert F, Trimble MR. Anticonvulsant drugs, cognitive function and memory. Acta Neurol. Scand., 1980; (S80): 75-80.
33. Thompson PJ, Huppert FA, Trimble MR. Phenytoin and cognitive functions: effects on normal volunteers and implications for epilepsy. British Journ. Clin. Psychol., 1981; 20: 155-162.
34. Thompson PJ, Trimble MR. Sodium valproate en cognitive functioning in normal volunteers. Br. Journ. Clin. Pharmacol., 1981; 12: 819-824.

35. Cohen AF, Crowley LAD, Land G, et al. Lamotrigine (BW430C), a potential anticonvulsant. Effects on the central nervous system in comparison with phenytoin and diazepam., Br. J. Clin. Pharmac., 1985; 20: 619-629.
36. Saletu B, Grunberger J, Linzmayer L, et al. Psychophysiological and psychometric studies after manipulating the GABA system by vigabatrin, a GABA-transaminase inhibitor. Int. J. Psychophysiol., 1986; 4: 63-80.
37. Curran HV, Java R. Memory and psychomotor effects of oxcarbazepine in healthy human volunteers. European Journal of Clinical Pharmacology, 1993; 44(6): 529-33.
38. Meador KJM, Loring DW, Abney OL, et al. Effects of carbamazepine and phenytoin on EEG and memory in healthy adults. Epilepsia, 1993; 34(1): 153-157.
39. Stevens JH, Schaffer JW, Brown CC. A controlled comparison of the effect of diphenylhydantoin and placebo on mood and psychomotor functioning in normal volunteers. Journal of clinical Pharmacol., 1974; 14: 543-551.
40. Tedeschi G, Casucci G, Allocca S, et al. Computer analysis of saccadic eye movements: assessment of two different carbamazepine formulations. Eur. J. Clin. Pharmacol, 1989; 37: 513-516.
41. MacLeod CM, Dekaban AS, Hunt E. Memory impairment in epileptic patients: selective effects of phenobarbital concentration. Science, 1978; 202: 1102-1104.
42. Camfield CS, Chaplin S, Doyle AB, et al. Side effects of phenobarbital in toddlers: behavior and cognitive aspects. J. Pediatri., 1979; 95: 361-365.
43. Wolf SM, Forsythe A, Stunden AA, et al. Long-term Effect of Phenobarbital on Cognitive Function in Children with Febrile Convulsions. Pediatrics, 1981; 68(6): 820-823.
44. Gallassi R, Morreale A, Lorusso S, et al. Cognitive effects of phenytoin during monotherapy and after withdrawal. Acta Neur. Scand., 1987; 75: 258-261.
45. O'Dougherty M, Wright FS, Cox S, et al. Carbamazepine plasma concentration. Relationship to cognitive impairment. Arch. Neurol. 1987; 44: 863-867.
46. Aldenkamp AP, Alpherts WCJ, Moerland MC, et al. Controlled Release Carbamazepine: Cognitive Side-effects in patients with Epilepsy. Epilepsia, 1987; 28: 507-514.
47. Aldenkamp AP, Alpherts WCJ, Blennow G, et al. Withdrawal of antiepileptic medication - effects on cognitive function in children - the results of the multicentre 'Holmfrid' study. Neurology, 1993; 43(1): 41-51.
48. Gallassi R, Morreale A, Lorusso S, et al. Cognitive effects of valproate. Epilepsy Research., 1990; 5: 160-164.
49. Dodrill CB, Temkin NR. Motor Speed is a contaminating Factor in evaluating the "cognitive" effects of Phenytoin. Epilepsia, 1989; 30: 453-457.
50. Macphee GJA, MacPhail EM, Butler E, et al. Controlled evaluation of a supplementary dose of carbamazepine on psychomotor function in epileptic patients. Eur J. Clin. Pharmacol., 1986; 31: 195-199.
51. Amman MG, Werry JS, Paxton JW, et al. Effects of carbamazepine on psychomotor performance in children as a function of drug concentration, seizure type and time of medication. Epilepsia, 1990; 31: 51-60.
52. Amman MG, Werry JS, Paxton JW, et al. Effect of sodium valproate on psychomotor performance in children as a function of dose, fluctuations in concentration and diagnosis. Epilepsia, 1987; 28: 115-124.
53. Gallassi R, Morreale A, Di Sarro R, et al. Cognitive effects of antiepileptic

drug discontinuation. Epilepsia, 1992; 33(S6): 41-44.
54. Calandre EP, Dominguez-Granados R, Gomez-Rubio M, et al. Cognitive effects of long-term treatment with phenobarbital and valproic acid in school children. Acta Neur. Scand., 1990; 81: 504-506.
55. Aikiae M, Kaelviaeinen R, Sivenius J, et al. Cognitive effects of oxcarbazepine and phenytoin monotherapy in newly diagnosed epilepsy: one year follow-up. Epilepsy Research, 1992; 11(3): 199-203.
56. Vining EP, Mellitis ED, Dorsen MM, et al. Psychologic and behavioral effects of antiepileptic drugs in children: a double-blind comparison between phenobarbital and valproic acid. Pediatrics, 1987; 80(2): 165-74.
57. Wilensky AJ, Friel PN, Ojemann LM, et al. Zonisamide in epilepsy: a pilot study. Epilepsia, 1985; 26: 212-220.

11 Current state of affairs; epilepsy surgery in children and adolescents

HERBERT SILFVENIUS
Department of Neurosurgery, University Hospital, Umeå, Sweden

Curative surgery for drug resistant partial epilepsy in children has a long tradition, but is performed less than in adults. Recently, palliative surgery has gained acceptance for epilepsy among children. In general, the surgical methods used in adults are also applicable to children and adolescents. The advantages of surgery should be weighed against the long-term negative consequences of pharmacologically resistant partial and generalized epilepsy. In this review on surgery for paediatric epilepsy (PES), the author shares his experience in the light of current literature. PES should be considered in every infant and young child with intractable seizures. For older children and adolescents, the indications for surgery are similar to those of adults. Early PES has better chances of favourable outcome. It facilitates clinical and psychosocial recovery and prevents future disability to a large extent. Children with moderate to severe drug resistant epilepsy need to be evaluated by epileptologists and correct diagnosis established without delay. The diagnostic armamentarium includes modern neuroimaging (CT-scan, MRI, DSA, SPECT, PET) and video EEG (surface and depth recording). They help to localise the epileptogenic focus more accurately. The presurgical study is usually brief as is the postoperative stay in hospital. The success of PES is related to etiology, site and size of lesion. About 50-80% become seizure-free or almost seizure-free after local excisions, and 65-85% after radical excisions. The complications are rare. Even a partial reduction in seizure tendency may provide significant improvement in cognitive, emotional, and psychosocial functions. Palliative PES, i.e. callosotomy, alleviates the care of multiple handicapped children. The referring paediatrician and the epilepsy surgical team share the responsibility to prevent long term disability in the children with surgically treatable epilepsy and alleviate the burden on their family. PES should neither be oversold nor underutilized. From a health economics point of view, PES is highly recommended. Presently, the demand for PES is beyond the infrastructure and professional expertise.

INTRODUCTION

At the beginning of this century, Paediatric Epilepsy Surgery (PES) was carried out as local cortical excisions.[1] In 1928 radical excision was performed, hemidecortication, or 'hemispherectomy', in an adult for a glioma.[2] This procedure was introduced as a treatment for epilepsy in

1938, and in 1950 for epileptic children with spastic cerebral hemiparesis.[3,4] Today a wide variety of surgical techniques are available to treat epilepsy in children. A survey of the literature on epilepsy surgery (ES) from 1900-1985 yielded 75 reports on local PES and 125 reports on hemispherectomy.[1] The corresponding numbers for the period 1985-1990 were 40 and 20. The ratio of local versus radical PES had apparently reversed in the last decade from 1:2 to 2:1. This tendency has further accelerated in the past three years. A Medline search for the period of 1991-1993 listed 85 reports on PES. Hemispherectomy accounted for only a fourth of PES. The number of PES reports has thus increased with time, but the trend has shifted from radical to localized PES, i.e. a greater number of children with focal epileptic lesions have recently undergone surgery. This reflects more liberal selection criteria in the currently expanded PES. Palliative PES for diffuse or multifocal epilepsy, i. e. corpus callosum section (CCS) was first performed in adults in 1940.[5] Before that another technique, callosal puncture had been described.[6] CCS has been performed in children.[7] About 180 reports of CCS have been published, mostly in adults from 1940 to 1990. During the subsequent three years there have been 10 such reports, which suggest a modest increase in utilization of this technique.[1] A common feature observed in these reports on curative as well as palliative PES is the presence of a structural lesion or an EEG pattern, consistent with the clinical symptoms. A recent trend is the identification of new surgical candidates, in children with neuronal migration disorders, Lennox-Gastaut syndrome, infantile spasms, and Landau-Kleffner syndrome, earlier not selected for PES.[9-12] Not too long ago the attitude towards PES was very cautious. The main concern was not to interfere with the maturation of brain, which was presumed to suppress partially or totally the epileptic activity. Currently, a different and more aggressive view is advocated. Apparently the functional plasticity of brain seems to be more resourceful after an early removal of epileptogenic lesion. As a result, early PES is likely to enhance clinical and psychosocial recovery. This review will focus attention on the current status of PES with particular emphasis on the promising surgical results and other related issues.

CANDIDATES FOR CORTICAL EXCISION

Despite accurate diagnosis and selection of appropriate Antiepileptic Drugs (AED's) partial epilepsy frequently proves refractory in infants, children and adolescents. It may vary in severity from epilepsia partialis continua to an occasional seizure. In the first situation, referral for PES is an easy choice. In the latter case (one or fewer seizures a month) the paediatrician, as well as the child's parents are hesitant to consider PES. In such cases the longterm negative consequences of seizures, neurological deficits and medications, such as educational, psychosocial problems should be carefully weighed against the risk of surgery. Rare partial seizures may therefore appear as 'unfounded criteria' for referral for PES evaluation. In such instances it can be helpful to request the parents to list the problems and inconveniences caused by epilepsy for the child and the family. One has to take also into account the secondary gains, such as extra love and empathy from the dear ones, to which the epileptic child is accustomed. The paediatrician may be inclined to assess the seizures as not seriously disabling for the child or his family or detrimental to his future. Frequent interaction between the paediatrician and the epilepsy team, while managing drug resistant cases would avoid the delay in offering surgical remedy. This would also prevent that only the worst cases come to the surgeon. It is worthwhile evaluating even milder cases from a surgical point of view. The epileptic child, as any healthy adult, has the right to a normal life. The undue fear of PES also contributes to the reluctance to refer children for evaluation. This does not usually weigh the negative consequences of the restrictions added by epilepsy to life. A common experience with ES for adults is that epilepsy often started in childhood, the opportunity to fully develop mental faculties was missed, the ES was unduly delayed, and the patient's quality of life deteriorated. Such valid arguments can be avoided if we consider PES early enough. Candidates for PES urge us to critically assess our diagnostic skills, timing and mode of therapy, investigations, and evaluation of the pros and cons of medical versus surgical therapy.[13] A family's desire for surgery on their child is relevant to the question about the epidemiological basis of PES. Expressed differently, how mild-moderate-severe drug resistant partial epilepsy are we ex cathedra entitled to consider acceptable for the child/family? Inventories on quality of life, life satisfaction and other psychosocial parameters should be included in the evaluation of drug resistant epilepsy in children. Unfortunately, these aspects are rarely investigated systematically.[7,15-19] Recent observations have favoured early evaluation.[14]

CANDIDATES FOR PALLIATIVE SURGERY

Is palliative PES warranted? If we consider only the frequency of seizures before and after surgery, as often done with resective PES, to measure the success rate of palliative PES, the results are not impressive. If the chances of freedom from seizures after palliative PES are after all so low, then why undertake it? However, the outcome scores applicable to resective surgery are not appropriate to measure the outcome of palliative PES, which is performed in an entirely different setting. A more realistic classification of outcome has been proposed for palliative PES.[7] Studies using such yardsticks have justified the role of palliative PES in improving the psychosocial capabilities and quality of life for the epileptic children and their families. We do not question medical treatment of poor responders just because their improvement is partial. Palliative PES also should be considered on same grounds. Some of the patients who would benefit from CCS are living outside the family, sometimes in special homes and centers. Even a partial improvement after palliative PES may reduce the demand for special attention and expensive medical resources required to support such children. CCS needs to be considered relatively early in children with drop attacks and bilateral diffuse or multifocal epileptic EEG abnormality. Besides CCS, cortical excisions and microsurgical undercutting (in eloquent areas where removal of a local epileptic region often cannot be done) are excellent palliative procedures. An improvement, even if subtle, may well be appreciated and prove to be vindicated. Many parents want to have their child evaluated for a possible PES, in order to reassure themselves that they have tried everything possible. It is the challenge and duty of the epileptologists to meet that concern with empathy, high standards and far-sightedness.

NEED FOR PES

The number of children expected to benefit from PES varies according to the selection criteria. Part I deals with the epidemiology in detail. Recent reviews on ES reveal that at least 4% of the general population will develop epilepsy. Five to ten percent of all incident cases will prove to have seizures refractory to AED's. About 60% of those could be partial seizures.[20] Another estimate states that 20-30% of the persons developing epilepsy will continue to have it despite AED treatment.[21] The estimate of PES candidates would increase if we include the negative secondary effects of epilepsy besides the seizure frequency, or otherwise expressed, 'there is more to intractability than a lack of remission of seizures'.[22] With the advent of diagnostic facilities more and more cases of neuronal migration disorders, Lennox-Gastaut Syndrome, and Infantile Spasms are subjected to PES, thereby broadening the epidemiological base. The epidemiological basis for palliative PES is obscure. The number of children undergoing CCS is less than that of resective PES.[20] It appears that the demand for PES is far beyond the capacity of the current expert care system. An optimal balance between 'overselling' and 'underutilization' needs to be developed.

ETIOLOGIES

Children with drug resistant epilepsy due to a variety of definable etiologies and syndromes are subjected to cortical excision (Table 1).

Table 1. Etiologies and syndromes of epilepsy remediable with PES. Modified from [23-26]

Developmental/genetic disorders
Tuberous sclerosis
Neurofibromatosis, Schwannoma
Sturge-Weber syndrome
Arteriovenous malformations, cavernous or other angioma
Glial, small/cryptic tumours, astrocytoma, oligodendroglioma, ganglioglioma
Dysembryoplastic neuroepithelial tumour
Porencephalic cyst
Cortical migration disorders, focal cortical dysplasia, microdysgeneses, polymicrogyria
Hemimegalencephaly
Hamartoma
Rasmussen's chronic encephalitis
Hemiconvulsions hemiplegia epilepsy (HHE) syndrome
Stroke
Trauma
Encephalitis/meningitis/abscess
Tumour
Mesial temporal sclerosis
Lennox-Gastaut syndrome
Infantile spasms
Landau-Kleffner syndrome

The onset of epilepsy may vary from at birth to a few months, at other times later. ES is particularly beneficial in epilepsy patients who have had febrile convulsions.[27] Cases subjected to palliative PES may have similar etiologies and chronobiology.[7]

AGE AT SURGERY

Recently ES has been extended to infants with uncontrolled seizures and/or status epilepticus with some promising results. Children with less severe but disturbing partial epilepsy have also been operated at an early age, i.e. less than 5 years. Some of these children had large unilateral hemispheric lesions and suffered from neurological deficits and mental handicaps. A similar reasoning applies to those with local epileptic lesions also.[28,29] The handicaps from large or local lesions may be; irritability, aggressive behavior, autism, mental retardation, dysphasia, spastic hemiparesis-plegia, etc. Appropriate AED trials for 1-2 years would suffice to establish drug resistance. The trial period could be reduced further by early accurate diagnosis. An exception to this rule may be a child with suspected chronic encephalitis (Rasmussen's type). Such patients who develop unilateral deficits may be kept under surveillance. Their parents may be eventually convinced of the need for radical PES.[30] Children over the age of 4-5 years and adolescents with drug resistant simple or complex partial epilepsies can be operated on similar criteria as the adults. The option of CCS can be withheld until 4-5 years of age, or until the seizures are fully characterized.

CLINICAL DIAGNOSIS

The presurgical evaluation in children is usually shorter and less demanding than in adults. The clinical assessment follows the knowledge gained from adult semiology, studied in great detail with invasive techniques. The clinical features such as aura, pattern of seizures, interictal behaviour and interictal deficits may well lateralize the lesion even though accurate localization may not be possible. In infants the localization of the epileptogenic lesion can be cumbersome, especially in the absence of a macroscopic structural lesion. In cases where the epileptic abnormality remains fairly stationary, it should be possible to home in on the focus based on the symptoms and signs. The reader may consult recent books on typical and probable seizure patterns from the two hemispheres, based on information from children and adults.[31-34] It is important to recognize the possibility of an intra- or interhemispheric projection of the seizure activity. The epileptic activity originating in a functionally silent brain region may express itself as a disturbance in the function of adjacent eloquent areas. This could manifest as disturbance of speech, language, motorsensory, visual, auditory functions, and be called 'fringe' phenomena. Detailed evaluation with invasive EEG can differentiate between fringe phenomena and seizure foci located in eloquent areas themselves. Particularly, simple partial sensorimotor seizures often have a 'fringe aspect' as they may reflect a lesion in the frontal, temporal, sensorimotor or parietal cortices. Skilful localization of the lesion and its relation to the vital cortical region in question will help to achieve best surgical cure. Furthermore, beneficial PES can be carried out within an eloquent cortical area with least neurological deficits. The merits of control of seizures and the accompanying improvement in quality of life often far outweigh the possible risk from surgery. Frequent shifts in the clinical symptomatology pose another challenge. At times the clinical features as well as the surface EEG may suggest independent epileptogenic foci. Some of these could in fact be selective propagation of the epileptic discharges to different areas of brain along the long association fibres and commissures that run between different lobes of the brain. If the underlying focus can be detected by invasive EEG or stimulation studies, surgical treatment can be offered. Neuronal migration disorders have recently been recognized as a cause for intractable seizures.[10,35-37] They may present with or without other neurological deficits besides epilepsy.

Some of these patients have presented as infantile spasms, Lennox-Gastaut syndrome or hypsarrhythmia alone. Excision of focal cortical dysplasias in some of these cases have proved beneficial.[10,38-42] Mental retardation in a child with epilepsy should not discourage against

PES. The possible improvement in epilepsy after surgery would facilitate habilitation of such children. The CCS candidates are easily recognized by their mental retardation, traumatizing generalized seizures or drop attacks and widely spread bilateral epileptic EEG disturbances and are distinct from children with other seizure patterns.

NEURORADIOLOGY AND IMAGING

Radiological and imaging techniques have improved dramatically in the recent past. Plain X-rays of skull may reveal calcifications characteristic of tuberous sclerosis, Sturge-Weber syndrome, cryptic astrocytoma, ganglioglioma and oligodendroglioma. Tell-tale evidence of hemimegalencephaly and atrophic lesions may also be seen. CT-scan will offer better visualization of these lesions. However, MRI scan with special views and Gadolinium enhancement is invaluable in the evaluation of intractable epilepsy in children for a number of reasons.[43] MRI offers the best structural delineation whenever a focal lesion can be detected. Mesial temporal lobe atrophy, can be visualized and volumetrically defined with special MRI protocols.[44] Unilateral mesial temporal atrophy, detected by MRI is usually congruent with the clinical and EEG findings. MRI has greatly improved the detection of neuronal migration disorders. They may be visualized as nodules or thickening of cortex. However, microscopic neuronal migration disorders may still be missed.[36] Furthermore, MRI is superior to CT-scan in detecting; cryptic glial tumours on the convexity or mesial temporal lobe (amygdala/hippocampus), small cavernous angiomas with hemosiderin 'wrapping', postvascular-inflammatory changes, gliosis and microgyria. MRI helps to differentiate between a fringe phenomenon and a lesion of the primary motor cortex. The lesion may be excised with good results.[45,46] Interictal SPECT (Tc 99m H-MPAO) is occasionally helpful to localise the epileptic focus as an area of hypometabolism particularly when there is no structural lesion detected on MRI. Ictal, or postictal SPECT may show regional changes over the focus.[47-50] But frequently the size of the lesion may be spuriously magnified on ictal SPECT. If the child needs to be sedated-anaesthetized for the investigation, the Tc 99m-HMPAO should be injected first. FDG-PET scan is an excellent but expensive tool in the evaluation of epilepsy and needs special facilities. FDG-PET may show characteristic metabolic changes according to various syndromes, like the Lennox-Gastaut syndrome, infantile spasms and hemimegalencephaly.[51-53] Carotid/vertebral angiography with Digital Subtraction technique, DSA, may be necessary in suspected vascular malformations and intracarotid Amytal procedure (IAP).[54-56]

ELECTROPHYSIOLOGICAL EVALUATION

The presurgical EEG analysis in children is less comprehensive and simpler than in adults, especially in infants and children with structural lesions. The epileptogenic region can be correctly localised in many children by repeated scalp videoEEG monitoring with or without pharmacologically induced sleep or low dose barbiturate activation.[57] Sphenoidal electrodes may prove beneficial in children with complex partial seizures. Lennox-Gastaut syndrome, infantile spasms or hypsarrhythmia, or bilateral, widespread ictal EEG abnormality may require high dose barbiturate administration to delineate the epileptogenic region.[58] Invasive videoEEG monitoring in children is necessary, when the suspected lesion is very close to or within the eloquent cortical areas, or other hidden parts (basal and mesial hemisphere cortices). The location of the implanted electrodes should be documented on skull X-ray, CT-scan or DSA for exact reference to functional and anatomic landmarks. Invasive chronic EEG monitoring after open craniotomy combined with electrical stimulation of cortex was described in children in 1978, and is a useful method.[59,60] If the placement of the electrodes is extradural, the dura must be completely cut and resutured, to avoid pain during electrical stimulation. By this technique we can study behavioral and subjective responses and EEG simultaneously. A simpler method will be subdural insertion of strip electrodes through burr holes.[61] Stereotaxic depthEEG monitoring can be performed in a child whose head can be fixed steadily to a stereotaxic apparatus. At times functional analysis with evoked potentials (sensory, auditory, visual) recorded with the implanted electrodes, or intraoperatively is helpful to improve the precision of surgical excision. Magnetoencephalography (MEG) is a new technique suitable for delineation of foci with nonradially oriented epileptic 'dipoles' hidden in cortical banks.[62,63]

NEUROPSYCHOLOGY/NEUROPSYCHIATRY

Formal neuropsychological tests may not be required in all children as it is often possible to make conclusions, based on history and clinical examination. Above the age of 5 years, discriminitative and invasive hemispheric tests can be performed as in adults with simplified test batteries.[64-67] Lateralization of speech and memory can be carried out with the IAP.[68] Children below five years may then require an introductory short general anaesthesia. Presence of mother during the IAP may encourage the child to participate in the test fearlessly. Language is tested with simple conversation. Memory is studied by asking the child to recognize tasks given during amobarbital effect, simple words, lines from songs or rhymes, shown pictures of common objects or real objects. It is arguable whether the IAP is essential when radical PES in the speech-dominant hemisphere is contemplated, since the plasticity of the young brain practically restores speech function (< age 6). IAP is mandatory when patients have anomalous cerebral circulation or when the epileptogenic lesion is adjacent to speech-language-memory areas. Knowledge of the hemisphere's speech-memory capacity facilitates the surgical procedure. It is also helpful in preparing the family to anticipate the postoperative outcome and development. Superselective intracarotid Amytal test via branches of the middle cerebral artery for sensorimotor (various cortical segments), or language capabilities (Broca's and Wernicke's areas) can further facilitate planning of surgery.[55] This technique requires correlation of behavioral changes with concomitant scalpEEG changes recorded over the area of interest. Documentation of the regional slow wave activity induced by the amobarbital injection will confirm the accuracy of the procedure. Children with behavioral problems, as those with large destructive hemisphere lesions, or with local lesions in the frontal or temporal lobes (baso-medial frontal, amygdala) should, if possible, have child psychiatric assessment in order to diagnose and assess the pre- and postoperative behaviour. A preliminary study on the effect of PES on autism with epilepsy was recently made.[69] Children referred for CCS usually do not need neuropsychological evaluation as most of them obviously are mentally retarded. In these children CCS will achieve a partial seizure relief, improve sociability and lessen care load. These results are obvious in daily life.

PSYCHOSOCIAL EVALUATION

A psychosocial evaluation of the epileptic child and his family is strongly recommended.[66,70] A social worker or coordinating epilepsy nurse can extract valuable information often neglected by physicians. The inquiry will specify the difficulties of the family with an epileptic child, assess the progress and shortcomings in home education, schooling, quality of life and the social and economical restrictions which epilepsy brings about. This type of assessment often ends up in an arrangement of personal or financial social help for the child and family. Knowledge of the broader life perspectives offers an extra dimension to the care of children with drug resistant epilepsy.

CONTRAINDICATIONS FOR PES

Resective PES is contraindicated in bilateral cortical lesions. Such patients may, however, benefit from CCS. Mental retardation (IQ <70) should not be considered a contraindication for PES, as it may ameliorate socially disturbing seizures or distressing behaviour.

PREPARATIONS FOR SURGERY

The family may require extensive discussions about the chances and risks of PES, supported by current statistics to make their decision regarding PES. Surgery should not be unduly delayed after a positive decision. The parents should be reassured about the absence of pain during and after PES. Most parents are relieved to learn that the postoperative stay in the hospital is brief, usually a week. Hemispherectomy cases may have a prolonged stay. The AED's have usually been reduced earlier during the presurgical evaluation, if not, they are halved or stopped on the evening before surgery to enhance the epileptic activity to be monitored intraoperatively. Steroids are also given then. In the operating room antibiotics are given. Mannitol is used to reduce brain edema. Children scheduled for CCS should not have their AED's changed in relation to the PES.

ANAESTHESIA AND OTHER MEASURES

The advances in neuroanaesthesia and invasive diagnostics present the option of local or special general anaesthesia during ES.[71,72] Traditionally children are operated under general anaesthesia. Special anaesthetic agents are given as induction (methohexital, pentothal, proprofol) followed by the inhalation of a nitrous oxide-oxygen or isoflurane-oxygen mixture with fentanyl as an adjunct analgesic. The purpose is not to suppress the epileptic focus when electrocorticography (ECoG) is used.[73] Before ECoG, the administration of nitrous-oxide is stopped in order to enhance the epileptic abnormality. Electrical stimulation as well as surface or depth ECoG (mesial temporal lobe) can also be done to define and delimit the epileptogenic region during PES. Hence short acting muscle relaxants are preferred. During electrical stimulation for functional mapping a common experience is that the motor cortex in young children (<2 years) may not respond as readily as an adult cortex. This may necessitate altered stimulation parameters.[74] If evoked somatic afferent responses are to be recorded, peripheral stimulation electrodes (face, hand, foot) are applied. The pre-resection ECoG may show ictal or vigorous interictal spiking during the 10-15 min of recording. If it is absent or sparse, iv methohexital can be used to enhance the activity. The ECoG findings are correlated with other diagnostic data and the type and extent of the surgery is tailored accordingly. Very rarely surgery remains an exploration only. Ideally there should not be any ECoG spiking from the remaining cortex after the excision. If additional surgical cleaning up is necessary its result can again be checked with ECoG, showing often broadened spikes reduced in number and amplitude. The relation between the clinical results and the post-resection ECoG finding is not a linear one. In some cases, such as selective amygdalo-hippocampectomy, there is no need at all for ECoG or special anaesthesia. A PES procedure takes 3-6 hours and the child is usually sent to the ward afterwards. Nursing in the intensive care unit may be required if the surgery was prolonged. The first postoperative dose of AED's may be given already in the OR. Early postoperative seizures are controlled with additional short acting AED's.

LOCAL CORTICAL EXCISIONS

The extent of the cortical excisions is tailored according to the size and nature of the epileptogenic lesion and its proximity to vital areas. The procedures include lesionectomy, (removal of any macroscopic structural lesion), topectomy, gyrectomy, selective amygdalo-hippocampectomy, partial/complete lobectomy, multilobectomy, and hemispherectomy.

Temporal lobe excision

The most frequent ES in adults is anterior temporal lobectomy (ATL), that is, in 70-80% of cases.[13] This also seems applicable to children and adolescents, although children have other lesions which demand radical PES.[75,76] There are different surgical techniques for the ATL. In the classical ATL the anterior 4-5 cm of the lateral temporal lobe and a variable mesial portion of amygdala, hippocampus and uncus are removed.[77] Exclusive anterolateral lobectomy or selective amygdalo-hippocampectomy are also currently used.[78] There are different variants of the latter procedure. Interference with language (Wernicke's area) is avoided if the ATL is limited to 4 cm from the tip on the speech-dominant side. A contralateral upper quadrantanopia follows in about 1/3 of patients after ATL. Most operated patients consider it of no practical significance. When removing the hippocampus, partially or completely, adequate functioning of memory in the contralateral hemisphere has to be ascertained preoperatively to avoid memory impairment. Epileptogenic lesions in the posterior temporal lobe can be excised by a posterior approach, or an enlarged ATL. On the non-dominant side the resection may then extend 8-9 cm posteriorly without causing deficits except for visual field cuts. On the speech-dominant side the inferior temporal gyrus may be excised avoiding the Wernicke area which is located in the mid and superior temporal gyri.[79]

Extratemporal excisions

Excisions in the extratemporal areas (frontal, sensorimotor, parietal and occipital lobes) are rarer than those in the temporal lobes. This partly reflects differences in location of epileptogenic lesions, referrals, prejudice about risks and lack of information of the benefits.

Frontal lobe excisions

Excisions in the frontal lobes in children are about 25% of those in the temporal lobes.[76] They can be performed on each side and must respect the Broca's area and motor cortices. This can be done in different ways. During presurgical evaluation a series of subdural electrodes can be implanted. Observation of the behavioral responses to stimulation of each of these electrodes provide helpful information about the topography of the area. Similar information can be gathered intra-operatively by direct stimulation of the motor cortex. The Broca's area in the latter situation remains undefined, but can be located indirectly in relation to the electrically defined motor cortex, the lateral frontal lobe and the Sylvian fissure. Postoperative dysphasia is practically nonexistent under these precautions.

Excisions in the motor cortex

Excisions can be safely carried out within or near the motor cortex. If the lesion occurred early in development, the functional plasticity of the brain may have displaced the 'normality' to the adjoining segments within the motor cortex. Its location, orientation, size and organization may vary from person to person. This provides a certain degree of safety for PES. These excisions, however, are rarer than those in the frontal lobe. A child with simple partial motor seizures (epilepsia partialis continua - status epilepticus - rare seizures) need not have an epileptic lesion in the motor cortex itself, as earlier pointed out in regard to the 'fringe phenomenon'. If, however, there is a need to remove a lesion within the face area of the motor cortex, it can be done without producing appreciable permanent deficits. The excision is then limited to the border of the thumb-finger motor region, thus contralateral hand paralysis can be avoided. Lesions in the premotor gyri can likewise be excised down to the depth of the precentral sulcus, without damaging the anterior bank of the motor cortex or the corticospinal fibres emerging from it. Intra-operative electrical stimulation can be used to control such excisions. If a lesion is located in the hand-arm-foot area and constitutes a serious disturbance to the child, its removal is recommended as the postoperative local paralysis is more acceptable and endurable than an uncontrollable extremity.

Excisions in the sensory cortex

Delicate removals in the sensory cortex require electrical stimulation, carried out either preoperatively or intraoperatively of the motor cortex, or somatosensory evoked potential (SEP) mapping of the sensory cortex itself. Simple partial sensory auras/seizures or mixed partial sensorimotor auras/seizures may also emerge from the clinically 'silent' regions of the parietal lobe. The posterior boundary of the sensory cortex can be exactly determined in relation to the motor cortex identified with electrical stimulation. Local excisions in the sensory cortex result in permanent proprio- and exteroceptive deficits, notable in the hand region and related to the size of the excision.[80]

Excisions in the parietal & occipital lobes

Local excisions in the parietal and occipital lobes are rarely carried out. As long as the language and visual cortices (and their afferent paths) are avoided the removal can be carried out on either side without postoperative neurological deficits.

Excisions in the language area

Restricted excisions in the language area (Wernicke) in cases with structural lesions (Sturge-Weber syndrome, tuberous sclerosis, cryptic tumours) can safely be carried out under general anaesthesia after preoperative functional mapping with electrical stimulation. The location of electrodes stimulation of which resulted in interference with language need to be charted on the plane X-ray films of skull, CT-scans and the stereotaxic landmarks to avoid deficits. They are also related to functional and diagnostic landmarks. The excisions can be made, whenever the lesion has been adequately delimited and microsurgical technique is used, respecting blood vessels. The neuronal plasticity has probably displaced the critical area aside from the lesion. Superselective intracarotid Amytal test may precede the surgery in such cases. A microsurgical cortical undercutting is preferred to excision in the Broca's area.[81]

RADICAL CORTICAL EXCISIONS AND HEMISPHERECTOMIES

There are different techniques for removing or functionally isolating an entirely epileptogenic hemisphere. The hemisphere may be removed/isolated completely whenever the child has impaired finger-hand motor performance, spastic hemiparesis, or debilitating simple partial sensorimotor seizures with finger-hand motor function. In the classical anatomic hemispherectomy the removal extends down to the subcortical nuclei.[4] Radical removal on the speech-dominant side should be done before the age of 5-6 years to allow for development of language. After radical surgery the child will have a contralateral hemiparesis/-plegia but can walk after some weeks-months with a limp. None of the author's hemispherectomized patients or families, have regretted the 'neurological price' paid in comparison with their preoperative problems. The postoperative hemianopsia is well compensated by directed head and eye movements. The large cavity left over by the surgery is initially filled with saline and later by CSF. This cavity accumulates deleterious and osmotically active degradation products of blood, hemosiderin and proteins.[82-84] As a result, superficial cerebral hemosiderosis and fatal intracranial hypertension (ICP) can develop even several years afterwards. Continuous drainage of the cavity or spinal drainage with a closed system during a week or so until the CSF protein is reduced, and/or a ventriculoperitoneal shunt operation will abolish the above risk of hemispherectomy. A way of reducing the volume of the cavity to minimize raised ICP was described in 1983. The cranial dura is mobilized to create a 'tent' with a subdural and an epidural compartment. The smaller subdural space reduces the longterm complications.[85] The risks of anatomical hemispherectomy can further be circumvented by the so called functional hemispherectomy.[86] Here the temporal and central cortices are excised and the frontal and parieto-occipital lobes are functionally isolated by cuts in their white matter, leaving them as vascularized absorbers in situ.[87] This method has further been modified with even less cortical removal, by strategic cuts in the white matter, and called a hemispherotomy.[88] The cuts prevent spread of the epileptic cortical activity beyond the isolation. A postoperative EEG will then show remaining epileptic abnormality of no clinical significance.[89] A subtotal hemispherectomy may be chosen when it is essential to preserve eloquent cortical areas or nonepileptic regions. The term multilobectomy is used when the removal for instance comprises a fronto-centro-temporal excision, a lesser version of subtotal hemispherectomy. Hemispherectomy candidates may show uni- or bilateral epileptic EEG abnormality. In the latter case, the child may still benefit from surgery done on the side with the predominant EEG abnormality, and signalling clinical signs and symptoms.

OTHER SURGICAL TECHNIQUES

Multiple subpial transsection

In addition to the surgical excisions mentioned above, multiple subpial transsection was described in 1969.[90] The epileptic discharges propagate along the horizontally oriented fibres that run along the crests of the gyri. The projection fibres that subserve the functions of these gyri travel vertically. The horizontal intracortical axons are cut at closely spaced intervals with the idea to limit the propagation of the seizure discharges. This technique does not sever the ascending and descending connections and hence spares the functions subserved by the gyri. This method has been used in eloquent cortical areas, with rare complications (Broca, Wernicke, motorsensory) and is thus a corollary to the other local excisions.[81]

Chronic invasive electrical stimulation

Chronic electrical stimulation of the thalamus via permanently implanted electrodes to control seizures has been advocated.[81] The success rate is lower than for resective surgery, and is hardly usable in children. Chronic cerebellar stimulation via implanted electrodes for the treatment of epilepsy was described in 1973 but abandoned as ineffective.[91] Recently, chronic Vagal Nerve electrical Stimulation (VNS) has been performed, mainly on adults with partial epilepsy, and on surgical failures.[92] The applicability of VNS on children remains to be tested. In brief, VNS reduces seizures by 50% in 30-50% of the cases.[81]

Stereotaxy functional surgery

Stereotaxic functional surgery in adults has for years been used for various forms of epilepsy. The seizure types described have varied greatly, as have the techniques.[81] This procedure is rarely practised in children. Stereotaxy may though be used in lesionectomies.

Non-invasive epilepsy surgery

The possibility of non-invasive ES is often brought up for discussion by parents of PES candidates. This discussion in a sense is a plead for a 'riskfree PES'. which unfortunately, does not exist. Patient selection, diagnostic work up, type of apparatus, targeting and dosimetry all contribute to its complexities and unsolved problems. Directed radiation therapy has been used in ES. The results from a small number of adults treated with a linear accelerator followed up for some years were recently published.[93,94] The results are not as good as from cortical excisions.

Palliative PES, corpus callosum sections

The corpus callosum (CC) may be sectioned partially or completely with microsurgical technique. It is preferable to perform an anterior 2/3 cleavage. Stereotaxic CCS has also been conducted.[95] A MRI (not

CT-scan) before the intervention determines the profile and shape of the CC. Another MRI after the surgery will show the extent of CCS. This method is applicable on children from the age of 3-4 years upwards, with the specific goal of achieving a reduction in severity of seizures, particularly the drop attacks. CCS should ideally be performed before the child turns into the apparition of a 'boxer' plagued by repeated skin cuts in the head and by dental/facial fractures.[96] The so called cerebral disconnection syndrome is not a major clinical problem even after a complete CCS.[7] In children the adverse longterm effects are minimal. In cases with bilateral speech representation some impairment of speech may ensue CCS.[97,98] CCS on epileptic children with psychiatric disorders has been reported.[99]

Re-operations

Cortical re-excisions should be considered in surgical failures, particularly in temporal lobe where an extended beneficial removal of mesial structures (hippocampus) can be carried out.[100-102] Removals in other cortical regions may also be performed, although those in the frontal lobe-sensorimotor regions are more risky because of postoperative cerebro-meningeal adhesions. An anterior CCS can also be made a complete section if clinical improvement was not as expected.

EVALUATING SEIZURE OUTCOME AFTER SURGERY

The effect of ES is methodologically evaluated with regard to seizure outcome after a postoperative follow up for at least 2 years. Different scorings are in use.[34,103-105] One of them uses four grades with subclassifications. Those who become seizure free (grade I) or almost seizure free (grade II) are accepted as surgical successes, while those who achieve a marked reduction in seizure 90% (grade III), or who benefit even less (grade IV) are considered as surgical failures. However, the partial improvement experienced in grades III-IV may get the label of true failures if other postoperative benefits are unspecified or ignored. Some data on a broader evaluation of PES will be presented after a strict listing of seizure outcome related to seizure patterns and PES procedure.

Outcome from temporal lobe excisions

Children operated on in the temporal lobes by current techniques have a 65-75% chance of becoming seizure free, or almost seizure free which is comparable to that of adults.[76] The partially improved constitute 15-20% (grade III), and those with less, or no benefit at all 10-15% of the cases.

Another recent report describes the good improvement in 8 out of eleven operated children.[106]

Extra-temporal excisions

The results in adults for excisions in the frontal lobe, sensorimotor cortex, parietal and occipital lobes are less impressive than those for the TL, 35-45% become seizure free, 15-20% almost seizure free, and 15-20% markedly improved, while 10-15% attain less or no surgical benefit. These results seem also to apply to children.[76]

Hemispherectomies

The overall success rate of radical excisions is high, but there are differences related to etiology and surgery. Children with anatomical hemispherectomy mostly for non-inflammatory etiology are likely to be seizure free in 40-70%, almost seizure free in 25-45%, markedly improved in 10%, and in 5-15% have little or no benefit.[107,108] For children treated with a functional hemispherectomy the chance for becoming seizure free is 78%. Another 16% have experienced more than a 80% reduction in their seizures, but about 6% benefit little or not at all. An anatomical or functional hemispherectomy for Rasmussen's encephalitis yields an outcome grade I-IV spectrum of 61%-16%-11%-11%.[104] Another report presents 90% grade I results in hemispherectomized children.[109] A widespread unilateral epileptic hemisphere lesion of devastating clinical nature can thus effectively be treated with radical surgery.

Multiple subpial transsection

Multiple cortical undercutting, often done in eloquent regions achieves results comparable to those after local extratemporal excisions, i.e. in a grade I-IV sequence of 55%-11%-16%-18%.[81] It is underscored that the children with surgery in speech language areas improve their speech/language capacities. Children with Landau-Kleffner syndrome have also been successfully operated with this technique.

Surgical results for Lennox-Gastaut syndrome and Infantile spasms

References have been made to Lennox-Gastaut syndrome and Infantile spasms. A recent report showed that 65% of 23 children became seizure free after excision, and another 13% achieved 90% seizure reduction.[38] Children of these categories have also benefited from palliative CCS (cf. text below).

Epilepsia partialis continua & status epilepticus, surgical results

Children with these serious clinical conditions should, if possible be operated. The seizure outcome varies, some report a fifty-fifty or higher chance of becoming seizure free, others a partial improvement.[110-113]

Results from re-operations

A re-operation is worthwhile in grade III-IV outcomes. About half of the re-operated cases have become seizure free, and another 1/3 have improved.[100,102]

Corpus callosum sections

The results of the ILAE 1980-1990 global survey of epilepsy surgery revealed that CCS achieved freedom from seizures in about 10% of all the children so operated.[76] According to another report of all cases treated achieved freedom from seizures in about 8%, 61% were markedly improved and 31% were not improved.[7] Besides abolishing generalized seizures, behavioral and psychiatric disturbances are improved in about 70-80% of patients.[96,99,114,115]

BROADER PERSPECTIVES ON PES OUTCOME

Other postoperative aspects should be incorporated in the evaluation of surgical outcome to make it more comprehensive and realistic. Significant improvement in mental capacities (IQ, memory, behavior), psychosocial function, quality of life of the child and the family need to be taken into account for assessing the surgical benefits, particularly in the grade III-IV outcome categories. Only recently have these broader psychosocial and rehabilitation aspects been more methodologically addressed.[14,116-118] A wider evaluation of these secondary ES gains is desirable. Additionally, the economic aspects of epilepsy and PES should be evaluated in order to balance the discussion on the 'high costs' of ES.[119] The high costs for the society from epilepsy are in fact mainly due to the absent or low participation in the production from the epilepsy population.[120] One recent report underscored the health economic benefits of ES on adults by calculating the gains achievable by establishing expert facilities for ES.[121] The gains are higher than the costs in expert medical and surgical treatment [122-123]. PES is likewise 'good business' both in regard to direct costs (AED's, care contacts, extra transportation, etc.) and indirect costs (i.e. loss of production). The economical savings are derived from reduced direct costs and increased production by the parents (reduced parental indirect costs) particularly by the mothers, reduced future disability pension costs for the treated children (reduced transferral costs) and future gains accomplished by the successfully operated and better educated grown up children taking part in production (reduced future indirect costs).[124] Further, a lowered mortality and prolonged survival (reduced loss of production) additionally contribute to the savings for the society.[125,126] Besides the medical and humane motives for PES these considerable economical benefits should encourage more PES and result in increased financial support for expert epilepsy care. Part IV elaborates more about the psychosocial aspects.

SURGICAL COMPLICATIONS

The complications of PES should be compared with those occurring during long term medical treatment, the clinical decline and psychosocial disabilities. In other words, a balanced input-output evaluation is necessary to avoid one from taking a too narrow view on surgical risks and disregarding other risks. Here, we shall only refer to the morbidity and mortality from presurgical evaluations and PES. The risks of invasive diagnostics are less than 0.5%.[76] The permanent operative morbidity ranges between 0-4.5%. Some risks were already briefly mentioned along with the individual surgical procedures. A comprehensive account of complications of ES is given elsewhere.[127] In addition to low frequency, infective complications postoperative haemorrhage has been reported.[76] The neurological morbidity by location of excision may be:
- Temporal excisions; contralateral upper quadrant hemianopsia, oculomotor nerve palsy, impaired memory, the latter in 1-4%,[127] dysphasia after speech-dominant excision, hemiparesis/hemiplegia, transient postoperative depression.
- Frontal excisions; Motor dysphasia (speech-dominant PES), hemiparesis, subtle mental impairments.
- Sensorimotor excisions; impaired proprio-/exteroceptive functions, mono-/hemi-, paresis/-plegia, Wernicke dysphasia (speech-dominant PES).
- Parieto-occipital excisions; deficits as from the sensory cortex, dysphasia, hemianopsia.
- Anatomical hemispherectomy; superficial cerebral hemosiderosis, late hydrocephalus, ventriculoperitoneal shunt problems. Functional hemispherectomy has rare complications. Partial cerebral disconnection syndrome occurs after complete CCS in about 15%.[127] The mean mortality after localized PES is around 1.5%.[76] The mortality after hemispherectomies is about 6%.

HISTOPATHOLOGY

The histopathological changes observed in cortical specimens excised from children are basically those previously listed in Table 1. Different histopathological findings are found under the clinical syndromes mentioned.

POSTOPERATIVE FOLLOW UP

The AED medication should be continued for 1-2 years after PES and then stopped or reduced if possible. The child can be followed up by the referring paediatrician after a few visits to the surgeon.

PERSONAL EXPERIENCE OF PES

Fifty-seven epileptic children, 25 girls and 22 boys, age 8 months-16 years, mean 9.8 years, have been operated on during 14 years at the Umeå University Hospital. The epilepsy onset ranged between birth and age 15 (mean 4.1 years), the follow up between 2 months 14 years. Children constitute about 1/3 of all cases operated. The investigations included clinical evaluation, in most cases scalp videoEEG monitoring, CT-scan, MRI, interictal SPECT, and at times the IAP test.[128] Children in the age group 8-16 years were tested neuropsychologically. Anterior temporal lobe excisions constituted 28%, extratemporal 44%, cases with anatomical/subtotal hemispherectomy combined with ventriculoperitoneal shunting 18%, and 10% had anterior CCS.

The proportion of seizure free or almost seizure free cases (grade I-II) was; for the hemispherectomy group 80%, the ATL group 81%, the extratemporal group 48%. All the six children with anterior CCS improved little (grades III-IV). There were no surgical complications. Two children (3.5%) died long after PES, one in relation to epilepsy, due to a generalized neuronal migration disorder. The histopathological changes were similar to those described in the literature, 51% had structural lesions (tuberous sclerosis, Sturge-Weber syndrome, gliosis, cavernous angioma, etc), 26% had small/cryptic glial tumours (astrocytoma, oligodendroglioma, ganglioglioma), and 23% had microscopic neuronal migration disorders alone, or together with other changes.[36]

A health economical analysis, based on a questionnaire, has been made on a part of this material, as well as on children operated by the author at another hospital.[128]

GENERAL COMMENTS

This review has discussed the merits and limitations of PES in the light of current literature. It must be emphasized that PES is an invaluable procedure in improving the quality of life for children with intractable or disabling epilepsy. Studies have also confirmed that there are strong economic benefits from PES. Recent advances in investigations have widened the scope of PES with minimal complications or deficits. It is up to the medical community to fully exploit these opportunities.

REFERENCES

1. A Bibliography on Epilepsy Surgery and Related Subjects 1900-1990. Commission on Neurosurgery of Epilepsy, The International League Against Epilepsy, ILAE. Manuscript.
2. Dandy WE. Removal of right cerebral hemisphere for certain tumours with hemiplegia. JAMA, 1928; 90: 823-825.
3. MacKenzie KG. The present status of the patient who had the right hemisphere removed. JAMA, 1938: 111-168.
4. Krynauw RW. Infantile hemiplegia treated by removing one cerebral hemisphere. J. Neurol. Neurosurg. Psychiatry, 1950; 13: 243-267.
5. Van Wagenen WP, Herren RY. Surgical division of commissural pathways in the corpus callosum; relation to spread of an epileptic attack. Arch. Neurol. Psychiatr., 1940; 44: 740-759.
6. Stieda A. Further experience with puncture of the corpus callosum, particularly in epilepsy, idiocy and related conditions (in German). Arch. klin. Chir., 1914; 105: 277-295.
7. Gates JR, Wada JA, Reeves AG, et al. Reevaluation of Corpus Callosotomy. In: Engel J, Jr. editor. Surgical treatment of the epilepsies. Second edition. New York: Raven Press, 1993: 637-648.
8. Carrazana EJ, Lombroso CT, Mikati M, et al. Facilitation of infantile spasms by partial seizures. Epilepsia, 1993; 34: 97-109.
9. Shields WD, Shewmon DA, Chugani HT, et al. Treatment of infantile spasms: medical or surgical? Epilepsia, 1992; 33 (S6): S26-S31.
10. Vinters HV, Fisher RS, Cornford ME, et al. Morphological substrates of infantile spasms: studies based on surgically resected cerebral tissue. Childs Nerv. Syst., 1992; 8: 8-17.
11. Solomon GE, Carson D, Pavlakis S, et al. Intracranial EEG monitoring in Landau-Kleffner syndrome associated with left temporal lobe astrocytoma. Epilepsia, 1993; 34: 557-60.
12. Cole AS, Anderman F, Taylor L, et al. The Landau-Kleffner Syndrome of acquired epileptic aphasia: Unusual clinical outcome, surgical experience and absence of encephalitis. Neurol., 1988; 38: 31-38.
13. Engel J. Jr, Shewmon DA. Overview: Who should be considered a surgical candidate? In: Engel J, Jr. editor. Surgical treatment of the epilepsies. Second edition. New York: Raven Press, 1993: 23-34.
14. Guldvog B. Evaluation of epilepsy surgery in Norway 1949-1988 (dissertation). Oslo: Oslo University, 1993.
15. Lesser RP, Fisher RS, Uematsu S. Assessment of surgical outcome. Epilepsy

Res., 1992; 5 (S5): 217-229.
16. Huttenlocher PR, Hapke RJ. A follow up study of intractable seizures in childhood. Ann. Neurol., 1990; 28 (5): 699-705.
17. Devinsky O, Cramer JA. Assessing Quality of Life in Epilepsy: Development of a New Inventory. Epilepsia, 1993; 34 (S4) S1-S44.
18. Vickery BG, Hays RD, Herman BP, et al. Outcomes with respect to quality of life. In: Engel J, Jr. editor. Surgical treatment of the epilepsies. Second edition. New York: Raven Press, 1993: 623-625.
19. Solomon LM, Shinnar S. Postscript Early intervention In: Engel J, Jr. editor. Surgical treatment of the epilepsies. Second edition. New York: Raven Press, 1993: 123-132.
20. Hauser WA. The natural history of epilepsy: Epidemiological considerations In: Theodore WH, editor. Surgical Treatment of Epilepsy Epilepsy Res., S5, 1992.
21. Sander JWAS. Some aspects of prognosis in the epilepsies: A review. Epilepsia, 1993; 34: 6, 1007-1016.
22. Dreifuss FE. Postscript Establishment of What constitutes medical failure In: Engel J, Jr. editor. Surgical treatment of the epilepsies. Second edition. New York: Raven Press, 1993: 133-136.
23. Shields WD, Duchowny MS, Holmes GL. Surgically remediable syndromes of infancy and early childhood. In: Engel J. Jr. editor. Surgical Treatment of the Epilepsies, Second Edition, New York: Raven Press, 1993: 35-48.
24. Cascino GD, Boon PAJM, Fish DR. Surgically remediable lesion syndromes. In: Engel J, Jr. editor. Surgical Treatment of the Epilepsies, Second Edition, Raven Press: New York, 1993: 77-86.
25. Daumas-Duport C, Scheithauer BW, Chodkiewicz J-P, et al. Dysembryoplastic tumour: a surgically curable tumour of young patients with intractable partial seizures: report of thirtynine cases. Neurosurg., 1988; 23: 545-556.
26. Vinters HV, Armstrong L, Babb TL, et al. The neuropathology of human symptomatic epilepsy. In: Engel J. Jr. editor. Surgical Treatment of the Epilepsies, Second Edition, New York: Raven Press, 1993: 593-608.
27. Abou Khalil B, Andermann E, Andermann F, et al. Temporal lobe epilepsy after prolonged febrile convulsions: excellent outcome after surgical treatment. Epilepsia, 1993; 34: 878-83.
28. Blume WT. Uncontrolled epilepsy in children. Epilepsy Res. (S5), 1992: 19-24.
29. Adler J, Erba G, Winston KR, et al. Results of surgery for extratemporal partial epilepsy that began in childhood. Arch. Neurol., 1991; 48: 133-140.
30. Andermann F. Chronic encephalitis and epilepsy, Rasmussens's Syndrome. Boston: Butterworth Heinemannn, 1991.
31. Theodore WH. Surgical Treatment of Epilepsy, editor. Epilepsy Res., (5S), Amsterdam: Elsevier, 1992.
32. Luders HO. Epilepsy Surgery. New York: Raven Press, 1992.
33. Chauvel P, Delgado-Escueta AV, Halgren, E. Frontal lobe seizures and epilepsies. Adv. Neurol., 57, New York: Raven Press, 1992.
34. Engel J. Jr. Surgical Treatment of the Epilepsies, second edition. New York: Raven Press, 1993.
35. Hirabayashi S, Binnie CD, Janota I, et al. Surgical treatment of epilepsy due to cortical dysplasia: clinical and EEG findings. J. Neurol. Neurosurg. Psychiatry, 1993; 56: 765-770.
36. Brännström, T, Silfvenius H, Olivecrona M. The range of disorders of corti-

cal organization in surgically resected epilepsy patients. A study dedicated to the memory of professor Patrick Sourander. In: Guerrini R, Andermann F, Canapicchi R, et al. editors. Dysplasias of Cerebral Cortex and Epilepsy. New York: Raven Press, 1993: 1-12.
37. Palmini A, Gambradella A, Andermann F, et al. Operative strategies for patients with cortical dysplastic lesions and intractable epilepsy. Epilepsia, 1994; 35 (6S): S57-S71.
38. Chugani HT, Shewmon DA, Shields WD, et al. Surgery for intractable infantile spasms: Neuroimaging perspectives. Epilepsia, 1993; 34: 764-771.
39. Burnstine TH, Vining EP, Uematsu S, et al. Multifocal independent epileptiform discharges in children: ictal correlates and surgical therapy. Neurology, 1991; 41: 1223-1228.
40 Vinters HV, De-Rosa MJ, Farrell MA. Neuropathologic study of resected cerebral tissue from patients with infantile spasms. Epilepsia, 1993; 34: 772-779.
41. Uthman BM, Reid SA, Wilder BJ, et al. Outcome for West syndrome following surgical treatment. Epilepsia, 1991; 32: 668-671.
42. Chugani HT, Shields WD, Shewmon DA, et al. Infantile spasms: I. PET identifies focal cortical dysgenesis in cryptogenic cases for surgical treatment. Ann. Neurol., 1990; 27: 406-413.
43. Kuzniecky R, Murro A, King D, et al. Magnetic resonance imaging in childhood intractable partial epilepsies: pathologic correlations. Neurology, 1993; 43: 681-687.
44. Jack Jr CR. MRI-based hippocampal valume measurements in Epilepsy. In: Wyllie, E. editor. Epilepsia, 1994; 35 (S6): S21-S29.
45. Dietrich RB, el-Saden S, Chugani HT, et al. Resective surgery for intractable epilepsy in children: radiologic evaluation. Am. J. Neuroradiol., 1991; 12: 1149-1158.
46. Olivier A. Extratemporal cortical resections: Principles and Methods. In: Luders HO, editor. Epilepsy Surgery. New York: Raven Press, 1992: 559-568.
47. Harvey AS, Bowe JM, Hopkins IJ, et al. Ictal 99m Tc-HMPAO single photon emission computed tomography in children with temporal lobe epilepsy. Epilepsia, 1993; 34 (5): 869-877.
48. Bercovic, SF, Newton, MR, Rowe, CC. Localization of epileptic foci using SPECT. In: Lüders HO, editor. Epilepsy Surgery. New York: Raven Press, 1992: 251-258.
49. Bercovic SF, Newton MR, Chiron C, et al. Single photon emission tomography. In: Engel J, Jr. editor. Surgical treatment of the epilepsies. Second edition. New York: Raven Press, 1993: 233-244.
50. Chiron C, Raynaud C, Jambaque I, et al. A serial study of regional cerebral blood flow before and after hemispherectomy in a child. Epilepsy Res., 1991; 8: 232-240.
51. Rintahaka PJ, Chugani HT, Messa C, et al. Hemimegalencephaly: evaluation with positron emission tomography. Paediatr. Neurol., 1993; 9: 21-28.
52. Chugani HT, Shewmon DA, Shields WD, et al. Surgery for intractable infantile spasms: neuroimaging perspectives. Epilepsia, 1993; 34: 764-771.
53. Olson DM, Chugani HT, Shewmon DA, et al. Electrocorticographic confirmation of focal positron emission tomographic abnormalities in children with intractable epilepsy. Epilepsia, 1990; 31: 731-739.

54. Rausch R, Silfvenius H, Wieser HG, et al. Intraarterial Amobarbital procedures. In: Engel J, Jr. editor. Surgical Treatment of the epilepsies. Second Edition New York: Raven Press, 1993: 341-357.
55. Wieser HG. Anterior Cerebral Artery Amobarbital Test In: Luders HO, editor. Epilepsy Surgery. New York: Raven Press, 1992: 515-524.
56. Peterson RC, Sharbrough FW. Posterior Cerebral Artery Amobarbital Test. In: Luders HO, editor. Epilepsy Surgery. New York: Raven Press, 1992: 525-529.
57. Lombroso CT. Use of intravenous shortacting barbiturates. A method of neurological investigation. Neurology, 1963; 13: 358.
58. Morrell F. Secondary epileptogenesis in man. Arch. Neurol., 1985; 42: 318-335.
59. Goldring SA. Method for surgical management of focal epilepsy, especially as it relates to children. J. Neurosurg., 1978; 49: 344-356.
60. Wyllie E. Invasive neurophysiologic techniques in the evaluation for epilepsy surgery in children. In: Luders HO, editor. Epilepsy Surgery. New York: Raven Press, 1992; 409-412.
61. Wyler AR, Ojemann GA, Lettich E, et al. Subdural strip electrodes for localizing epileptogenic foci. J. Neurosurg., 1984; 60: 1195-1200.
62. Hämäläinen M, Hari R, Ilmoniemi RJ, et al. Magnetoencephalography theory, instrumentation, and applications to noninvasive studies of the working human brain. Reviews of Modern Physics. The American Physical Society, 1993; 65: 413-498.
63. Paetau, R. Neuromagnetic signals: Developmental aspects and applications to epileptic disorders (dissertation). Helsinki: University of Helsinki, 1994.
64. Caplan R, Guthrie D, Shields WD, et al. Communication deficits in children undergoing temporal lobectomy. J. Am. Acad. Child Adolesc. Psychiatry 1993; 32: 604-611.
65. Berent S, Giordani B. Neuropsychological considerations in the presurgical examination of children and mentally retarded patients. In: Lüders HO, editor. Epilepsy Surgery. New York: Raven Press, 1992: 495-500.
66. Dodrill CB, Hermann BP, Rausch R, et al. Neuropsychological Testing for assessing prognosis following surgery for epilepsy. In: Engel J, Jr. editor. Surgical Treatment of the Epilepsies, Second Edition. New York: Raven Press, 1993: 263-271.
67. Fenwich PBC, Blumer DP, Caplan R, et al. Presurgical psychiatric assessment. In: Luders HO, editor. Epilepsy Surgery. New York: Raven Press, 1992: 273-290.
68. Williams J, Rausch R. Factors in children that predict performance on the intracarotid amobarbital procedures. Epilepsia, 1992; 33: 1036-1041.
69. Gillberg C. Uvebrant P, Carlsson C, et al. Case report: Autism and epilepsy. Neuropsychiatric aspects before and after epilepsy surgery. In Press.
70. Jones-Gotman M. Presurgical neuropsychological evaluation for localization and lateralization of seizure focus. In: Luders HO, editor. Epilepsy Surgery. New York: Raven Press, 1992: 469-475.
71. Trop D. Conscious-sedation analgesia during neurosurgical treatment of epilepsies, practice at the Montreal Neurological Institute. Int. Anesthesiol. Clin., 1986; 24: 175-184.
72. Kraemer DL, Spencer DD. Anaesthesia in Epilepsy surgery. In: Engel J, Jr. editor. Surgical Treatment of the Epilepsies, Second Edition. New York: Raven Press, 1993: 527-538.

73. Fiol ME, Boening JA, Cruz-Rodriguez R, et al. Effect of isoflurane (Forane) on intraoperative electrocorticogram. Epilepsia, 1993; 34 (5): 897-900.
74. Jayakar P, Alvarez LA, Duchowny MS, et al. A safe and effective paradigm to functionally map the cortex in childhood. J. Clin. Neurophysiol., 1992; 9: 288-293.
75. Duchowny M, Levin B, Jayakar P, et al. Temporal lobectomy in early childhood. Epilepsia, 1992; 33: 298-303.
76. A Global Survey on Epilepsy Surgery 1980-1990. The Commission on Neurosurgery of Epilepsy, The International League Against Epilepsy, ILAE. Silfvenius H, editor. Epilepsia, In Press.
77. Penfield W, Baldwin M. Temporal lobe seizures and the technic of subtotal temporal lobectomy. Ann. Surg., 1952; 136: 625-634.
78. Wieser HG. Selective amygdalohippocampectomy: Indications, investigative techniques and results. In: Symon L, et al. editors. Adv Techn Stand Neurosurg Vienna: Springer-Verlag, 1986; 13: 39-133.
79. Penfield W, Roberts L. Speech and Brain Mechanisms, New York: Princeton University Press, 1959.
80. Olivier A. Extratemporal Cortical Resections: Principles and Methods. In: Luders HO, editor. Epilepsy Surgery. New York: Raven Press, 1992: 559-568.
81. Fisher RS, Uthman BM, Ramsay RE, et al. Alternative Surgical Techniques for Epilepsy. In: Engel J, Jr. editor. Surgical Treatment of the Epilepsies, Second Edition. New York: Raven Press, 1993: 549-564.
82. Oppenheimer DR, Griffith HB. Persistent intracranial bleeding as a complication of hemispherectomy. J. Neurol. Neurosurg. Psychiatry, 1966; 9: 239-240.
83. Rasmussen T. Postoperative superficial hemosiderosis of the brain, its diagnosis, treatment and prevention. Trans Amer. Neurol. Ass., 1973; 98: 133-137.
84. Wilson PJE. Cerebral hemispherectomy in the treatment of infantile hemiplegia. A report of 50 cases. Brain, 1970; 93: 147-180.
85. Adams CBT. Hemispherectomy - a modification. J. Neurol. Neurosurg. Psychiatr., 1983; 46: 617-619.
86. Villemure JG, Rasmussen T. Functional hemispherectomy in Children. Neuropaediatrics, 1993; 24: 53-55.
87. Tinuper P, Andermann F, Villemur JG, et al. Functional hemispherectomy for treatment of epilepsy associated with hemiplegia: rationale, indications, results, and comparison with callosotomy. Ann. Neurol., 1988; 24: 27-34.
88. Villemure J. Hemispherectomy Techniques. In: Luders HO, editor. Epilepsy Surgery. New York: Raven Press, 1992: 569-578.
89. Smith SJ, Andermann F, Villemure JG, et al. Functional hemispherectomy: EEG findings, spiking from isolated brain postoperatively, and prediction of outcome. Neurology, 1991; 41: 1790-1794.
90. Morrell F, Hanbery JW. A new surgical technique for the treatment of focal cortical epilepsy. Electroencephalograph Clin. Neurophysiol., 1969; 28: 120.
91. Cooper IS. Chronic stimulation of cerebellar cortex in epilepsy and generalized myoclonus in man. New York. Amer. Epil. Soc. Meeting, 1973.
92. Penry JK, Dean JC. Prevention of intractable partial seizures by intermittent vagal stimulation in human. Preliminary results. Epilepsia, 1990; 32: (S2), S40-S43.
93. Bardia-Salorio JL, Vanaclocha V, Roldan P. Radiosurgical treatment of epilepsy. 9th Meeting World Soc. Stereot. and Funct. Neurosurg., 1985; July

4-7, Toronto.
94. Bardia-Salorio JL, Barcia JA, Hernandez G, et al. Radiosurgical treatment of epilepsy. Acta Neurochir., 1992; 117: 109.
95. Marino R, Ragazzo PC. Selective criteria and results of selective partial callosotomy. In: Reeves AG, editor. Epilepsy and the Corpus Callosum. New York: Plenum Press, 1985: 281-301.
96. Nordgren RE, Reeves AG, Viguera AC, et al. Corpus callosotomy for intractable seizures in the paediatric age group. Arch. Neurol., 1991; 48: 364-372.
97. Sass KJ, Spencer SS, Novelly RA, et al. Amnestic and attention impairments following corpus callosum section for epilepsy. J. Epilepsy, 1988; 1: 61-66.
98. Sass K, Novelly R, Spencer D, et al. Post callosotomy language impairments in patients with crossed cerebral dominance. J. Neurosurg., 1990; 72: 85-90.
99. Septien L, Giroud M, Sautreaux JL, et al. Effects of callosotomy in the treatment of intractable epilepsies in children on psychiatric disorders (in French). Encephale, 1992; 18: 199-202.
100. Awad IA, Wingkun EC, Nayel MH, et al. Surgical Failures and Reoperations. In: Luders HO, editor. Epilepsy Surgery. New York: Raven Press, 1992; 679-685.
101. Olivier A, Tanaka T, Andermann F. Reoperations in temporal lobe epilepsy. Epilepsy, 1988; 29: 678.
102. Polkey CE, Awad IA, Tanaka T, et al. The place of reoperation In: Engel J, Jr. editor. Surgical Treatment of the Epilepsies, Second Edition. New York: Raven Press, 1993: 663-667.
103. Rasmussen T. Surgery of epilepsy arising in regions other than the temporal and frontal lobes. In: Purpura DP, et al. editors. Adv. Neurol. New York: Raven Press, 1975: 207-226.
104. Villemure JG, Andermann F, Rasmussen TB. Hemispherectomy for the treatment of epilepsy due to chronic encephalitis. In: Andermann F, editor. Chronic Encephalitis and Epilepsy Rasmussen's Syndrome. Boston: Butterwork-Heinemann, 1991: 235-244.
105. Engel J. Jr. Outcome with respect to epileptic seizures. In: Engel J, Jr. editor. Surgical Treatment of the Epilepsies. New York: Raven Press, 1987: 553-571.
106. Hopkins IJ, Klug GL. Temporal lobectomy for the treatment of intractable complex partial seizures of temporal lobe origin in early childhood. Dev. Med. Child Neurol., 1991; 33: 26-31.
107. Andermann F, Rasmussen TB, Villemure JG. Hemispherectomy: Results for control of seizures in patients with hemiparesis. In: Luders HO, editor. Epilepsy Surgery. New York: Raven Press, 1992: 625-632.
108. Villemure JG, Adams CBT, Hoffman HJ, et al. Hemispherectomy. In: Engel J, Jr. editor. Surgical Treatment of the Epilepsies, Second Edition. New York: Raven Press, 1993: 511-518.
109. Honavar M, Janota I, Polkey CE. Rasmussen's encephalitis in surgery for epilepsy. Dev. Med. Child Neurol., 1992; 34: 3-14.
110. Desbiens R, Berkovic SF, Dubeau F, et al. Life threatening focal status epilepticus due to occult cortical dysplasia. Arch. Neurol., 1993; 50: 695-700.
111. Ribaric II, Nagulic M, Djurovic B. Surgical treatment of epilepsy: our experiences with 34 children. Childs Nerv. Syst., 1991; 7: 402-404.
112. Gorman DG, Shields WD, Shewmon DA, et al. Neurosurgical treatment of refractory status epilepticus. Epilepsia, 1992; 33: 546-549.

113. Fusco L, Vigevano F. Reversible operculum syndrome caused by progressive epilepsia partialis continua in a child with left hemimegalencephaly. J. Neurol. Neurosurg, Psychiatry, 1991; 54: 556-558.
114. Cendes F, Ragazzo PC, da Costa V, et al. Corpus Callosotomy in treatment of medically resistant epilepsy: preliminary results in a paediatric population. Epilepsia, 1993; 34: 910-917.
115. Nakatani S, Nii Y, Ikejiri Y, et al. Partial callosotomy for Lennox Gastaut Syndrome first cases in Japan. Neurol. Med. Chir. Tokyo, 1990; 30: 930-939.
116. Girvin JP. Complications of Epilepsy Surgery. In: Luders HO, editor. Epilepsy Surgery. New York: Raven Press, 1992: 653-660.
117. Lindsay J, Ounsted C, Richard P. Long-term outcome in children with temporal lobe seizures. V. Indications and contraindications for neurosurgery. Dev. Med. Child Neurol., 1984; 26: 25-32.
118. Fraser RT, Gumnit RJ, Thorbecke R, et al. Psychosocial Rehabilitation: A pre and post operative perspective. In: Engel J. Jr. Surgical Treatment of the Epilepsies Second Edition. New York: Raven Press, 1993: 669-677.
119. National Institutes of Health. Plan of Nationwide Action on Epilepsy. DHEW publication No (NIH) 79-1115, Department of Health, Education and Welfare, Washington, 1978.
120. Silfvenius H, Dahlgren H, Jonsson E, et al. Epilepsy Surgery (in Swedish). The Swedish Council on Technology Assessment in Health Care. Stockholm, 1991.
121. Wieser HG, Siegel AM, editors. Guidelines for comprehensive epilepsy centers Proceedings of the Round Table held during the Second International Zurich Epilepsy Symposium April 9-11, Zurich, 1992 .
122. Kriedel T. Cost benefit analysis of epilepsy clinics. Soc. Sci. and Med., 1980; 14: 35-39.
123. Silfvenius H, Lindholm L, Olivecrona M. Epilepsy Surgery: Cost/benefit calculation in Sweden. In: Wieser HG, Siegel AM, editors. Guidelines for comprehensive epilepsy center. Proceedings of the round table held during the 2nd International Zurich Epilepsy Symposium. Zurich, 1992: 79-85.
124. Silfvenius H., Lindholm L. Säisä J, et al. Costs of and Savings from Pediatric Epilepsy Surgery; A Swedish Study, in press.
125. Olivecrona M, Silfvenius H. Mortality in Epilepsy in Sweden 1969-1986; As Revealed by Death Certificates. Epilepsia, submitted.
126. Harvey AS, Nolan T, Carlin JB. Community-based study on mortality in children with epilepsy. Epilepsia, 1993; 34 (4): 597-603.
127. Pilcher WH, Roberts DW, Flanigin HF, et al. Complications of Epilepsy Surgery. In: Engel J, Jr. Surgical Treatment of the Epilepsies, Second Edition. New York: Raven Press, 1993: 565-581.
128. Christianson S-Å, Heijbel J, Olivecrona M, et al. Procedures in pediatric epilepsy surgery. In: Pedersen B, Dam M, editors. Proceedings of the Northern European Epilepsy Symposium. Acta Neurol. Scand., 1990; 82 (S133): 19.

12 Neuropsychological aspects of epilepsy surgery

MICHAEL WESTERVELD
KIMBERLEE J. SASS

Yale University School of Medicine, Departments of Surgery (Neurosurgery), Pediatrics, and Child Study Center, New Haven CT, USA

It is widely assumed that successful surgery for epilepsy at a younger age (e.g., school age, prior to late adolescence) will minimize or eliminate the cognitive, behavioral, and psychosocial impairments commonly observed in adults with chronic seizures. This assumption is based upon the belief that sparing the developing brain from frequent seizures and long-term exposure to antiepileptic drugs will facilitate normal development. Although there is ample documentation of the neuropsychological deficits associated with chronic seizures, there are few studies which provide adequate pre- and post-operative evaluation of cognitive, social, and behavioral functioning in children undergoing surgery. This is due, in part, to the fact that only children with more severe epilepsy and accompanying cognitive impairments have been considered candidates for surgery. Although this practice is changing, empirical verification of the neuropsychological benefits of surgery is still lacking. Prospective, longitudinal studies of patients undergoing surgery is necessary to demonstrate the absence of significant cognitive morbidity and improved behavioral and psychosocial functioning in patients undergoing surgery. This chapter will summarize existing literature on outcome of epilepsy surgery performed in childhood and adolescence. The emphasis will be on identifying areas in need of further study and the role of the neuropsychologist in a paediatric epilepsy program.

EPILEPSY SURGERY IN CHILDREN

It is clear that seizures exert a deleterious effect on development of cognitive and psychosocial skills. It is also clear that these deficits arise fairly early after the onset of seizures, are associated with seizure frequency and duration, and that decline in cognitive functioning is an ominous prognostic sign. Demonstration of improved function in the absence of cognitive morbidity associated with surgery for epilepsy would indicate that there is no compelling reason to delay intervention for those patients deemed to be refractory to antiepileptic drug (AED) therapy and with little hope of remission. It should be emphasized that, even among adult patients with epilepsy these cases represent a minority, however when appropriately selected they benefit greatly from surgery.

The first report of successful surgery for epilepsy appeared in 1886 when Horsly reported that seizures could be eliminated following focal resection of cortex which, when stimulated electrically, gave rise to an aura or seizure.[1] The evolution of technology since that time has resulted in improved selection of candidates and reduced morbidity, with greater understanding of the pathology underlying many types of epilepsy. However, the improvement has been primarily in the identification of adult surgical candidates. There remains a need to improve **early** identification of appropriate surgical candidates, which may greatly reduce the neuropsychological and psychosocial sequelae of chronic epilepsy.

Falconer[2-4] and colleagues[5,6] provided the earliest reports of seizure outcome following surgical intervention focusing explicitly on children and adolescents. These authors argued that "any child who has had several epileptic fits" should be considered a potential candidate for surgery.[5] It was believed that, while many forms of childhood epilepsy may remit without intervention, epilepsy of temporal lobe origin is more likely to persist into adulthood.[2] Furthermore, they believed that the outcome of operation during school age is more favorable than similar operations performed in later decades.[4] Their experience led them to conclude that epilepsy may be more appropriately performed before adulthood, around the age of 10-12 years.[3] The outcome of surgery in the series was favorable, with seizure control results comparable to similar adult series. Although the initial evaluations of surgical success concerned seizure reduction, an added benefit of early intervention was that behavior was 'improved' or 'much improved', and that a general trend of improved IQ and learning was observed.[6] Despite the reported successful outcome of surgery in this and other series which followed[1,7-12], epilepsy surgery is not often considered an option for children and young adolescents.

EARLY SURGICAL INTERVENTION: CONTROVERSIES AND RATIONALE

Controversies

Referrals to surgical centers may be delayed in the hope that recurrent seizures will remit. However, the natural history of many seizure types is poorly understood, and estimates of spontaneous remission rates vary considerably. For complex partial seizures, the rate of remission is low,[13] and may be as low as 10 to 30%[14,15] When a structural abnormality is present in the temporal lobe, the likelihood is that seizures will persist.[14] Referrals may also be delayed in the hopes that AED treatment will be successful, and that seizures will not recur after medication withdrawal. However, epidemiological studies attempting to identify risk factors associated with seizure recurrence estimate recurrence rates as high as 70%.[16] A recent prospective study [17] found that when AED's are discontinued, even after a mean seizure free interval of 2.9 years, the risk of seizure recurrence was 36%[17] Thus, a significant portion of children who suffer repeated episodes of unprovoked complex partial seizures are likely to require long term AED treatment, and may eventually become refractory to pharmacologic treatments. More accurate means of identifying, at an early age, those who will not derive long-term benefits from AED treatment are needed. Although cognition has been cited as being an important prognostic variable, most of the studies involving prediction of seizure recurrence do not include neuropsychological findings in their risk assessment.

Availability of surgery as a viable option has also limited its widespread application. Evaluation of seizures in children involves numerous steps to adequately characterize the electroclinical syndrome.[18] Continuous EEG/video monitoring is necessary to precisely document focality of ictal and interictal epileptic activity, with some centers requiring a minimum of 2 separate admissions to document consistency of the information obtained.[13] When localization using scalp EEG is not precise, invasive studies such as the Intracarotid Amytal Procedure[19-21] and intracranial electrodes[18,22] may be required. When seizures are found to originate in or near functional cortex, electrocortical stimulation to map neuropsychological functions may also be required. However, many tertiary care facilities do not have the specialized equipment or staff necessary to provide such sophisticated and comprehensive evaluation.

Criteria for surgery in children, as well as adults, requires failure to adequately treat seizures with exhaustive trials of antiepileptic drugs (AED's) despite adequate compliance.[23] There are numerous AED's, and it may require protracted trials in order to truly evaluate the efficacy of a particular regimen. Several years may be required to adequately exhaust medication trials in children.[24] In addition to the numerous AED's presently available, as many as 22 new AED's have undergone clinical trials[25], adding to the non-surgical treatment options which need to be exhausted.

There is also a concern about the potential for cognitive morbidity following epilepsy surgery in children. The material specific memory deficits which occur following temporal lobectomy in adults are well described.[26] In addition, focal cortical resections may result in specific deficits for information processing or sensorimotor skill related to the zone of cortical resection. Although there may be deficits associated with the function of a resected epileptogenic focus, there may also be improvement in cognitive functions typically mediated by cortical areas not involved in seizure propagation or surgical resection. [27,28]

These concepts are derived from studies in adult patients, in whom surgical treatment of seizures was undertaken. However, extrapolation of these results to children with epilepsy may not be direct. Neurological maturation is not complete until late adolescence. Paroxysmal bombardment by epileptogenic lesions may be sufficient to disrupt normal functional maturation but insufficient to alter its pattern, the result being the commonly reported pattern of deficits observed in adult epileptic patients. However, there is some indirect evidence that the window for developmental plasticity which allows for at least partial language acquisition by the nondominant hemisphere extends into adolescence.[29] A surgically induced lesion at an earlier maturational stage may provoke compensatory mechanisms, resulting in fewer cognitive deficits. Thus, it could be argued that while the potential for cognitive morbidity certainly exists, there is evidence that compensation and recovery of function occur, and theoretically could be greater in younger patients.

Rationale for Surgical Intervention

The outcome of surgery during childhood and adolescence with respect to seizures is not substantially different than in adults. That is, the majority of appropriately selected candidates may expect to gain substantial relief from recurrent seizures. The percent with favorable outcome range from 48%[30] to 94%,[31] with the majority of patients enjoying complete relief from seizures with minimal surgical morbidity. Delaying surgery until adulthood does not result in improved seizure control,[31] but may result in greater neuropsychological impairments. Conversely, early intervention which spares the brain from recurrent seizures during critical periods of development without exposure to the potentially deleterious effects of AED's may reduce the long term cognitive impairments.

Effects of seizures on cognitive functioning
Investigators have established a relationship between the severity and duration of epilepsy and the extent of neuropsychological impairments. Farwell et al.[32] compared a matched control group to a sample of 118 children with seizures on a battery of neuropsychological tests. Children with seizures performed significantly worse than the controls. Furthermore, intelligence was found to be significantly related to duration of the seizure disorder; children having a longer duration of seizures obtained lower IQ scores. Other authors have noted a strong association between intellectual impairment and early age of seizure onset[33,34] and seizure frequency.[32,35] Although these results document the dramatic effect of seizures on IQ scores, the Full Scale IQ represents a composite of many divergent skills. It is also important to include more detailed assessment of cognitive ability so as to identify potential cognitive liabilities which are related to the lateralization and focality of the epileptogenic cortex.

Fedio & Mirsky[36] found a dissociation in neuropsychological abilities among children with temporal lobe epilepsy versus centrencephalic seizures. Children with centrencephalic epilepsy tended to demonstrate greater impairments in sustained concentration tasks versus children with temporal lobe epilepsy (TLE). In contrast, patients with TLE demonstrated material specific cognitive deficits in the expected direction (VIQ<PIQ for Left TLE; VIQ>PIQ for Right TLE patients), while no such effect was found for controls or centrencephalic patients. In addition, children with temporal lobe epilepsy did more poorly on material specific measures of memory than their extratemporal and control counterparts. Other authors have confirmed that the lateralization of epileptic discharges exerts a disruptive influence on performance of material specific cognitive tasks, with verbal tasks susceptible to interference from dominant hemisphere discharges while nondominant hemisphere discharges disrupt performance on nonverbal-spatial tasks.[37-39]

Effects of seizures on academic performance
Even among those children with seizures who do not demonstrate obvious deficits in IQ or memory variables, learning and academic performance is often well below that of their peers. Children with epilepsy are over-represented in special education classes,[40] and tend to demonstrate achievement well below grade placement in reading[41-43] and arithmetic.[44] Disease variables such as the frequency of seizures, age of onset of habitual seizures, and location of the epileptogenic focus are important factors in determining the type and severity of the learning disability. However, diminished academic performance may also be related to iatrogenic effects of treatment with antiepileptic drugs.

Effects of AED's on cognitive functioning
Although many types of seizures may be adequately controlled with AED's, these may themselves contribute to the diminished cognitive efficiency observed in children with epilepsy. Although it is sometimes difficult to separate the deleterious effects of medication from the effects of the seizures themselves, numerous studies have documented the impact of AED's on cognition.[45] Polytherapy tends to have a greater impact on cognition than treatment with individual drugs, however AED's vary in their effects on performance.[45] Farwell[46] noted that phenobarbital depressed IQ by an average of 8.4 points compared with a control group. Furthermore, they noted that diminished cognitive performance persisted for several months following discontinuation of treatment. Phenytoin has also been associated with lower cognitive performance, even when taking into account seizure activity.[45] Although not all AED's exert such dramatic effects on global cognitive functioning, there are few which can be said to be entirely benign with respect to cognition or psychomotor performance.

Effects of seizures on psychosocial and vocational functioning
Much of the investigation into quality of life issues in epilepsy has focused on adults. However, many of the difficulties cited by adults with epilepsy begin in childhood and evolve concomitantly with the seizure disorder. Individuals with epilepsy often fail to develop appropriate social skills, and as a result have few, or poor, relationships outside of immediate family. This may infrequently be due to mental illness associated with seizures, but more commonly is a function of social isolation.[47] In addition to

social dysfunction, vocational adaptation is often poor in patients with epilepsy. Although this is primarily an adult issue, unemployment or underemployment frequently is a result of failure to attain educational or job skills during earlier school years.[47]

Ironically, although one of the goals of early surgery is to minimize the deficits associated with chronic seizures, there are few reports which adequately document the benefits or risks to cognition associated with early surgical intervention. Although nearly all of the reported series refer to neuropsychological testing as an important component of the preoperative assessment, none present detailed outcome data. Extrapolation of the adult literature to paediatric populations is not straightforward because of developmental plasticity and the varying developmental issues at different ages.

SURGICAL PROCEDURES

Temporal Lobectomy

The most commonly performed surgical procedure for complex partial epilepsy is resection of the anterior temporal lobe,[48] yet there are few reports of neuropsychological status following temporal lobectomy in children and adolescents. The focus of early reports was primarily on seizure outcome, with improvement in cognition or behavior considered an added benefit.[6] Many reports which include outcome aside from seizure control focus on global measures of cognition such as IQ[12,49] and/or qualitative evaluation of behavioral and academic improvement.[12,14,15,31] These reports are often retrospective, base their conclusions on comparison of early versus late operated patients, and may include patients who undergo surgery as adults, but had seizure onset in childhood.[12,50]

In one of the few studies which included prospective collection of neuropsychological data, Meyer et al.[9] report on 37 of their 50 patients who underwent neuropsychological assessment before and after temporal lobectomy. They found no change in IQ scores as a function of side of surgery. However, there was an inverse relationship between duration of seizures prior to intervention and both Verbal and Performance IQ scores, with longer duration of seizures prior to surgery associated with less favorable outcome. These authors also evaluated memory outcome, and found no significant association between side of surgery or duration of seizures and changes in memory. Although there was no significant decline in memory scores, these authors used the Wechsler Memory Scale (WMS) composite score to evaluate memory function in their sample of children. The WMS composite score may not be sufficiently sensitive to detect important material specific memory problems, and may not be sensitive to developmental memory questions because it is primarily intended for use with adults.

Dennis et al.[51] found memory deficits following temporal lobectomy to be related to side of surgery and extent of posterior temporal cortex resection. Patients in their sample who underwent left temporal lobectomy demonstrated greater difficulty for verbal recognition and judgement of context than patients undergoing right temporal lobectomy. Even though the authors found greater difficulty to be associated with larger resections on the left, it is impossible to conclude that the deficit was caused by the resection in the absence of testing before and after surgery. Larger resections may have been associated with more severe epilepsy, or at least larger areas of involvement in seizure onset, resulting in greater pre-operative difficulty. Only 3 of their patients underwent testing before and after surgery; two of these patients improved.

Prospective studies which evaluate individual as well as group outcomes in specific neuropsychological domains (e.g., memory, language) are clearly needed to determine if the risks to cognitive functioning in children are similar to those in adults, and to affirm the presumed benefits of early surgery.

Neuropsychology of Hemispherectomy

Functional hemispherectomy has replaced anatomically complete removal of a cerebral hemisphere due to the complication of late hemosiderosis which may follow the latter procedure. However, the neuropsychological indications and sequelae are essentially the same. The procedure is typically considered for patients who have hemiplegia or dense hemiparesis,[52] and for whom no additional deficit in motor functions would be anticipated following removal of the diseased hemisphere. Most of these patients suffered perinatal injury (infantile hemiplegia), although functional hemispherectomy is increasingly performed for patients with progressive unilateral encephalopathy[53] (i.e., Rasmussen's encephalitis). Neuropsychological testing performed in patients who are candidates for functional hemispherectomy generally reveals below average functioning, with behavior management problems and aggression also common. Complete relief of seizures is achieved in 80-85% of cases,[52] with concomitant improvement in behavior.

In patients with Rasmussen's syndrome, initial functioning may be normal but progressive deterioration prior to surgery is the rule. In patients with initially normal ability and evidence of left hemisphere disease, there may be a legitimate concern for significant language morbidity depending on the age at surgery. Patients with evidence of early left hemisphere injury typically have acquired right hemisphere speech dominance and therefore constitute less of a risk for post-operative aphasia, regardless of age at surgery. However, in some patients with left hemisphere disease, language deterioration is part of the clinical presentation suggesting left hemisphere dominance for speech. These patients constitute a much greater risk for post-operative language deficits. Vining et al.[53] report on their experience with several patients undergoing left hemispherectomy with varying degrees of aphasia persisting following surgery, with the degree of language recovery apparently relating to age at surgery. The oldest patient undergoing dominant hemispherectomy in this series was nearly 13 years of age. This patient reportedly has recovered some language, and at 6 years following the surgery is reported to use clear short phrases although language and intellectual deficits remain.

Neuropsychology of Disconnection Procedures

Corpus callosum section is a palliative, rather than curative, procedure. It is considered an option primarily when secondarily generalized seizures are debilitating, and no single resectable focus can be found. Callosotomy effectively disconnects the two hemispheres, preventing the ictal discharge from involving the hemisphere contralateral to seizure onset. However, by disconnecting the hemispheres to eliminate interhemispheric spread of epileptic activity, interhemispheric communication of sensory and perceptual information is also disrupted. The study of adult patients undergoing this procedure has identified only infrequent debilitating impairments related to disconnection phenomena. The most striking of these include the 'alien hand', and intermanual conflict. More subtle deficits are common, although they are often difficult to detect outside of specialized testing conditions and are rarely disabling.[54] These include anomia for information presented unilaterally to the nondominant hemisphere (e.g., tachistoscopic presentation, or placing objects in the nondominant hand), and difficulty with bilateral perceptual-motor integration.

Clinically evident deficits following callosotomy in adults are likely to occur under one of two conditions. When seizures result in atypical hemispheric representation of speech, post callosotomy language deficits may occur.[55] Neuropsychological impairments may also occur when surgical disconnection interrupts callosally mediated interhemispheric compensation mechanisms.[56] It is possible that the age at which callosotomy is performed, relative to certain developmental milestones, will alter the pattern of cortical reorganization. The extent to which the corpus callosum plays a role in facilitating recovery from certain developmental injury is also unknown, therefore it is unknown whether callosotomy performed at an early age will alter the manifestation of deficits.

The study of neuropsychological functioning following cerebral disconnection has been largely limited to adult patients. Geoffroy et al.[57] reported on a series of 9 children who underwent callosotomy. In their series, no patient suffered a decline in mental age or IQ. All but one patient in their series underwent language evaluation, each of them demonstrating improvements relative to pre-operative assessment. The three patients who underwent memory testing also demonstrated improved functioning. There were no reported incidents of significant deficits following callosotomy in this series, although one patient did not benefit from surgery either in terms of seizures or developmental outcome. This series, in conjunction with adult series, suggests that callosotomy may be performed in children as well as adults with low risk to global cognition in appropriately selected candidates. However, there is an absence of information regarding the influence of early callosotomy on the development of specific areas of cognitive function such as language, memory, attention, and executive function.

ROLE OF CHILD NEUROPSYCHOLOGY IN AN EPILEPSY SURGERY PROGRAM

Patient Selection

Most surgical evaluation protocols include neuropsychological assessment as an integral component. The neuropsychologist contributes to the selection of appropriate candidates by documenting focal cognitive disturbances which are convergent with seizure onset as established by other noninvasive means, by determining hemispheric dominance for speech and memory capacity via intracarotid amytal testing, and by identifying patients in whom surgery might produce unacceptable cognitive morbidity. However, developmental and practical issues complicate these assessments in children. Performance on standardized psychometric measures is effected not only by the disease process, but also by plasticity in the presence of neurological dysfunction and the natural emergence of cognitive abilities (e.g., language, mnemonic strategies) which occurs with maturation. Greater understanding of the interaction of these factors in children is needed in order to refine the neuropsychological contribution to patient selection.

Until recently, the role of assessing psychosocial factors in epilepsy surgery was given minimal attention. However, part of selecting patients is identifying those who are handicapped by their seizures, which includes assessment of individual emotional, psychosocial, and family factors.[24] Neuropsychology is uniquely positioned to provide standardized assessment of these variables in addition to more 'traditional' neuropsychological assessment.

Outcome Assessment

Neuropsychology also provides a unique contribution to outcome assessment. There are several series reporting seizure outcome following surgery.[1-12] However, seizure improvement without concomitant improvement in quality of life issues such as cognition, and social/emotional adjustment is of questionable value. There is ample preliminary evidence that, with respect to global cognition, there is little risk. However, as more and more children are considered for surgical intervention, greater understanding of the impact of an early surgical lesion upon development of specific cognitive domains is needed. There are no studies which address language, attention, memory, perceptual and perceptual-motor skills, or social/emotional development following surgery using standardized psychometric instruments designed to assess development of these abilities in children. Studies which utilize appropriately matched control groups can effectively evaluate the relative risks vs. benefits of early surgery.

Developmental Research

Much of what we presently understand about the neuropsychology of higher cortical functioning is owed to the study of epilepsy and epileptic patients.[58] The necessary evaluation of children's neuropsychological functioning prior to undertaking surgery for epilepsy provides a natural laboratory for study of cognitive development and plasticity of brain behavior relationships. Application of appropriate neuropsychological study to children undergoing epilepsy surgery may provide a similar window of increased understanding to concepts such as developmental plasticity and emergence of higher cortical functions in children.

Acknowledgements
The authors wish to thank Hugh G. Henry for his bibliographic assistance in preparation of this manuscript.

REFERENCES

1. Green JR. Surgical treatment of Epilepsy during Childhood and Adolescence Surg. Neurol., 1977; 8: 71-80.
2. Falconer MA. Significance of surgery for temporal lobe epilepsy in childhood and adolescence. J. Neurosurg., 1970; 33: 233-252.
3. Falconer MA. Temporal lobe epilepsy in children and its surgical treatment. Medical J. Australia, 1972; 1: 1117-1121.
4. Falconer MA. Place of Surgery for Temporal Lobe Epilepsy during Childhood. British Medical Journal, 1972; 2: 631-635.
5. Davidson S, Falconer MA, Stroud CE. The Place of Surgery in the Treatment of Epilepsy in Childhood and Adolescence: A Preliminary Report in 13 Cases. Dev. Med. Child Neurol., 1972; 14: 796-803.
6. Davidson S, Falconer MA. Outcome of Surgery in 40 Children with

Temporal Lobe Epilepsy. Lancet, 1975: 1260-1263.
7. Green JR, Pootrakul A. Surgical aspects of the treatment of epilepsy during childhood and adolescence. Arizona Medicine, 1982; 39: 35-38.
8. Goldring S. A method for surgical management of focal epilepsy, especially as it relates to children. J. Neurosurg., 1978; 49: 344-356.
9. Meyer FB, Marsh WR, Laws ER, et al. Temporal lobectomy in children with epilepsy. J. Neurosurg., 1986; 64: 371-376.
10. Drake J, Hoffman HJ, Kobayashi J, et al. Surgical management of children with temporal lobe epilepsy and mass lesions. Neurosurgery, 1987; 21: 792-797.
11. Jay V, Becker LE, Otsubo H, et al. Pathology of temporal lobectomy for refractory seizures in children. J. Neurosurg., 1993; 79: 53-61.
12. Erba G, Winston KR, Adler JR, et al. Temporal lobectomy for complex partial seizures that began in childhood. Surg. Neurol., 1992; 38: 242-243.
13. Luders H, Dinner DS, Morris HH, et al. EEG Evaluation for epilepsy surgery in children. Cleveland Clinic Journal of Medicine, 1989; 56(suppl): S53-S61.
14. Hopkins IJ, Klug GL. Temporal lobectomy for the treatment of intractable complex partial seizures of temporal lobe origin in early childhood. Dev. Med. Child Neurol., 1991; 33: 26-31.
15. Kotogal P, Rothner AD, Erenberg G, et al. Complex partial seizures of childhood onset: A five year follow-up study Arch. Neurol., 1987; 44: 1177-1180.
16. Berg AT, Shinnar S. Relapse following discontinuation of antiepileptic drugs: A Meta-Analysis Neurology, 1994; 44: 601-608.
17. Shinnar S, Berg AT, Moshe SL, et al. Discontinuing antiepileptic drugs in children with epilepsy: A prospective study. Ann. Neurol., 1994; 35: 534-545.
18. Wheless J. Evaluation of Children for Epilepsy Surgery. Pediatric Annals, 1991; 20: 41-49.
19. Williams J, Rausch R. Factors in children that predict performance on the Intracarotid Amobarbital Procedure Epilepsia, 1992; 33: 1036-1041.
20. Szabo CA, Wyllie E. Intracarotid Amobarbital testing for language and memory dominance in children. Epilepsy Res., 1993; 15: 239-246.
21. Westerveld M, Zawacki TM, Sass KJ, et al. (In Press) Intracarotid Amytal Procedure Evaluation of Hemispheric Speech and Memory Function in Children and Adolescents J, Epilepsy.
22. Wyllie E, Luders H, Morris HH, et al. Subdural electrodes in the evaluation for epilepsy surgery in children and adults. Neuropediatrics, 1988; 19: 80-86.
23. Holmes GL, King DW. Epilepsy Surgery in Children Wiener klinische Wochenschrift, 1990; 102: 189-197.
24. Wyllie E, Luders H. Complex partial seizures in children: Clinical manifestations and identification of surgical candidates. Cleveland Clinic Journal of Medicine, 1989; 56(suppl): S43-S52.
25. Fisher RS. Emerging Antiepileptic drugs. Neurology 1993; 43(suppl 5): S12-S20.
26. Jones-Gotman M. Localization of lesions by neuropsychological testing. Epilepsia, 1991; 32(suppl 5): S41-S52.
27. Novelly RA, Augustine EA, Mattson RH, et al. Selective memory improvement and impairment in temporal lobectomy for epilepsy. Ann. Neurology, 1984; 15: 64-67.

28. Cavazutti V, Winston K, Bakaer R, et al. Psychological changes following surgery for tumors in the temporal lobe. J. Neurosurg., 1980; 53: 618-626.
29. Strauss E, Wada J, Goldwater B. Sex differences in interhemispheric reorganization of speech. Neuropsychologia, 1992; 30: 353-359.
30. Fish DR, Smith SJ, Quesney LF, et al. Surgical treatment of Children with Medically Intractable Frontal or Temporal Lobe Epilepsy: Results and Highlights of 40 Years' Experience Epilepsia, 1993; 34(2): 244-247.
31. Duchowny M, Levin B, Jayakar P, et al. Temporal Lobectomy in Early Childhood Epilepsia, 1992; 33(2): 298-303.
32. Farwell JR, Dodrill CB, Batzel LW. Neuropsychological abilities of children with epilepsy. Epilepsia, 1985; 26(5): 395-400.
33. Huttenlocher PR, Hapke RJ. A follow-up study of intractable seizures in childhood. Ann. Neurol., 1990; 28: 699-705.
34. O'Leary DS, Seidenberg M, Berent S, et al. Effects of age of onset of tonic-clonic seizures on neuropsychological performance in Children. Epilepsia, 1981; 22: 197-204.
35. Dodrill CB, Troupin AS. Seizures and adaptive abilities. Arch. of Neurol., 1976; 33: 604-607.
36. Fedio P, Mirsky A. Selective Intellectual Deficits in children with temporal lobe or centrencephalic epilepsy. Neuropsychologia, 1969; 7: 287-300.
37. Kasteleijn-Nolst Trenite DGA, Sibelink BM, Berrends SGC, et al. Lateralized effects of subclinical epileptiform EEG discharges on Scholastic performance in children. Epilepsia, 1990; 31(6): 740-746.
38. Kasteleijn-Nolst Trenite DGA, Bakker DJ, Binnie CD, et al. Psychological effects of epileptiform EEG discharges I: Scholastic skills. Epilepsy Research, 1988; 2: 111-116.
39. Binnie CD, Channon S, Marston D. Learning disabilities in Epilepsy: Neurophysiological aspects. Epilepsia, 1990 31(suppl 4): S2-S8.
40. Aldenkamp AP, Alpherts WCJ, Dekker MJA, et al. Neuropschological Aspects of Learning Disabilities in Epilepsy. Epilepsia, 1990; 31(suppl 4): S9-S20.
41. Rutter M, Graham P, Yule WA. A neuropsychiatric study in childhood. 1970 London: Heinemann Medical.
42. Stores G, Hart J. Reading skills of children with generalized and focal epilepsy attending ordinary school. Dev. Med. Child Neurol., 1976; 18: 705-716.
43. Stores G. Schoolchildren with epilepsy at risk for learning and behavior problems. Dev. Med. Child Neurol., 1978; 20: 502-508.
44. Aldenkamp AP. Epilepsy and Learning Behavior In: Parsonage M, Grant RHE, Craig AG, Ward AA, (eds): Advances in epileptology: the XIVth International Epilepsy Symposium New York: Raven Press, 1983: 221-229.
45. Trimble MR. Antiepileptic drugs and cognition. Epilepsia, 1990; 31(suppl 4): S30-S34.
46. Farwell JR, Lee YJ, Hirtz DG, et al. Phenobarbital for febrile seizures-effects on intelligence and seizure recurrence. New England Journal of Medicine, 1990; 322: 364-369.
47. Vickrey BG, Hays RD, Hermann BP, et al. Outcomes with Respect to Quality of Life In: Engel J.(ed) Surgical treatment of the Epilepsies New York: Raven Press 1993: 623-636.
48. Cascino GD, Jack CR, Parisi JE, et al. MRI in the presurgical evaluation of patients with frontal lobe epilepsy and children with temporal lobe epilepsy:

Pathologic correlation and prognostic significance. Epilepsy Research, 1992; 11: 51-59.
49. Harbord MG, Manson JI. Temporal lobe epilepsy in childhood: Reappraisal of etiology and outcome. Ped. Neurol., 1987; 3: 263-268.
50. Mizrahi EM, Kellaway P, Grossman RG, et al. Anterior Temporal Lobectomy and Medically Refractory Temporal Lobe Epilepsy of Childhood. Epilepsia, 1990; 31(3): 302-312.
51. Dennis M, Farrell K, Hoffman HJ, et al. Recognition memory of item, associative, and serial order information after temporal lobectomy for seizure disorder. Neuropsychologia, 1988; 26: 53-65.
52. Rasmussen T, Villemure JG. Cerebral Hemispherectomy for seizures with hemiplegia. Cleveland Clinic Journal of Medicine, 1989; 56(Suppl 1): S62-S68.
53. Vining EPG, Freeman JM, Brandt J, et al. Progressive unilateral encephalopathy of childhood (Rasmussen's Syndrome): A reappraisal. Epilepsia, 1993; 34(4): 639-650.
54. Westerveld M, Sass KJ, Spencer SS, et al. Neuropsychological function following corpus callosotomy for epilepsy. In Devinsky O, Theodore WH, (eds.) Epilepsy and Behavior, 1991; 203-211.
55. Sass KJ, Novelly RA, Spencer SS, et al. Post-callosotomy language deficits in patients with crossed cerebral dominance. J. Neurosurg., 1990; 72: 85-90.
56. Novelly RA, Lifrak MD. Forebrain commissurotomy reinstates effects of pre-existing hemisphere lesions: An examination of the hypothesis In: Reeves AG, (ed). Epilepsy and the Corpus Callosum New York: Plenum Press, 1985: 467-500.
57. Geoffroy G, Lassonde M, Delisle F, et al. Corpus callosotomy for control of intractible epilepsy in children. Neurology, 1983; 33: 891-897.
58. Novelly, R.A. (1992) The debt of neuropsychology to the epilepsies. American Psychologist, 47: 1126-1129.

13 The impact of epilepsy on cognitive development and learning behaviour

ALBERT P. ALDENKAMP

Department of Neuropsychology "Meer en Bosch", Epilepsy Center, Heemstede, The Netherlands

Although exact prevalence estimates are lacking, literature shows that children with epilepsy, as a group, have a greater risk for developing problems in learning behaviour. Moreover some studies report that children with epilepsy have other 'types' of learning disability, than observed in the general 'non-neurological' population. Especially the more frequent problems with arithmetic suggests such specific impairment. In this chapter the complex relationships between learning disabilities, underlying cognitive impairment and epileptic conditions are analyzed. Factors such as the localization of the epileptogenic focus in specific (in particular temporal) areas of the brain, seizure activity (i.e. seizure type, seizure severity and seizure frequency), and central side-effects of the antiepileptic medication are discussed in subsequent paragraphs. The final paragraphs address the long-term prognosis and issues of assessment and treatment.

INTRODUCTION

Children with epilepsy, as a group, are at risk for developing learning problems.[1-6] However, 'learning problems' is a rather ill-defined category[7] and any uniformity in methods to assess learning problems is still lacking.[8] Consequently, prevalence estimates of learning problems in children with epilepsy vary widely, percentages mentioned in the literature ranging from 5% to 50%.[3] Nevertheless ample -although circumstantial evidence exists showing that epilepsy may impair the abilities to learn. Approximately one-third of children with refractory epilepsies are receiving some form of special educational support.[3,5,9-11] Moreover, academic underachievement in children with epilepsy, relative to their own abilities, has been noted by several authors.[6,12]

As yet, it is unknown which factors contribute to the development of learning impairment. Parental and peer group attitudes may be expected to influence the child's learning behaviour. Parents may e.g. worry about the seizures, about the side-effects of antiepileptic medication, or about possible future social handicaps.[13-15] Consequently, parents tend to have other expectations of their children with epilepsy relative to their healthy children, such as more emotional problems, poorer concentration, lower academic achievement and fewer employment opportunities.[15,16] Because of such special concerns and expectations, parents may behave differently towards their children with epilepsy. However, there is considerable variation among parents and other members of the family in their reactions to a child with epilepsy, which may range from overprotection to rejection and scapegoating. Various developmental problems may occur due to such extreme reactions. Examples are low self-esteem, lack of social skills[17] feelings of guilt, or the adoption of a sick role.[18] All these reactions may have effects on learning behaviour in school.

Specific cognitive deficits may also be responsible for learning problems and educational underachievement. Slowing on speeded tasks, involving complex information processing, slowed decision making, attention and concentration difficulties, all are well established phenomena in epilepsy.[19,20] Factors such as the localization of the epileptogenic focus, the seizure activity, and the central side-effects of antiepileptic medication may underlie such cognitive deficits and thus interfere with learning processes.[21]

TYPE OF LEARNING DISABILITY

Generally, 'learning disabilities' is seen as a homogeneous disorder with a single cause. In this unitary concept, all children are supposed to have the same 'type' of learning disability. In contrast, multiple syndrome research successfully hypothesized the existence of independent 'types' of learning disabled children[7,22-24] based on underlying patterns of cognitive weaknesses and strengths. For every-day practice in school, this is important because it may open possibilities for differential treatment. As yet, such differential treatment is lacking. Remarkably, e.g. none of the special schools for epilepsy use special learning techniques. Yet, it is emphasized repeatedly that children with epilepsy must be seen as a special group with respect to their educational needs.[25,26]

Research still has to reveal which type of school problem dominates in children with epilepsy. Several studies have demonstrated specific problems in the area of reading in children with epilepsy. The Isle of Wight study[1] showed specific under-achievement in reading in children with epilepsy, with an average retardation of approximately 12 months. These results were corroborated in several studies.[2,16,27] Other studies suggested arithmetic[4,5,9,12,28,29] or spelling and writing skills[30] to be the impaired academic skill in children with epilepsy .

If we accept that at least a subgroup may have specific arithmetical impairment and that difficulties may occur in several academic skills simultaneously, then children with epilepsy may have other 'types' of learning disability, than observed in the general 'non-neurological' population, where, by and large the problems are restricted to impaired reading ability.[1,7,23] Studies have found some evidence for at least two 'epileptic' subtypes of learning disability. Firstly, a 'mental slowing subtype' exhibiting slowing of information processing especially in complex task configurations.[21] These children also show pronounced retardation of their arithmetical skills. A correlation was found between this subtype and type of epilepsy (all these children had refractory localization related epilepsies) and the use of polytherapy (in particular long-term treatment with phenytoin). Secondly, a 'problem-solving disorder type' was found[4,5] representing disorders in central cognitive processing, such as logical thinking and verbal reasoning. This disorder could not be related to epileptic factors directly, but with psychosocial factors, such as the personal cognitive style. Children, exhibiting this subtype showed impairment on arithmetic, reading and other aspects of the school programme simultaneously.

RESEARCH ON COGNITIVE FACTORS

A considerable body of research exists with regard to the general issue of cognitive functioning in epilepsy. Speed of information processing, memory, vigilance, alertness, sustained and focused attention and motor fluency appear to be particularly vulnerable areas of functioning in epilepsy.[6,27,31,32] In children with epilepsy, learning problems may well have a different cognitive basis than in their non-epileptic counterparts, due to the specific influence of the epilepsy. This may contribute to the observed types of learning problems that are specific for epilepsy and that are different from those in other learning disabled children.

Cognitive impairment may thus be considered as a factor, mediating between epileptic conditions and the inability to learn in school, although this relationship is far from simple. Reading problems for example may be caused by defects in several levels of the information processing system such as visual scanning, the ability to transform information from one modality to an other, linguistic reasoning, and comprehension.[7,23,24]

Current research in children with epilepsy aims at analyzing the complex relationships between learning disabilities, underlying cognitive impairment and epileptic conditions. Factors such as the localization of the epileptogenic focus in specific (in particular temporal) areas of the brain, seizure activity (i.e. seizure type, seizure severity and seizure frequency), and central side-effects of the antiepileptic medication may underlie cognitive impairment and thus interfere with learning processes. These factors will be discussed here, in subsequent paragraphs.

The impact of seizure activity on cognitive function and learning

Direct effects of diurnal seizures are reported on several aspects of the information processing system: alertness, short-term learning and abstraction.[31] Prolonged post-ictal effects on cognitive functioning exist, but are more difficult to detect and therefore to accept for the child and his family. Nevertheless, a tonic-clonic seizure in the preceding 30 days may have a negative effect on test scores.[31,32] Cognitive impairment and lower performance on intelligence tests is found up to five hours after short periods (10s or less) with subtle seizures.[33] Conversely, Rodin et al.[34] and Seidenberg et al.[35] have found a steady rise in performance on intelligence tests for patients who became seizure free.

Nocturnal seizures are thought to have detrimental effects on language functions, on memory[36] and on alertness, possibly through the effects of disturbed sleep patterns.[37]

Seizure type comes out as an important risk factor, with most effects on the cognitive functions from tonic-clonic and complex partial seizures.[31,35,38,39] Nonetheless the effect of, e.g., absence seizures on cognitive function must not be ignored. An early onset of epilepsy with frequent seizures can interfere with brain development and may consequently have a long-term impact on cognition, by inhibiting of mitotic cell activity, affecting myelinisation and reducing cell numbers and cell size.[36] This factor may explain the poor prognosis for cognitive development in age-dependent encephalopathies, such as the West syndrome, as these developmental defects cannot be compensated for in later life-phases.[40] Apart from these effects, an early onset may also be equivalent with a greater number of seizures during life time that is suggested to be a risk factor for cognitive deterioration in a number of studies.[31,34,41] These findings suggest that seizure activity has a substantial impact on cognitive function. When evaluating separate cognitive factors, especially the attentional factors appear to be vulnerable for seizure activity. This is also in line with factor-analytical studies on Wechsler intelligence tests, performed on several groups of patients with intractable seizures.[42,43] These studies show the attentional subtest 'digit symbol' to be the most consistent predictor of neuropsychological impairment and show a relative prominence of the subtests: 'digit symbol', 'digit span' and 'arithmetic', all tests that are sensitive for distractibility.

An indirect approach to the relationship between seizure activity and cognition examines the effect of characteristic ictal events in the electro-encephalogram (EEG). Prolonged seizure discharge in the form of an electrical status or non-convulsive status epilepticus may result in a broad spectrum of impairment, reaching from total lethargy to decreased awareness. Repeated non-convulsive status may result in significant intellectual deterioration.[44] The extent of cognitive impairment during brief bursts of epileptiform discharges in the EEG varies with the number of spike components and the involvement of frontocentral regions.[45] Moreover impairment is more common during prolonged (> 3s) generalized spike-wave discharges than during focal activity.[46,47] An important additional factor appeared to be the level of task complexity, with more impairment in the mentally demanding type of tasks.[48] Especially the acquisition phase of memory (i.e., 'learning new words') is vulnerable if these discharges occur.

Brief episodes with epileptiform discharges are sometimes divided into activity accompanied by clinical symptomatology and 'subclinical' activity. Early research by, e.g., Gibs in 1936[49] and Schwab in 1939[50] has shown that 'subclinical activity' may not always be completely without impairment. Binnie[46] and Stores[44] have successfully demonstrated cognitive impairment during such periods with 'subclinical activity'. Even specific asymmetrical effects have been established with localized discharges.[46] For these events, the concept of transient cognitive impairment has been proposed,[44,46] which has proven to be of value in explaining episodic learning difficulties and fluctuations in cognitive function. In some investigations a positive association between atypical test profiles or impaired learning on the one hand and subclinical discharges in the EEG on the other hand have been established.[33] Transient cognitive impairment may also account for the test-retest variability that has been reported in IQ-testing of children with epilepsy.[35,41,51] Impaired learning may especially occur in tasks which are sensitive to the effects of subclinical discharges.

Localization of epileptogenic foci - effects on cognition and learning

The temporal lobe, and the neurophysiological-neurochemical mechanisms involved in the limbic system are known to be important in the mediation of cognitive and learning behaviour. One might thus expect that epilepsies originating in the temporal lobe are a special risk for cognitive development and learning behaviour. Indeed, a high incidence of cognitive impairment in children with temporal lobe epilepsy has been reported since the 1950's.[52] A very large body of literature now exists concerning the etiological importance of temporal lobe dysfunction in the psychological and social difficulties of children with epilepsy. Memory deficits, e.g., are seen as an inevitable consequence of temporal lobe dysfunction.[6,9,31,34,35] This may induce difficulties in the elementary reading process, as it may obstruct the development of word recollection or word recognition. The relationship between temporal lobe dysfunction and cognitive impairment may, however, be obscured by other factors. Many intractable epilepsies are characterized by epileptic foci of temporal lobe origin; seizure frequency is also higher in these patients; phenytoin, as a factor of drug treatment is also a common factor in these patients[34] and several reports suggest a detrimental effect of this drug on cognitive function. Hence, temporal lobe dysfunction, seizure frequency and antiepileptic medication may all be correlated factors in the same patients.

Our knowledge about the relationship between cognition and learning on the one hand and the localization of epileptogenic activity on the other hand is increased by our experiences with functional surgery for epilepsy. During preoperative WADA-testing, testing during depth-electrode stimulation or functional mapping during Penfield operations, relations between hemispheric dominance, localization of brain dysfunctioning and the interictal effects on learning and cognition can be studied in much greater detail. Nonetheless, we must take into account that, as yet, most of these studies involve adults, whereas cerebral organization does not necessarily follow the same patterns in children.

Functional brain imaging techniques, such as the Positron Emission Tomography (PET), Single Photon Emission Computerized Tomography (SPECT) and functional Magnetic Resonance Imaging (MRI) have demonstrated that epileptogenic foci are often characterized by lesions, based on functional (e.g., regional cerebral blood flow) or structural pathological changes of specific parts of the brain. The rCBF-SPECT displays foci as 'cold' lesions with interictal hypoperfusion and ictal hyperperfusion.[53,54] In accord with neuropsychological studies of, e.g., CVA-lesions, asymmetrical effects have been obtained. Epileptogenic foci in the left hemisphere appeared to be associated with impaired verbal learning, verbal memory, and serial information processing. Naming problems are more prevalent with left posterior EEG-abnormalities. In contrast, right hemispheric foci produce alterations of performance in visual-spatial tasks, directed attention, modulation of affect, paralinguistic aspects of communication and parallel information processing. The finding of deficits in attention and general activation in generalized epilepsies is sometimes interpreted as an effect of the non-specific function of the thalamic system.[6,32,36,,38,39,41,46,53,55] Current experimental studies show that, whereas memory problems mostly have been attributed to bilateral lesions in the hippocampus, severe amnesia may be caused by unilateral impairment of specific structures outside the hippocampus.[56] Kertesz,[55] e.g., points to the involvement of the prefrontal cortex. The findings with rCBF-SPECT in several studies also indicate that much larger areas than the hippocampal-limbic circuits are involved in memory dysfunction, situated particularly in the frontal structures.

Cognitive side-effects of antiepileptic treatment

During the last decade our understanding of the impact of antiepileptic drugs (AED) on cognitive functioning has increased considerably.[57] A large number of studies suggest that AED treatment has a much greater impact on cognitive function than had hitherto been suspected.[58,59] In addition to the deteriorating effects of polypharmacy, as opposed to monotherapy, differential effects of specific drugs on higher cerebral functions have been obtained. Phenytoin for instance shows an overall trend towards poorer performance when compared to carbamazepine or to some of the newer antiepileptic drugs.[57-59] Especially slowing of information processing is mentioned as a central effect of phenytoin, possibly caused by slowed central conduction, as measured with visual evoked responses.[60]

Nevertheless, substantial effort will be needed to obtain a more complete model of the relationships between specific drugs and cognitive impairment. A review of our current knowledge of cognitive effects of AED treatment in children is given elsewhere in this volume.

Additional issues have been advanced. Recently, some attempts have been made to investigate the significance of pharmacokinetic properties of antiepileptic drugs for cognitive functioning. In carbamazepine, transient cognitive deficits have been detected in relation to high peak serum levels. The pharmacokinetic profile of this drug, characterized by rapid and marked fluctuations in serum levels, may differentially affect test performance across short periods during the day. A more stable pattern of cognitive functioning was found with a controlled release preparation.[61] The clinical relevance of this finding was illustrated by the choice of patients to use the controlled release preparation after the experiment, even though some of them had a higher seizure frequency.

To explain this, we may point to the impact of drug-treatment on the subjective evaluation of well-being or 'quality of life' by the patient himself. Up to now, our understanding of the impact of drug treatment on behaviour is mainly derived from experimental studies. Occasionally, the patients' subjective well-being is taken into consideration. Yet, it seems quite important to explain the differential extent to which patients accept certain side-effects of their treatment in exchange for more seizure control. This issue is further discussed in other chapters.

ISSUES OF ASSESSMENT AND TREATMENT

Progress in understanding the learning problems of children with epilepsy is achieved with the development of adequate assessment techniques. The integration of microcomputer technology and the use of combined EEG-recording and cognitive testing may be crucial factors. This is especially important as computerized cognitive assessment supports a more accurate evaluation of epileptiform EEG-correlates for fluctuations in learning behaviour. Several earlier mentioned issues require some form of EEG-recording during psychological testing or during monitoring of learning behaviour. Stores[44] and Binnie[46] recommended EEG monitoring during psychological testing in any child, who shows inconsistent performance in his/her learning behaviour. The possibility of connecting the computer with EEG equipment permits on-line monitoring of both parameters without time-delay, using feedback impulses from the computer. This system can be coupled to brain mapping facilities, e.g. to evaluated event-related evoked responses, such as the 'cognitive' generated P-300, which may enhance our understanding of cognitive impairment in relation to lateralized discharges in children with learning impairment. On-line and combined computerized cognitive testing and EEG-recording may also provide us the accuracy, necessary for simultaneously testing with depth electrode stimulation[62] or with functional brain imaging.[63] Recently several systems for computerized cognitive testing in epilepsy became available (see the review of these systems by Dodson and Kinsbourne[64]).

Nonetheless, we are still in the process of implementing these new techniques in the assessment of children with epilepsy. An early warning was provided by Stores[2] reporting that teachers' rating of inattentiveness in the class room did not correlate with any of the cognitive tests for measuring attention. This illustrates the need for further evaluation of the relationships between cognitive deficits and the actual problems that occur in the class room.

There is limited experience with special learning techniques for the learning impaired child with epilepsy, although some findings clearly indicate that subgroups of children with epilepsy may have special needs with regard to the type of educational support.[26] Stores, e.g., found improvements in attention with sensory stimulation in children with epilepsy. Some of the children in his study seemed to be chronically 'underaroused'. Stimulation, through the introduction of extra noise by earphone had an alerting effect on their performance.[2] This may be related to the finding that a high level of arousal generally reduces epileptiform activity. Likewise, some evidence exists that extra stimulation may also be beneficial in drug-induced slowing of information processing.

On the basis of these and subsequent observations, specific training programs have to be developed, further tailored to the needs of specific 'types' of learning disabilities, such as discussed in the first paragraphs of this chapter. Up to now, there is limited experience with the development of such learning techniques, and to our knowledge, systematic trials are completely absent. Consequently, the special schools for children with epilepsy provide programs developed for the general population of learning disabled. This may explain the decrease in referral rates to special schools for epilepsy. The number of children referred to special schools for epilepsy in the U.K., e.g., decreased with 80% in approximately 30 years.[25,65] At this moment mental handicaps and behaviour disorders are the most frequent reason for referral,[66] whereas 'seizures' account for only 21% of the cases. Besag[26] recommends that these schools specialize for a selected group and play an important role as a research and resource facility. Another option is to offer specialized educational help to regular schools, where most children with epilepsy are educated. In both the U.K. and The Netherlands, mobile epilepsy teams provide special help for learning disabled children with epilepsy who are in regular schools. This decentral approach is also more in accord with recent legislation and is encouraged by governmental institutions.

Some studies give directions for the development of specific learning techniques. Firstly, when organizing educational support we come to the heterogeneous nature of the condition immediately. Therefore, single factor theories lack essential information and methods of treatment require integration of several disciplines. Rarely such attempts of multidisciplinary approach have been made.[25,26,65] Even more seldom are proposals for practical treatment with a multidisciplinary approach. Possibly, the concept of the case-conference, as suggested by Strang[8,22] may be used for further experiments. Programs for function training have been developed, especially for disorders in problem solving, cognitive style[3,4] and for memory impairment.[67] Although some elements of these programs are now operative, they are still in the process of long-term evaluation and implementation in school practice.

The information on differential effects of AED treatment may enlarge therapeutical possibilities, as both efficacy and the expected impact on cognitive development of a specific drug may be equally valued during treatment. Accordingly, the first choice of drug may sometimes be replaced by an alternative drug with less side-effects. Where this is not possible, we may focus on adaptations in drug administration, such as the use of controlled release preparations. In some trials, add-on treatment, using drugs with assumed beneficial effects are tested. Among others, experiments have been initiated with amphetamines[68] with nootropic drugs (both in learning disabled children[69] and in memory impaired patients with epilepsy[70]) and with antihypoxic compounds.[71] The agents of the nootropic class are thought to enhance neuronal efficiency, especially when this is impaired by hypoxia or other noxious influences to which especially the telencephalic structures are sensitive.[69]

SOCIOECONOMIC STATUS

Although somewhat beyond the scope of this chapter, we may consider briefly the 'endpoint' of several critical influences and focus on the social-economic status of the adolescent and young adult with epilepsy. Being able to obtain and maintain a satisfactory job and an income is obviously relevant to an individual's psychosocial functioning, if only because unemployment introduces economic pressure, and may reduce the opportunities for social interaction and leisure activities. Unfortunately, unemployment and underemployment of people with epilepsy are much more frequent than in the general population. The unemployment rate for people with epilepsy is two times the national average in the United States.[72] Many young adults with epilepsy (estimated up to 50%[73]) experience problems finding work and it is by no means uncommon that people loose their job because of seizures. Not surprisingly, people with epilepsy generally have lower than average income.

The relationship between epilepsy and lower socioeconomic status is, however, complex. Characteristics of the seizures may be such that they limit an individual's employment opportunities. Bothersome personality or behavioral characteristics may contribute to difficulties with employment. Cognitive functioning may be a significant factor in determining successful or unsuccessful employment status. Neuropsychological investigations of epilepsy have found, for example, that measures of higher cortical function predict vocational status and adequacy of psychosocial functioning.[74,75] Also, epilepsy continues to be associated with considerable stigma and ignorance, manifesting itself in various forms of social discrimination, e.g. difficulties in obtaining a driver's license, discrimination in obtaining employment,[73] difficulty in obtaining all types of insurance. The National Commission for the Control of Epilepsy and its Consequences (1978) has outlined the societal sanctions in greater detail. Such sanctions are conducive to social exclusion and ostracism, which may result in limited opportunities for extended social contact. Public attitudes towards epilepsy and misconceptions about this condition may go a long way in accounting for the difficulties experienced in getting employment.

REFERENCES

1. Utter M, Graham P, Yule WA. A neuropsychiatric study in childhood. Clinics in Developmental Medicine. London: Heinemann, 1970.
2. Stores G. Schoolchildren with epilepsy at risk for learning and behaviour problems. Devel. Medicine and Child Neurology, 1978; 20: 502-508.
3. Thompson PJ. Educational attainment in children and young people with epilepsy. In: Oxley J, Stores G, editors. Epilepsy and Education. London: The Medical Tribune Group, 1987: 15-24.
4. Aldenkamp AP. Epilepsy and learning behavior. In: Parsonage M, Grant RHE, Graig AG, et al., editors. Advances in Epileptology. New York: Raven Press, 1983: 221-229.
5. Aldenkamp AP. Learning disabilities in epilepsy. In: Aldenkamp AP, Alpherts WCJ, Meinardi H, et al., editors. Education and Epilepsy. Berwyn: Swets & Zeitlinger, 1987: 21-38.
6. Aldenkamp AP, Alpherts WCJ, Dekker MCA, et al. Neuropsychological aspects of Learning disabilities in Epilepsy. Epilepsia, 1990; 31(S4): 9-20.
7. Rourke BP. Neuropsychology of learning disabilities. Essentials of subtype analysis. New York: Guilford, 1985.
8. Strang J. Educational and related treatment considerations concerning the child with epilepsy: a developmental neuropsychological approach. In: Aldenkamp AP, Alpherts WCJ, Meinardi H, et al., editors. Education and Epilepsy. Berwyn: Swets & Zeitlinger, 1987: 118-135.
9. Green JB, Hartlage LC. Comparative performance of epileptic and non-epileptic children and adolescents. Dis. Nerv. Syst., 1971; 32: 418-421.
10. Holdsworth L, Whitmore K. A study of children with epilepsy attending ordinary schools. Devel. Med. and Child Neur., 1974; 16: 746-758.
11. Ross EM, Tookey P. Educational needs and epilepsy in childhood. In: Trimble MR, Reynolds EH, editors. Epilepsy, Behaviour and Cognitive Function. Chistester: John Wiley & Sons, 1988: 87-97.
12. Seidenberg M, Beck N, Geisser M, et al. Academic achievement of children with epilepsy. Epilepsia, 1986; 29: 753-759.
13. West P. An investigation into the social construction and consequences of the label epilepsy. Sociological Review, 1979; 27: 719-741.

14. Ward F, Bower BD. A study of certain social aspects of epilepsy in childhood. Developmental Medicine and Child Neurology, 1978; 39: 1-50.
15. Suurmeijer TPBM, Dam A van, Blijham M. Socialization of the child with epilepsy and school achievement. In Meinardi H, Rowan J, editors. Advances in Epileptology. Lisse: Swets & Zeitlinger. 1978: 46.51.
16. Long CG, Moore JL. Parental expectations for their epileptic children. Journal of Child Psychology and Psychiatry, 1979; 20: 299-312.
17. Fenton GW. Personality and behavioral disorders in adults with epilepsy. In: Reynolds EH, Trimble MR, editors. Epilepsy and Psychiatry. Edinburgh: Churchill Livingstone, 1981: 77-91.
18. Lechtenberg R. Epilepsy and the family. Boston: Harvard University Press, 1984.
19. Aldenkamp AP, Vermeulen J, Alpherts WCJ, et al. Validity of computerized testing: patient dysfunction and complaints versus measured changes. In: Dodson WE, Kinsbourne M. Assessment of cognitive function. Demos, New York, 1992: 51-68.
20. Alpherts WCJ, Aldenkamp AP. Computerized neuropsychological assessment in children with epilepsy. Epilepsia, 1990; 31(S4): 35-40.
21. Vermeulen J, Kortstee SWAT, Alpherts WCJ, et al. Cognitive performance in learning disabled children with and without epilepsy. Seizure, 1994; 3: 13-21.
22. Strang JD. Cognitive deficits in children: adaptive behavior and treatment techniques. Epilepsia, 1990; 31(S4): 54-59.
23. Ceci SJ. Handbook of cognitive, social and neuropsychological aspects of learning disabilities; Volume 1 and 2. London: Lawrence Erlbaum Associates Publishers, 1986.
24. Strang J, Rourke BP. Adaptive behavior of children who exhibit specific arithmetic disabilities and associated neuropsychological abilities and deficits. In: Rourke BP, editor. Neuropsychology of learning disabilities; essentials of subtype analysis. New York: Guilford, 1985: 302-331.
25. Radley R. The Educational needs of children with epilepsy. In: Oxley J, Stores G, editors. Epilepsy and Education. London: The Medical Tribune Group, 1987: 9-15.
26. Besag F. The role of a special school for children with epilepsy. In: Oxley J, Stores G, editors. Epilepsy and Education. London: The Medical Tribune Group, 1987: 65-73.
27. Stores G, Hart J. Reading skills of children with generalised and focal epilepsy attending ordinary school. Developm. Med. Child Neur., 1976; 18: 705-716.
28. Bagley CR. The educational performance of children with epilepsy. British Journal for Educational Psychology, 1970; 40: 82-83.
29. Ross EM, West PB. Achievement and problems of British eleven year olds with epilepsy. In: Meinardi H, Rowan, J. editors. Advances in Epileptology. Lisse: Swets and Zeitlinger, 1978: 24-34.
30. Jennekens-Schinkel A, Linschooten-Duikersloot EMEM, Bouma PAD, et al. Spelling errors made by children with mild epilepsy: writing-to-dictation. Epilepsia, 1987; 28: 555-563.
31. Dodrill CB. Correlates of generalized tonic-clonic seizures with intellectual, neuropsychological, emotional and social function in patients with epilepsy. Epilepsia, 1986; 27: 399-411.
32. Baird HW, John ER, Ahn H, et al. Neurometric evaluation of epileptic

children who do well and poorly in school. Electroencephalography and Clinical Neurophysiology, 1980; 48: 683-693.
33. Aldenkamp AP, Gutter T, Beun AM. The effect of seizure activity and paroxysmal electroencephalographic discharge on cognition. Acta Neurologica Scandinavica, 1992; 86: 111-122.
34. Rodin EA, Schmaltz S, Twitty G. Intellectual functions of patients with childhood-onset epilepsy. Developmental Medicine & Child Neurology, 1986; 28: 25-33.
35. Seidenberg M, O'Leary DS, Giordani B, et al. Test-retest changes of epilepsy patients: assessing the influence of practice effects. Journal of Clinical Neuropsychology, 1981; 3: 237-255.
36. Renier WO. Restrictive factors in the education of children with epilepsy from a medical point of view. In: Aldenkamp AP, Alpherts WCJ, Meinardi H, et al., editors. Education and Epilepsy. Berwyn: Swets & Zeitlinger, 1987: 3-14.
37. Declerck AC, Linden I van, Oei LT, et al. Are learning problems in children with epilepsy due to a disturbed sleep? Abstract 18th Intern. Epilepsy congress (book of abstracts), New Delhi, 1989.
38. Brittain H. Epilepsy and intellectual functions. In: Kulig BM, Meinardi H, Stores G, editors. Epilepsy and Behavior. Lisse: Swets and Zeitlinger, 1980: 2-13.
39. Klove H, Matthews CG. Neuropsychological studies of patients with epilepsy. In: Reitan RM, Davison LA, editors. Clinical Neuropsychology. New York: John Wiley, 1974: 237-267.
40. Aicardi J. Epileptic syndromes in childhood. Epilepsia, 1988, 20 (Supl) 1-5.
41. Bourgeois BFD, Presky AL, Palkes HS, et al. Intelligence in epilepsy: a prospective study in children. Annals of Neurology, 1983; 14: 438-444.
42. Kupke T, Lewis R. WAIS and neuropsychological tests: common and unique variance within an epileptic population. Journal of Clinical and Experimental Neuropsychology, 1985; 7: 353-366.
43. Bornstein RA, Drake jr ME, Pakalnis A. Wais-R factor structure in epileptic patients. Epilepsia, 1988; 29: 14-18.
44. Stores G. Effects on learning of 'subclinical' seizure discharge. In: Aldenkamp AP, Alpherts WCJ, Meinardi H, et al., editors. Education and Epilepsy, Berwyn: Swets & Zeitlinger, 1987: 14-21.
45. Mirsky AF. Studies of paroxysmal EEG phenomena and background EEG in relation to impaired attention. In: Evans CR, Mulholland TB, editors. Attention in Neurophysiology, London: Buttersworths, 1969: 310-322.
46. Binnie CD. Seizures, EEG discharges and cognition. In: Trimble MR, Reynolds EH, editors. Epilepsy, Behaviour and cognitive function, New York: John Wiley & Sons, 1987: 45-51.
47. Binnie CD. Seizures, EEG discharges and cognition. In: Trimble MR, Reynolds EH, editors. Epilepsy, Behaviour and cognitive function, New York: John Wiley & Sons, 1987: 45-51.
48. Aarts JHP, Binnie CD, Smit AM, et al. Selective cognitive impairment during focal and generalised epileptiform EEG activity. Brain, 1984; 107: 293-308.
49. Gibbs FA, Lennox WG, Gibbs EL. The electroencephalogram in diagnosis and in localisation of epileptic seizures. Arch. Neurol. Psychiatr., 1936; 36: 1225-1235.
50. Schwab RS. A method of measuring consciousness in petit mall epilepsy.

Journ. Nerv. Ment. Dis., 1939; 89: 690-691.
51. Aldenkamp AP, Alpherts WCJ, De Bruïne D, et al. Test-retest variability in children with epilepsy - a comparison of Wisc-R subtest profiles. Epilepsy Research, 1990; 7: 165-172.
52. Gibbs FA, Stamps FW. Epilepsy Handbook. Springfield Ill: Charles C. Thomas, 1953.
53. Verhoeff NPLG, Weinstein H, Aldenkamp AP, et al. Focus Localization in patients with partial epilepsy with ^{99}TC-^{m} HMPAO SPECT under continuous surface EEG monitoring. Nuclear Medicine Communications, 1992; 13: 127-136.
54. Sperling MR, Sutherling WW, Nuwer MR. New techniques for evaluating patients for epilepsy surgery. In: Engel J Jr, editor. Surgical treatment of the epilepsies, New York: Raven Press, 1987: 235-257.
55. Kertesz A. Localization in Neuropsychology. New York: Academic Press, 1983.
56. Mellanby J, Hawkins C, Wilks L. The relationship between seizures and amnesia in experimental epilepsy. Acta Neurol. Scand., 1984; 99: 119-124.
57. Aldenkamp AP, Alpherts WCJ, Blennow G, et al. Withdrawal of antiepileptic medication - effects on cognitive function in children - the results of the multicentre 'Holmfrid' study. Neurology, 1993; 43(1): 41-51.
58. Trimble M. Anticonvulsant drugs: mood and cognitive function. In: Trimble MR, Reynolds EH, editors. Epilepsy, Behaviour and Cognitive Function, Chichester: John Wiley & Sons, 1988: 135-145.
59. Trimble M, Thompson PJ. Anticonvulsant drugs, cognitive function and behavior. Epilepsia, 1983; 24: 55-63.
60. Green JB, Walcoff M, Lucke JF. Phenytoin prolongs far-field somatosensory and auditory evoked potential interpeak latencies. Neurology, 1982; 32: 58-88.
61. Aldenkamp AP, Alpherts WCJ, Moerland MC, et al. Controlled Release Carbamazepine: Cognitive Side-effects in Patients with Epilepsy. Epilepsia, 1987; 28(5): 507-514.
62. Fenwick PBC. Seizures, EEG discharges and behaviour. In: Trimble MR, Reynolds EH, editors. Epilepsy, Behaviour and Cognitive Function, Chichester: John Wiley & Sons, 1988: 51-67.
63. Lassen NA, Roland PE. Localization of cognitive function with cerebral blood flow. In: Kertesz A, editor. Localization in Neuropsychology, New York: Academic Press, 1983: 141-153.
64. Dodson WE, Kinsbourne M, editors. Assessment of cognitive function. New York: Demos, 1992.
65. Aldenkamp AP, Das Gupta A, Saxena VS, editors. Epilepsy and Education; educational policies for the child with epilepsy. Bombay: Reckitt & Colman of India Limited, 1990.
66. Corbett J. Epilepsy as part of a handicapping condition. In: Ross E, Reynolds EH, editors. Paediatric Perspectives in Epilepsy, Chichester: John Wiley & Sons, 1985: 78-79.
67. Aldenkamp AP, Vermeulen J. Neuropsychological rehabilitation of memory function in epilepsy. Journal of Neuropsychological Rehabilitation, 1992; 1(3): 199-214.
68. Henriksen O. Education and Epilepsy: assessment and remediation. Epilepsia, 1990; 31(S4): 21-26.
69. Ashton H. Brian systems, disorders and psychotropic drugs. Oxford: Oxford

University Press, 1987.
70. Aldenkamp AP, Van Wieringen A, Alpherts WCJ, et al. Double-blind placebo-controlled, neuropsychological and neurophysiological investigations with oxiracetam (CGP 21690E) in memory-impaired patients with epilepsy. Neuropsychobiology, 1991; 24(2): 90-102.
71. Aldenkamp AP, Overweg J, Smakman J, et al. The effect of R 58 735 (Sabeluzole) on memory functions in patients with epilepsy. Neuropsychobiology, in press.
72. So EL, Penry JK. Epilepsy in adults. Annals of Neurology, 1981; 9: 3-16.
73. Fraser RT. Vocational aspects of epilepsy. In Hermann BP, editor. A Multidisciplinary Handbook of Epilepsy. Springfield Ill: Charles C. Thomas, 1980: 74-105.
74. Dikmen S, Morgan SF. Neuropsychological factors related to employability and occupational status in persons with epilepsy. Journal of Nervous and Mental Disease, 1980; 168: 236-240.
75. Dodrill CB. Interrelationships between neuropsychological data and social problems in epilepsy. In Canger R, Angeleri F, Penry JK, editors. Advances in epileptology. New York: Raven Press, 1980: 191-197.

14 The impact of epilepsy on behaviour and emotional development

PAMELA J. THOMPSON
Chalfont centre for epilepsy, Chalfont St. Peter, U.K.

Chronic illness in childhood can have a damaging impact on other aspects of a child's life as well as health. Epilepsy brings with it additional characteristics which threaten the stability of the family and the emotional development of the child. The impact of epilepsy will depend on a number of factors related to the seizures, their treatment, the social environment of the child and the developmental stage when the seizures begin. Of all the mediating variables reviewed, the most potent for emotional development are environmental factors, namely the approach and attitude of the family, the school and the wider public. Approaches which focus on educating and supporting the child's family and others in the child's social environment are of vital importance. To conceive of epilepsy as a largely medical health problem with scant concern for its psychological impact may seriously compromise the emotional development of the child.

INTRODUCTION

Children require a stable upbringing to develop emotionally and anything which threatens that stability may have negative consequences for subsequent development and psychological health. Epilepsy, with the unpredictable and at times intense nature of its symptoms, poses such a threat. Available evidence provides some support for this hypothesis, with studies reporting elevated levels of emotional problems and behavioural disturbance in children and adolescents with epilepsy in comparison with other chronic conditions and healthy controls.[1-4] Specific problems highlighted include increased dependency,[5-8] low self-esteem and self-concept,[7,9-11] social isolation,[12] neurosis,[13] aggression and overactivity.[14,15] It would be unwise and misleading to make sweeping statements about the impact of epilepsy on emotional and behavioral development on the basis of these findings. Clearly, as this review will demonstrate, mediating factors exist which can exert a positive or negative influence on childhood development.

SEIZURE RELATED FACTORS

Aetiology

Epilepsy can arise as a consequence of a wide range of brain pathologies.

There exists a relationship between brain damage and brain immaturity and behavioral disturbance in childhood. In many instances behavioral difficulties in childhood epilepsy arising in the context of brain damage are more a consequence of the underlying cerebral disturbance than of the seizure disorder.[16] This may be a direct effect or secondary to an impact on cognitive development. For example, children who sustain damage to the left cerebral hemisphere may as a consequence have delayed or arrested language development. Language problems are associated with an increased risk of behavioral problems.[17] Regrettably, too often behavioral difficulties in the child with epilepsy are attributed by the family and often the physician to the occurrence of seizures rather than the existing brain damage. When seizures become well controlled but the behaviour difficulties persist then the medication may be unfairly blamed.

Seizure frequency

Intuitively frequent seizures might be expected to negatively impinge upon development. However the research evidence in support of this is far from overwhelming. Some authors have reported a relationship although not always a strong one[2,4,18] but others have found no statistically significant association.[1,19] Even if a relationship exists, it is unclear whether this is a primary effect or secondary to increased exposure to psychosocial factors such as stigma and social isolation.[18]

Seizure type

Nuffield[20] proposed children with typical absences tended to present with neurotic type behaviours whereas those with temporal lobe discharges presented with more antisocial behaviours including aggression. Rutter et al.[1] reported an association between elevated rates of behaviour problems and complex partial seizures. Stores and Piran[6] found that boys with left temporal spike discharges showed significantly greater need for emotional and physical contact with their mothers than children with generalised epilepsy or those with seizure activity localised to the right hemisphere.

Whitman et al.[21] assessed a sample of 123 children with epilepsy attending a tertiary referral centre. When the sample was classified into temporal lobe or generalised cases, no differences emerged between the groups on parental or teachers ratings of behaviour. The authors did report an association between anterior temporal lobe spike discharges arising from either hemisphere in males and aggressive behaviour. The authors however rightly insert a note of caution in their discussion since this was one of only three significant findings out of a total of almost ninety correlations performed.

TREATMENT

Antiepileptic drugs

Attention to the effects of antiepileptic medication has tended to focus on cognitive rather than emotional development.[22] However some evidence does exist that antiepileptic drugs can influence other behaviours adversely. The drug most implicated in this regard is phenobarbitone. Studies have suggested an association between this treatment and restlessness, irritability, sleep difficulties, low self-esteem and other depressive symptomatology.[23,24] There is some support for the view that carbamazepine and sodium valproate are associated with a low incidence of behavioral side effects.[22] Studies, however, are few particularly those of the more recently introduced medications lamotrogine and vigabatrin.

Hermann et al.[18] reported an association between polypharmacy and increased rates of behavioral problems in children. Cull[25] provides additional evidence for the adverse behavioral effects of multiple drug therapy. The study reported followed up children either having drugs added or withdrawn from their regimens. A decrease in the number of drugs prescribed was associated with an improvement six months later in behaviours rated as inappropriate and anti-social. Antiepileptic drugs may have an impact on emotional development secondary to a negative effect on cognitive development or via adverse cosmetic changes.

Surgery

Total or markedly improved seizure control is the main aim of surgical programmes. Parents and physicians, however, often assume surgery will result *"in more normal subsequent development with greater opportunities for academic and vocational growth than would have been the case otherwise"*.[4] While the available evidence provides some support for this belief, the studies are few, often retrospective with limited measurement of behaviour beyond the anecdotal. Vaernet[26] presents findings from 33 temporal lobectomy cases aged between 4 and 21 at the time of surgery. He highlights substantial improvements in behaviour postoperatively particularly for those children with conduct disorders. The observed behavioral improvements were not entirely related to improved seizure control.

Corpus callosotomy and hemispherectomy are two other surgical procedures undertaken in childhood epilepsy although much less frequently than temporal lobe operations. Even less systematic research has been undertaken with these cases. The majority of children have preexisting emotional problems. Available evidence suggests some individuals do benefit in behavioral terms.[27,28]

ENVIRONMENTAL

The family

The family is the first and most important environmental influence over the child's behaviour. Where the child has epilepsy perhaps there is a greater need for a stable family life given the unstable nature of the condition and the uncertainty surrounding the prognosis.

"Families are miniature societies in which children make their first attempts at adapting to others and in which they learn patterns of behaviour which tend to persist throughout life. The child's family should facilitate development from a state of complete dependency in infancy to one of independence from the family in adult life".[16]

Experts in child development are in general agreement that rates of emotional disturbance are elevated in the context of a disturbed family background. This also seems to be the case for childhood epilepsy.[29,30] Hermann et al. found rates of behavioral problems were higher in the presence of marital disharmony and divorce.[18] Upon reflection, a relationship between childhood epilepsy and family instability is not surprising. A diagnosis of epilepsy can have a negative impact on the emotional security of the family. Adjustment to the diagnosis and the loss of the 'perfect child' may develop in time. The family, however, also has to live with the threat of ongoing seizures. Living with this threat can be a source of major stress which can serve to undermine normal family functioning. Many families articulate feelings of helplessness and concern that the endpoint of an attack may be injury or death. Such feelings are expressed even where seizures are under good control.[31] Understandably, if parents experience high levels of anxiety, this will influence their approach to the child with epilepsy.

Reports in the literature exist of relationships between the parenting style adopted and measures of emotional adjustment in the child. Overprotection fostering increased dependency in the child has been cited as a common parenting style in epilepsy.[6,7,32] Overprotection on the part of the parents can result in the curtailing of experiences both social and emotional essential for normal development. This includes positive experiences such as playing with peers and participating in sporting activities and negative experiences such as appropriate discipline following behavioral misdemeanours. The child needs to be exposed to a range of emotions and situations as they will in adult life. The effects of overprotection may vary but can lead to increased self-doubt, lack of confidence, selfishness, rebelliousness and manipulativeness. Professionals are often critical of the overprotective parent. The distinction, however, between being protective and overprotective is by no means clear cut.[32]

Emotional skewing is another parenting style reported in the literature. This results in a disproportionate amount of time being devoted to the child with epilepsy. Resentment from siblings may follow, which in itself can have a disruptive impact on family stability. Mulder and Suurmeijer[12] report the findings from in depth interviews with 13 families. The parents expressed a need to spend more time with the child with epilepsy at the expense of time spent with their other children. Ward and Bower[33] reported jealousy in siblings arising from perceived neglect and resentment for extra responsibilities assigned to them involving overseeing their brother or sister with epilepsy. If too much is asked of a child in terms of self-control and responsibility, this can result in increased rates of behavioral difficulties.

Some families attempt to conceal the epilepsy, often as a consequence of the perceived stigma attached to the condition. The shame engendered by having epilepsy within the family may be so great that even mention of it becomes taboo. Shame about epilepsy may easily be internalised by the child with negative consequences for the development of self-identity and self- esteem.

Expectations of the parents about the ability of the child may similarly be internalised and influence self-esteem. In a study of nineteen families, parental expectations of children with epilepsy were lower than for healthy siblings.[7] Parents expected the affected child to develop emotional problems, to do less well at school academically and at sports. Lowered goals have been reported by other investigators.[5]

More recent studies in the literature focus on both positive and negative approaches, adopted by families. These are to be welcomed as they draw attention to the fact that many families cope well and have emotionally well adjusted children. Austin and McDermott[34] studied 27 parents with children with epilepsy ranging 6-16 years with a mean seizure frequency of 6.7 per month. These authors found positive parental attitudes and coping styles existed and these correlated with good family adjustment. Conversely negative attitudes correlated with poor family adjustment. A relationship was reported between a longer duration of epilepsy and positive attitudes and coping styles. This was interpreted as indicating that over time it may be possible for parents to develop more positive reactions to their child with epilepsy.

Hermann and Austin[4] report the findings of a study which explored factors associated with good and poor adaptation in a sample of 128 children, 8-12 year old. All children had epilepsy of at least one years duration. Approximately one third of the sample were classified as good adaptors and one third as poor adaptors. The strongest discriminators between these two subgroups were family-esteem and communication, extended social support, family adaptation ,financial well-being and family stress. High ratings for the first four variables and low ratings for the fifth were associated with good emotional adjustment.

Using behavioral observations of mother and child dyads, Lothman and colleagues[35] explored the relationship between family dynamics and behavioral adjustment. Twenty children with epilepsy, aged from 8-12 years, were observed undertaking a card sorting and a constructional task in the presence of their mothers. All children had experienced seizures for at least five years. The mothers behaviour was rated globally with regard to provision of emotional support, respect for autonomy and quality of instruction and at a more discrete behavioral level with regards to amount of eye contact and the giving of feedback and hints. Maternal approaches and behaviours rated as positive were associated with high ratings of confidence, flexibility and independent problem solving skills in the child and good behaviour in the classroom. On the other hand, maternal approaches considered undermining were associated with passive ill-directed problem solving behaviours in the child and teachers ratings of problem behaviours in the classroom. The authors suggest that mothers who become overinvolved with a child impede the ability of the child *"to internalise a sense of themselves as effective, active agents in coping with situations and demands"*.

The school

A child's school environment will also exert a significant influence over subsequent development. Less work has been undertaken on the role of the school in relation to epilepsy, but some associations have been noted, in particular in relation to the attitudes of teachers and classmates. Children who had seizures at school were reported in one study to be more prone to significant depression compared to children who experienced seizures outside school hours.[36] The main cause of such depressive reactions was proposed to be rejection by classmates and their parents, due to a fear of epilepsy. Ross and Peckham[38] present findings from an epidemiological project, the National Child Development Study, and propose school refusal and truancy amongst an epilepsy sample attending normal school to be the result of low self-esteem, arising from bullying or teasing. Kato and co-workers[39] presented four cases of school refusal in children with epilepsy. Like the above authors they cite being shunned by classmates as a contributory factor.

In a survey of practising teachers in the UK, lack of knowledge about epilepsy was noted by 70% of the respondents. The teachers reported insufficient documentation and communication regarding specific children. Most also reported that their previous training in the area was lacking. Indeed in a separate survey by the same author of teacher training colleges, the amount of input about epilepsy was upwards of five minutes.[40]

Society

The wider social environment is also important. Stigma about epilepsy will have a negative impact on the attitudes of the family, school and individuals in the child's local neighbourhood.

DEVELOPMENTAL STAGE

The stage of development during which epilepsy begins might be expected to exert a dominant influence on subsequent development. Research studies have not adequately addressed this issue.

During adolescence, children normally spend increasingly more time with their friends and less time with their parents and siblings. This process may be delayed or even suspended as a consequence of parental anxieties when a young person has epilepsy. Furthermore, there are those children whose social lives already have been restricted, who enter adolescence without a peer group to provide a mechanism of separating from parents. Adverse effects of antiepileptic drugs, most notably cosmetic effects, may present an additional burden. Drug induced weight gain or gum hypertrophy may negatively influence an individuals' self-image at a stage of life when there is much peer group and media pressure for the *perfect body*. The formation of self-identity at this time will also be affected by the chosen career path. Moreover, the adolescent with epilepsy may have to face significantly reduced prospects. The inability to obtain a drivers licence may be a particularly bitter blow for many young people.

Viberg[10] studied sixteen young people with epilepsy and sixteen healthy controls using interviews, questionnaires and the Thermatic Apperception Test (TAT), a projective test. All participants attended mainstream schools. The epilepsy group had a range of seizure types but they did not represent a particularly intractable group as ten had been seizure free in the year prior to the study. The authors emphasised a high level of emotional adjustment difficulties including low self-esteem and raised stress in their epilepsy group. This was more evident in their response to the TAT than in their interview or questionnaire responses.

Kaminer and associates report a study which compared rates of psychological disturbance in adolescents with either epilepsy or bronchial asthma.[41] All participants were interviewed by a child psychiatrist. There were no differences between the groups on the measures of psychological adjustment employed. The authors however note that within each group there were individuals with signs of emotional disturbance.

Clement and Wallace[42] interviewed 58 children, 12-17 years old. All attended a tertiary referral centre and thus were representative of the severer end of the epilepsy spectrum. Low rates of difficulties were reported. Westbrook and colleagues[43] report a study of adolescents with relatively well controlled uncomplicated epilepsy, representatives of the other end of the severity spectrum. The interview responses of 13-19 year old children with epilepsy were compared with similarly aged groups of children with other chronic illnesses and healthy controls. No major differences were reported with the exception of raised levels of perceived stigma in the epilepsy group. The findings of these last two studies must be viewed cautiously as assessment of emotional adjustment was limited and superficial in comparison with the earlier studies cited.

POSITIVE APPROACHES

Aetiology

A good understanding of the consequences of the aetiology of epilepsy, when this is known or suspected, is important. Implications for behavioral and cognitive development can then be anticipated and intervention instigated at the earliest opportunity. Cognitive difficulties are hidden deficits which can easily be overlooked and adversely effect development.

Treatment

For all children optimal treatment needs to afford maximal seizure control with minimal side effects. Such a balance is not always easy to maintain. Where surgical treatment is a viable option families should have access to counselling which can explore their expectations. Clinically it is not uncommon to find widely unrealistic hopes regarding expected post-operative behavioral change. Where drug or surgical treatment is successful in controlling seizures problems of emotional adjustment may still persist. Attention to the wider impact of having epilepsy must also be addressed.

The family

The family need to be well informed about the child's epilepsy and the physician has a key role to play in this respect. Families will require information concerning prognosis and the treatment plan but also guidance on wider issues such as sports activities, lifestyle and at later stages implications for driving and employment. [32,44] The imparting of information should not be viewed as a one off exercise but will need reviewing and updating over time. As early as possible the child should be made an active participant in this process and during late adolescence the young person should be provided with the opportunity for appointments without the parents present. Austin and McDermott[35] suggests provision of information in the way outlined above will influence parenting styles positively and promote better adjustment of the child with epilepsy.

Most families cope very well with the strain of living with epilepsy but it is vitally important to explore this aspect and acknowledge the difficulties families may be facing. Enquiry should be made as to the psychological well-being of the parents and also any siblings. Schooling difficulties or conduct problems with the other children may be a reaction to the overindulgence of the child with seizures. Where there are signs of family stress referral for counselling or other forms of psychological input may be warranted.[32,44]

Parents may also benefit from time spent away from their *caring role*. Such breaks should be supported as parents can feel guilty about doing this. Engel[45] writes *"parents should be encouraged to spend time with each other, away from the patient in order to strengthen their own relationship and thereby create a supportive family environment"*. A supportive family is one of the most important factors in promoting healthy psychological development for any child. The valuable role of respite for parents is by no means a new proposal.[46]

The school

The potential role of the school in influencing the psychosocial outcome of the child has long been recognised in reviews of services for people with epilepsy. The need for effective communication between teachers, family and physician is invariably stressed. The teacher is in a much better position to deal with a seizure in the classroom when they are fully informed about what to expect and the acute management required.

The teacher following discussions with the family may also feel it advisable to inform the class. Such a step may help to reduce teasing and bullying. Considerable success has been reported in promoting positive attitudes toward epilepsy from educational groups for potential classmates.[47] Teachers must be encouraged not to uphold unnecessary restrictions over the child with epilepsy. A wide range of experiences are essential for the development of a positive self-image. Careful monitoring of the child's progress at school is also needed so that any difficulties can be identified early and addressed.

Society

Educating the general public is a major undertaking. The voluntary agencies have made great strides in this area. Much, however, can be accomplished at a local level for instance by writing supporting letters to accompany applications to join leisure clubs or college courses when unjustified opposition is raised.

REFERENCES

1. Rutter M, Graham P, Yule, W. A neuropsychiatric study in childhood. Clin. Dev. Med., 1970; 35/36: 175-185.
2. Holdsworth L, Whitmore K. A study of children with epilepsy attending ordinary schools, I: their seizure patterns, progress and behaviour in school. Dev. Med. Child Neurol., 1974; 16: 746-758.
3. Hoare P. The development of psychiatric disorder among school children with epilepsy. Dev. Med. Child Neurol., 1984; 26: 3-24.
4. Hermann B, Austin J. Psychosocial status of children with epilepsy and the effects of epilepsy surgery. In: Wylie E. editor. The treatment of epilepsy: principles and practices. Philadelphia: Lea & Febiger, 1993: 1141-1148.
5. Hartlage LC, Green JB, Offutt L. Dependency in epileptic children. Epilepsia, 1972; 13: 27-30.
6. Stores G, Piran, N. Dependency of different types in school children with epilepsy. Psychol. Med., 1978; 8: 441-445.
7. Long CG, Moore JR. Parental expectations for their epileptic children.J Child Psychol. Psychiat., 1979; 20: 299-312.
8. Hoare P. Does illness foster dependency? A study of epileptic and diabetic children. Dev. Med. Child Neurol., 1984; 26: 20-24.
9. Matthews WS, Barabas G, Ferrari M. Emotional concomitants of childhood epilepsy. Epilepsia, 1982; 23: 671-681.
10. Viberg M, Blennow G, Polski B. Epilepsy in adolescence: implications for the development of personality. Epilepsia, 1987; 28: 542-546.
11. Austin J. Childhood epilepsy: child adaptation and family resources. J. Child Adolesc. Psychiatr. Ment. Health Nurs., 1988; 1: 18-24.
12. Mulder HC, Suurmeijer ThPBM. Families with a child with epilepsy. J. Biosoc. Sci., 1977; 9: 13-24.
13. Hoare P, Kerley S. Psychosocial adjustment of children with chronic epilepsy and their families. Dev. Med. Child Neurol., 1991; 33: 201-215.
14. Henderson P. Epilepsy in school children. Br. J. Prev. Soc. Medicine, 1953; 7: 9-13.
15. Ounsted C. Aggression and epilepsy: rage in children with temporal lobe epilepsy. J. Psychosom. Res., 1969; 13: 237-242.
16. Barker P. Basic child psychiatry. London: Granada, 1983.
17. Davey D, Thompson P. Interictal language functioning in chronic epilepsy. J. Neurolinguistics, 1991; 6: 381-399.
18. Hermann BP, Whitman S, Dell J. Correlates of behaviour problems and social competence in children with epilepsy aged 6-11. In: Hermann BP, Seidenberg M. editors. Childhood epilepsies: Neuropsychological, Psychosocial and Intervention Aspects. Chichester: John Wiley and Sons, 1989: 143-158.
19. Stores G. Schoolchildren with epilepsy at risk for learning and behaviour problems. Dev. Med. Child Neurol., 1978; 20: 502-508.
20. Nuffield EJA. Neurophysiology and behaviour disorders in epileptic children. J. Ment. Sci., 1961; 107: 438-458.
21. Whitman S, Hermann BP, Black RB, et al. Psychopathology and seizure type in children with epilepsy. Psychol. Med., 1982; 12: 843-853.
22. Trimble MR. Antiepileptic drugs, cognitive function and behaviour in children: evidence from recent studies. Epilepsia, 1990; 31 (4S): 30s-34s.

23. Brent DA, Crumrine PK, Varma RR, et al. Phenobarbital treatment and major depressive disorder in children with epilepsy. Pediatrics, 1987; 80: 909-917.
24. Vining EPG, Mellits ED, Dorsen MM, et al. Psychological and behavioural effects of antiepileptic drugs in children: a double-blind comparison between phenobarbital and valproic acid. Pediatrics, 1987; 80: 165-174.
25. Cull CA, Trimble MR, Wilson J. Changes in antiepileptic drug regimen and behaviour in children with epilepsy. J. Epilepsy, 1992; 5: 1-9.
26. Vaernet K. Temporal lobotomy in children and young adults. In: Parsonage M, Grant RHE, Craig AG, Ward AA. Advances in epileptology XIVth Epilepsy International Symposium. New York: Raven Press, 1983: 255-261.
27. Nordgren RE, Reeves AG, Viguera AC, et al. Corpus callosotomy for intractable seizures in the paediatric age group. Arch. Neurol., 1991; 48: 364-372.
28. Beardsworth ED, Adams CBT. Modified hemispherectomy for epilepsy: early results in 10 cases. Br. J. Neurosurg., 1988; 2: 73-84.
30. Bagley C. The social psychology of the child with epilepsy. London: Routledge and Kegan Paul, 1971.
31. Hoare P. Psychiatric disturbance in the families of epileptic children. Dev. Med. Child Neurol., 1984; 26: 14-19.
32. Thompson PJ, Upton D. Quality of life in family members of persons with epilepsy. In: Trimble MR, Dodson WE, editors. Epilepsy and Quality of Life. New York: Raven Press, 1994: 19-33.
33. Maj M, Del Vecchio M, Tata MR, et al. Perceived parental rearing behaviour and psychopathology in epileptic patients: a controlled study. Psychopathology, 1987; 20: 196-202.
34. Ward F, Bower BD. A study of certain social aspects of epilepsy in childhood. Dev. Med. Child Neurol., 1978; 20 (39S): 145s-148s.
35. Austin JK, McDermott N. Parental attitude and coping behaviours in families of children with epilepsy. Journal of Neuroscience Nursing, 1988; 20: 174-179.
36. Lothman DJ, Pianta RC, Clarson, SM. Mother-child interaction in children with epilepsy: relations with child competence. J. Epilepsy, 1990; 3: 157-163.
37. Pazzaglia P, Frank-Pazzaglia L. Record in grade school of pupils with epilepsy: an epidemiological study. Epilepsia, 1976; 17: 361-366.
38. Ross EM, Peckham CS. School children with epilepsy. In: Parsonage M, Grant RHE, Craig A, Ward AA, editors. Advances in Epileptology: XIVth Epilepsy Symposium. New York: Raven Press, 1983: 215-220.
39. Kato H, Mori T, Moriuchi, et al. Psychosocial aspects of schoolchildren with epilepsy;schoolchildren who fell into school phobia. Folia Psychiatrica et Neurologia Japonica, 1979; 33: 437-439.
40. Corbridge P. Teacher training-attitudes towards epilepsy. In: Oxley J, Stores G, editors, Epilepsy and Education. London: Medical Tribune Group, 1987: 31-34.
41. Kaminer Y, Apter A, Aviv A, et al. Psychopathology and temporal lobe epilepsy in adolescents. Acta psychiatr. Scand., 1988; 77: 640-644.
42. Clement MJ, Wallace SJ. A survey of adolescents with epilepsy. Dev. Med. Child Neurol., 1990; 32: 849-857.
43. Westbrook LE, Silver EJ, Coupey SM, et al. Social charateristics of adolescents with idiopathic epilepsy: A comparison to chronically ill and

nonchronically ill peers. J. Epilepsy, 1991; 4: 87-94.
44. Santilli N. Psychosocial aspects of epilepsy:education and counseling for patients and families. In: Wylie E. editor. The treatment of epilepsy: principles and practices. Philadelphia: Lea and Febiger, 1993: 1163-1167.
45. Engel J. Seizures and epilepsy. Philadelphia: FA Davis Company, 1989.
46. Lennox WG. Science and seizures. New York: Harper and Brothers, 1941.
47. Gibson PA. Psychosocial interventions for adolescents with epilepsy. Epilepsia, 1993; 34 (2S): 31s.

15 The impact of epilepsy on social integration and 'quality of life': family, peers, and education

THEO P.B.M. SUURMEIJER
Department of Sociology, Department of Health Sciences
Northern Centre for Health care Research (NCH)
University of Groningen, The Netherlands

The impact of a chronic condition such as epilepsy is not well understood if one merely considers the physical aspects of the condition. Often, human behavior is seen as 'something innate'. That is particularly assumed to be the case if a child suffers from epilepsy, a 'nervous disease' that would not only influence the child's physical but also its behavioral, social and mental functioning. Something 'in the brain' is assumed to be responsible for daily functioning of the child with epilepsy, sometimes too easily defined as deviant and 'incurable' and consequently excluding the child from social interactions and opportunities. However, the social environment often plays a very dominant role in the way children 'behave'. A more comprehensive approach, in which also the role of the social environment is taken into consideration, is therefore recommended. Both the family's and the child's quality of life is at stake when a child is suffering from epilepsy. Familial and particularly parental reactions towards the epilepsy and the child with epilepsy, that is, parental attitudes and child rearing patterns, influence the daily functioning of the child with epilepsy. As a result, the child's position inside and outside the family may be sometimes jeopardized, affecting its social integration in the family, peer group and, via educational achievements, in the 'larger society'. All these factors may affect the quality of life of the child. Improving the quality of care in terms of informational and relational skills of doctors may add to the quality of life of the patient and their family, as well as social support provided by 'significant others'.

INTRODUCTION

Chronic disorders cannot be seen as a homogeneous entity and may, according to the medical characteristics of the disorder, have differential effects.[1,2,3] Nevertheless, from a psychological and sociological point of view, a distinction of chronic diseases according to dimensions such as visibility/invisibility, acute/chronic, life threatening/not life threatening, or paroxysmal/not paroxysmal, may be more useful than a distinction of chronic diseases in terms of medical labels or diagnoses.[2,4] All illnesses have psychosocial consequences and may thus interfere with daily life. This is especially true for chronic conditions. A chronic condition does not only affect physical aspects but also a series of areas of social and physic aspects of functioning. In spite of the type of chronic condition, patients may have to face '...*separation from family, friends, and other sources of gratification, loss of key roles, disruption of plans for the future, assaults on self-image and self-esteem, uncertain and unpredictable futures, distressing emotions and anxiety, depression, resentment and helplessness, as well as ... changes in physical appearance or in bodily functioning'*.[5] Moreover, a chronic condition not only affects the patients but also their social 'environment'. This social environment consists of: their families (1), their peers, and their leisure groups (2), their school and schoolmates (3). According to their position in each of these domains, the social integration of the child into his social world can be described. The impact of a chronic disease on these three social domains will, in turn, effect the course and consequences of the disorder.

FAMILY AND ILLNESS

Usually, depending on the age of the child, the relative importance of the three mentioned areas (family, peers, school) will vary and change over time. Consequently, the relative impact of these social areas will also vary during life time. A chronic disorder may further affect the relative importance of a particular social group. For example, the parents of a chronic ill child will take a more dominant position in his life than otherwise would have been the case, while the health care professionals are added as important sources of information and regulations. It is increasingly recognized that the family is a 'basic unit' in health, illness, (medical) care, and rehabilitation.[1,6-9] The interactions between this unit and a disease are summarized in figure 1.

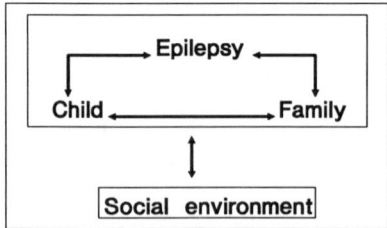

Figure 1. Relationships between epilepsy, child, family, and the social environment (including the health care system).

Since Hamilton's statement in 1945 that 'patients have families', the interest of the health care professional for the family in matters of health and illness is gradually expanding. Doherty and Campbell[10] refer to several factors, which may explain this growing interest. Among these:

psychosocial factors both as 'etiological agents' and in terms of 'consequences of a chronic illness', and treatment considerations ('compliance'). Both are very closely related to the family unit. *"The family affects the individual's health, and the individual's health affects the family"*. In this context, Doherty and Campbell refer to the 'biopsychosocial model' in which biological, mental or psychological and social issues are taken into consideration. Since 'health' is used here according to the 1948 WHO definition of health,[1] this approach is in line with the so called 'quality of life' and, as far as 'health related quality of life' is concerned, with 'quality of care'. Both constitute the two sides of the same coin.[12,13] The nature of the care provided will influence (the quality of) the lives of patients, parents and partners.

MULTI-DIMENSIONALITY OF CHRONIC ILLNESS, MULTIDIMENSIONALITY OF PROBLEMS

From the literature, it appears clearly that illnesses, in particular chronic illnesses, are multidimensional phenomena. This is shown in figure 2.

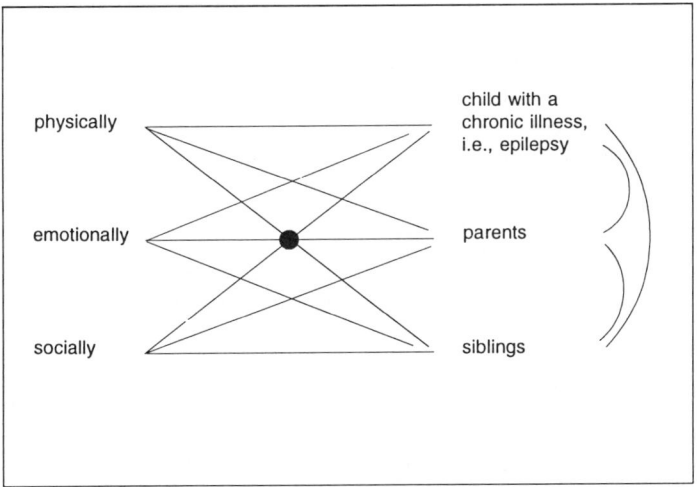

Figure 2. Multidimensionality of chronic illness and 'problems'.

Although this figure seems very complex, real life is even more complex. The figure illustrates the complexity of influences of a chronic illness on psychosocial functioning of both individuals and families. One aspect of the 'stressor illness' has to do with the physical health of the ill child, and an adaptive coping strategy is, of course, to seek and receive adequate medical treatment. But another problem has to do with emotional health and social life of the affected child, and how disturbances in these domains can be prevented or minimized. A third group of aspects has to do with the emotional, social and physical impact a chronic illness may have both on the parents and the siblings of the chronically ill child.[8,9] Their is an additional complicating factor. Not only family members but also physical, emotional and social aspects are often highly interrelated. Consequently, the problem solving strategy that is adaptive in one area or for one family member, does not necessarily need to be adaptive in another area or for another family member. For example, parental overconcentration of attention on the child with epilepsy (or with another chronic condition) is frequently reported and child-rearing practices are distorted. This may be adaptive for parents to reduce, for example, their feelings of anxiety. In some respects, it may also be adaptive for the ill child, for example for the prevention of serious accidents. It will, nonetheless, disturb a healthy development of the ill child. This overconcentration on one child may lead to a certain neglect of the affective and social needs of the other children in the family. The following quotations summarize the foregoing: "..*In epilepsy as in other chronic conditions, total eradication of the physical symptoms remains the goal of the physician. Yet, this is not enough, for the family and the community are also involved*".[14] "*Epilepsy is more than just having fits*',[15,16] "...*Possibly the least understood and most neglected aspects of epilepsy are the social, psychological, and behavioural problems that are so common*". "...*In terms of the psychological effect on the child, the illness itself is less menacing than familial response to that illness*".[9] Much of the literature focuses on the emotional impact epilepsy has on the child.[17] Since this aspect is extensively dealt with in other chapters of this volume, we will concentrate here in the social consequences of epilepsy: social integration in terms of 'family functioning', 'educational achievements' and 'leisure activities'.

FAMILY FUNCTIONING AND THE CHILD WITH EPILEPSY

Roughly estimated, a quarter of the children, aged 0-18, have 'some' chronic condition[18] while one in every five or six households has at least one member with a chronic somatic condition.[19] With a prevalence rate of epilepsy of about 0.75%, about one in every 33 families consisting of about four persons, will be confronted with epilepsy.[20-26] Thus, many families have to cope with the problem of 'chronic illness' and in our case with epilepsy and must find solutions for the primary and secondary problems that may arise from the epilepsy. 'Secondary' in this context does *not* imply that these problems are less relevant than the 'primary problems' of epilepsy itself; it only refers to a time sequence. The way parents handle the epilepsy, is based on their understanding of the epilepsy. These parental reactions will, in turn, strongly influence the child's development.
Parental reactions to the epilepsy will in itself be influenced by the

'societal context', that is by the interpretations attitudes and behaviors of people in the society. Scambler and Hopkins[27] speak of 'parental training in stigma of epilepsy'. By transfer of such attitudes, the children with epilepsy will consider their epilepsy as something 'wrong', 'bad', or 'disgracing' that have to be concealed as much as possible; gradually they may come to see themselves as 'worthless', excluding themselves or being excluded from normal social experiences. Through denial and self-denial of opportunities, a slow downward spiral is set in motion, which ultimately may end in low self-esteem, low social participation and low educational and occupational achievements.[28,29] These 'societal reactions' will certainly aggravate the burden of epilepsy for parents and child. The way parents perceive the epilepsy will further be influenced by the emotional impact of witnessing in particular larger seizures, by the dangers the seizures may have for the safety of the child and by the uncertainty about the medical and social prognosis: course of the disease, side-effects of antiepileptic drugs and social prospects (i.e. mental development, educational and occupational attainments). To compensate for feelings of helplessness, anxiety, uncertainty, guilt or shame, parents may develop socialization patterns that, from the point of view of a healthy personality development, may be labelled as 'inadequate': patronizing, overprotective, overindulgent, too strict or inconsistent disciplining; scapegoating, rejecting and other negative emotional pedagogic attitudes and behaviors like annoyance and frustration. All these types of responses from parents have been mentioned in the existing literature on the socialization of children with epilepsy (as with other chronic conditions).[8,9,17] Evidently, this will profoundly affect the development of the child suffering from epilepsy. A number of studies have pointed to the increased interpersonal distance that some families show, when coping with a chronic illness in one of their members. Beside an increase in marital distress, both higher levels of familial stress and sibling rivalry and reduced levels of family closeness and interactions have been reported.[8-10,17] According to Ferrari et al.[17] children with epilepsy and their families differed in a number of ways from diabetic children and healthy controls. According to their parents, children with epilepsy less often initiated activities within the family and judged their families as 'less close'. Also, more periods of 'emotional upset' occurred in families with a child with epilepsy than in the other families. These findings on lower family integration and more family stress in families with a child with epilepsy are corroborated by several other studies.[30-40] Fathers and particularly mothers of a child with epilepsy are more overprotective and indulgent, are more worried and have a more passive attitude about the future of their child, and considered education less important for their child, compared to the parents of healthy controls. Moreover, they feel more annoyed, frustrated and disappointed by the behavior of their child. On average, the differences between the parents of the children with epilepsy and the healthy controls vary from 10-30%.[39]

Family distress, overprotection and parental worries about the future are illustrated in figure 3.

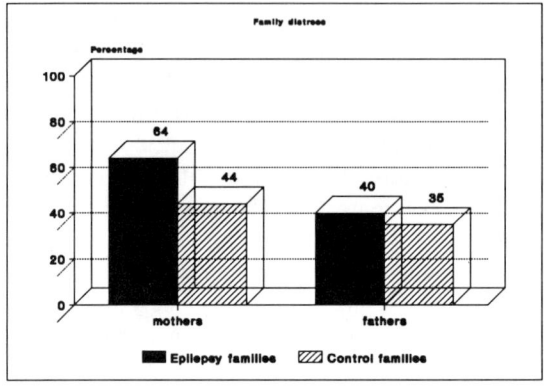

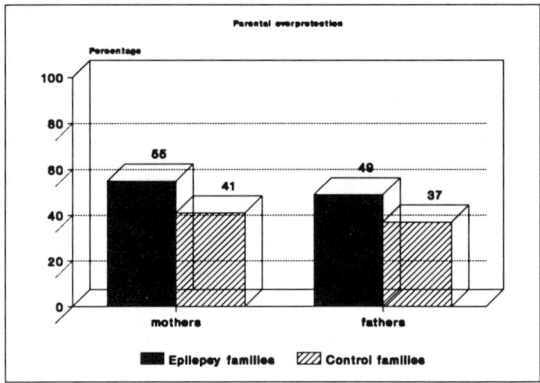

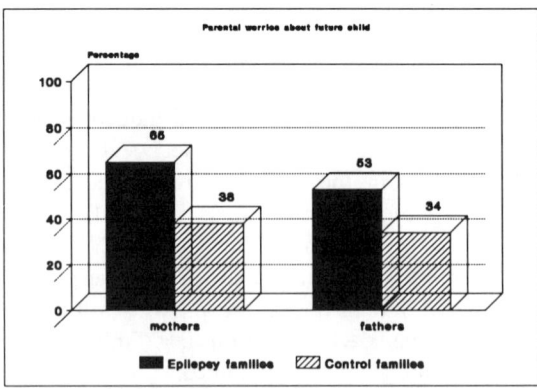

Figure 3. Differences between mothers and fathers of a child with epilepsy and healthy children. Scores are dichotomized in 'low and high' around mean scores. The percentages refer to 'high scoring parents'.

Factors such as seizure type, type of epilepsy, seizure-frequency, duration of the disease or hospitalization of the child had none or hardly any influence on these differences. Most often feelings of fear for the occurence of seizures had a significant influence on the differences between the two groups of parents. It is often assumed that patient education about the condition, about the treatment and about the consequences of the disease and treatment, will lead to more adaptive coping behavior. This concurs with Szasz and Hollender's doctor-patient role relational model.[41] In case of a chronic condition the most relevant or adjusted doctor-patient relation is one of mutual participation at least after an initial diagnostic phase. Within such a relationship the patient or his/her parents become actively engaged in the treatment program and attention is given to the experiences and the practical knowledge of the patient and his family. In practice, however, an 'activity/passivity role relation' is often observed. In such a treatment-relation insufficient attention is given to the experience and feelings of the patients and his parents. Our own research showed that when parents have the feeling that they are well informed about the epilepsy of their child, this will facilitate practical management of the epilepsy and of their child[42] (Table 1).

Table 1. Information and parental attitudes towards the child with epilepsy

		(Strongly) agree	
		Mothers	Fathers
If you know what is wrong with your child:			
1.	You also will more easily accept that something is wrong with your child	87%	95%
2.	You will be less frightened as well	60%	76%
3.	You also know better how to deal with your child	72%	98%

Table 1 shows that parents attach much value to being well informed which, in their perception, is beneficial for the relationship with their child with epilepsy. There is a need for information about the disease itself, its treatment, the prognosis, but also about raising their child, educational and occupational choices. This is in line with the results of a study among over 1100 adults with epilepsy, their partners or the parents of a child with epilepsy.[12] With 'being informed' clearly not only a technical explanation is meant but also information about psychosocial aspects. The translation of the original physical complaints of the child by the doctor in medical categories should be followed by a translation in social categories: what does epilepsy mean for daily life.[42,43] A good quality of care not only implies a total eradication of physical symptoms[14] but also paying attention to possible consequences of the epilepsy for daily life and the future of the child and its parents. Besides medical expertise, social skills and competence are required from physicians as well. It contributes, among other things, to adequate help seeking behavior, to compliance, and to a more adequate parent-child relationship. In turn, this will facilitate a healthy social-emotional development of the child with epilepsy. Furthermore, Table 1 shows that fathers and mothers have somewhat different perceptions of the efficacy of information. These differences may point at the different positions fathers and mothers (still) have with respect to child-rearing, leading to a more realistic appraisal of the situation by mothers as compared to fathers. In addition, it may refer to a greater social-emotional burden for mothers than for fathers in raising their child with epilepsy (see Figure 3).[31,39,44] One of the child-rearing behaviors, directly influencing the behavior of the child is the number of restrictions, imposed on the child with epilepsy. Not only parents but also doctors and others differ as to the types and degree of restrictions to be imposed on the child with epilepsy.[17,39,45] Of course, the number and types of restrictions will depend on the age of the child. But partly associated with the parental fear for the occurrence of seizures, it probably also indicates the degree of parental overprotection and overconcernedness.[39] Children with epilepsy are often imposed more restrictions by their parents than healthy controls. We asked the parents of a child with epilepsy and those of healthy controls how many restrictions they imposed on their child. These restrictions concerned the following ten activities: physical training; using gymnastics apparatus; practising a sport such as playing football, hockey, handball; joining a school camp of one day or longer; going to a fair; going to the movies; swimming; taking a bath; going to school alone; cycling. For each of these activities we asked the parents whether it was 'allowed to their child', 'only allowed when supervised' or 'never allowed'. The results are presented in Figure 4.[46]

This study shows furthermore that, when parents are both 'medically' and 'psychosocially' well informed by their doctors, this will reduce the number of restrictions imposed on the child: on average one to three restrictions less. Confidence and reassurance strengthens this effect.[42,46]
Concerning the restrictions to be imposed on the child, one should always ask: what are really dangerous situations for this child with this type of epilepsy and this seizure type and within this neighbourhood. How large is the risk for a certain activity to perform, and which risks are acceptable. Often both parents and doctors are placed in a dilemma.

Although a difficult subject to counsel parents, one always should think about the consequences of restrictions. Restricting the daily activities and thus restricting the lives of the child with epilepsy, will exclude them from normal social interactions and consequently, from the development of 'normal' social skills and experiences. As a result it will affect the child's position inside and outside the family.[47]

One may argue that the GP, may very well be the most proper person to advise the parents in these matters, because he is best informed

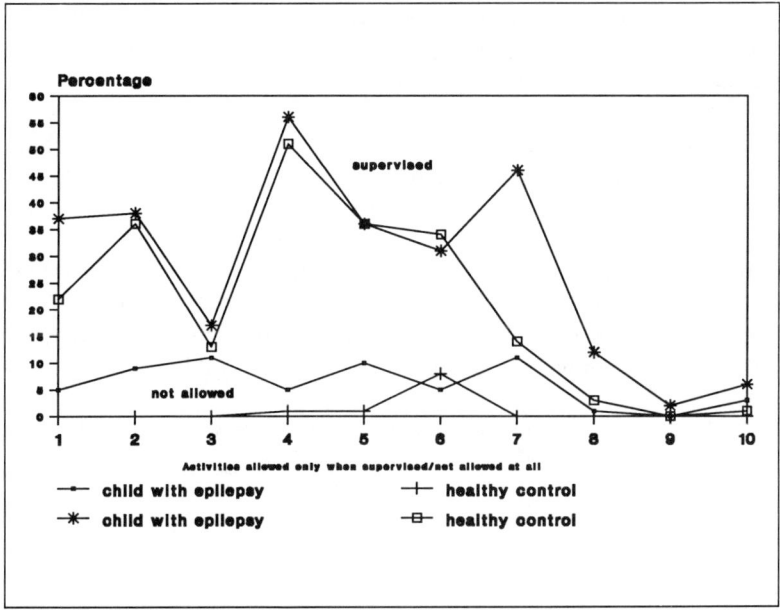

Figure 4. Number of restrictions imposed on the child with epilepsy as compared to 'comparable healthy controls'.

about the child, its parents and the neighbourhood. In this respect the GP can play a vital role in the quality of the child's as well as of the family's life, sharing the risks the parents are willing to take by guiding and supporting.[48]

CHILD AND EPILEPSY: PEERS AND EDUCATION

This section will focus on the social participation of the child: the interactions of the child with its peers and memberships of sports and other clubs or associations. Moreover, attention is given to the educational achievements of the child with epilepsy.

Social participation

The way a child is treated by its peers will be influenced by all kinds of characteristics of the child itself, those of his peers and those of the context in which interactions take place. As far as children with epilepsy are concerned, an important additional characteristic may be made up by his illness. It may be a 'starting point' to treat the child differently. Ignorance and prejudice may lead to an overemphasizing of the epilepsy: the epilepsy may be seen as an important or even central characteristic of the child's personality, in extreme cases completely determining the reactions of others towards the child. In this latter case one speaks of 'stigmatization'. Often normal characteristics and behaviors of the child are now ascribed to the epilepsy of the child and defined as 'deviant', leading to a negative treatment of the child by his peers.[49,52] As a result, children with epilepsy may feel excluded or may exclude themselves (to avoid negative treatment) from 'normal interactions': having friends and being a member of clubs. From our own studies among children with epilepsy, it appeared that these children were much less often member of sports clubs than the Dutch youth of the same age (11 years): 46% versus 32% were no member. More generally, one can say that they were somewhat more socially isolated then comparable healthy controls[39] (Table 2).

Table 2. Social participation of children with epilepsy and health controls (about 11 years of age)

	children with epilepsy	healthy controls
- no friends, no club memberships	9%	2%
- friends, no club memberships	26%	15%
- no friends, club memberships	20%	17%
- friends and club memberships	45%	66%

They also were more often negatively treated (teasing, excluding, quarrelling, etc.) by their peers at school and in the neighbourhood: 26% of the children with epilepsy, as compared to 13% of the healthy controls. To be afraid to have a seizure 'somewhere' and 'sometime' and to be 'out of control', in order to avoid to show that 'you need medicines' and more generally, 'to deviate from what is normal' for or allowed may all, separately and in combination, be reasons for their relatively larger social isolation. Similar results were also found or mentioned in other studies. [17,33,37,51-56] In this context, it should be pointed out that self-denial of opportunities, for example because of feelings or expectations of rejection and a tendency to conceal the epilepsy, may be as important as or perhaps even more important than actual stigmatization and discrimination.[17,57-59]

What happens inside the family is not independent of what happens outside. It has been found that parental attitudes, frustration, annoyance and overprotection were related on the one hand with behavioral restrictions imposed on the child and on the other hand with being negatively treated and socially isolated.[39,42,46] The interpretation of these relationships is shown in figure 5.

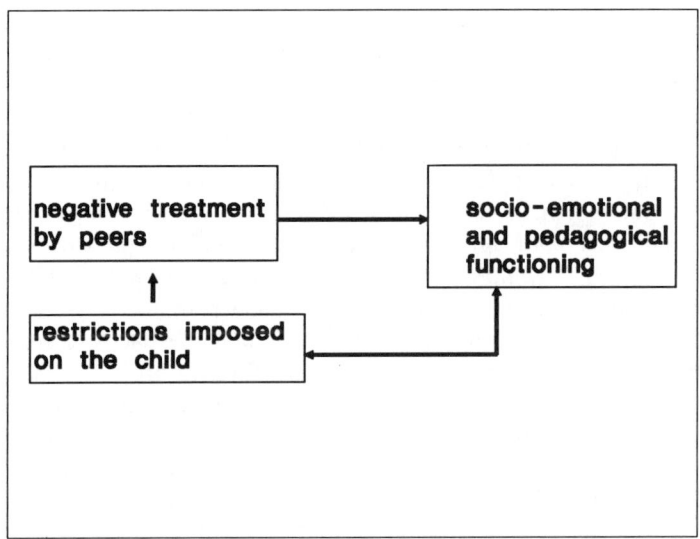

Figure 5. Relations between 'behavioral restriction', 'reactions social environment' and 'parental behavior and attitudes'.

Imposing behavioral restrictions upon the child may be interpreted as a tendency on the side of the parents to protect their child. This, in turn, will place the child in separate position which will negatively effect its relationships with peers. And that again, will be of influence on other aspects of the parent-child relationship. As several authors have indicated, this may lead to emotional and behavioral symptoms of disturbance and a (further) decrease in self-esteem.[28,29,37] Subsequently, fathers and particularly mothers of the child with epilepsy, who see how their child behaves and is treated by its peers, are being more frustrated and annoyed, are feeling themselves more disappointed and worried about their child and respond by protecting their child still more.

Education

Many studies have been carried out to study the effect of epilepsy on cognitive functioning and education. This emphasis on intelligence and education, as well as employment with regard to people with epilepsy is not a coincident. In modern industrialized societies the attitude of the population toward physical or mental conditions will be more negative, the more essential capacities are supposed to be affected. This is especially the case with regard to such highly valued capacities as intelligence and self control which are essential for efficiency, productivity and smooth interpersonal relations. Consequently, their is very little tolerance of behavioural and mental deviations that tend to disrupt the smooth functioning and easy flow of interpersonal relations; and it is people with epilepsy who are, or were, supposed to be impaired on these capacities'.[60] Both neurological and psychosocial factors are assumed to be responsible for the way people with epilepsy function. *"Evaluation of the relative contribution of each of these factors showed that neurological conditions, i.e. brain dysfunction, leads to specific defects in cognitive functioning and concern relatively small groups, whereas psychosocial factors have a more general effect on cognition and affect more patients in our sample"*.[61] In addition, neurological factors are always modulated by the way parents and others cope with the condition.[30,37,62-65] Although the data are sometimes contradictory, it is often assumed that in the absence of additional neurological deficits, and controlling for 'ability', the 'environmental factors' (attitudes of and treatment by parents, siblings, peers and teachers) are a significant factor in explaining underachievement of children with epilepsy. Using the 'Home Observation for Measurement of the Environment' (HOME), Mitchell et al.[66] found that children with epilepsy academically underachieved even after controlling for intelligence and that 'environmental factors' were mainly responsible for these findings. Children with epilepsy were doing worse on reading, comprehension, spelling, mathematics, and general knowledge. Neither medication nor seizure-related variables appeared to be a risk factor for underachievement. The authors state that this *"re-emphasizes the importance of socioeconomic and cultural variables as potential confounders in studies of ill populations"*. Recently, children with epilepsy were found to experience significantly lower quality of life in the psychological, social and schooldomains ('schoolprogress' and 'schoolachievement') than children with asthma.[32,33] Among about 11-years old children with epilepsy (all at normal schools, no additional neurological deficits), it appeared that children with epilepsy did worse at school as compared to healthy controls: lower marks on reading, grammar and arithmetics, and they more often remained below the age-appropriate grade level. As compared to the parents of the healthy control children, expectations of the fathers and mothers concerning the education to be followed by their child after primary school, were significantly lower.[39,60] These expectations of parents and other 'significant others' such as teachers about the intelligence and educational abilities and, more generally, about the future of the child with epilepsy, may be of utmost importance for the achievements of the child. Lower expectations and being worried about the future of the child may lead to lower educational choices and achievements.[39,60,67-72] This is confirmed by the results of or own study (Figure 6).

These findings illustrate that there is no clear evidence of correction of the educational careers of the children with epilepsy closing the initial gap between these children and comparable healthy controls: a low(er) academic start will finally result in a low(er) educational end level. These results were the more remarkable, since the 'medical situation' of many of these children had very much improved since the onset of the epilepsy.[7-3,74] Of course, 'education' is neither the only important thing in life nor will it fully determine happiness in life but, it is an important method to facilitate the access to other 'goods' and 'markets' which are considered to be worthwhile to strive for in our western world. It will influence, for example, one's access to the labour market and marriage-market, one's income level, health and health care utilization, and so on. Educational expectations and choices will thus affect the type and degree of social integration of people with epilepsy as compared to other people. Therefore, *'it seems very desirable for parents, teachers, and other relevant others to be in frequent contact with each other. Low stimulation and estimation of the intellectual capabilities and, consequently, (too) low academic choices and results of the children with epilepsy may then perhaps be prevented'*.[73] Also here, it is true that *'many of his or her successes or failures in situations outside the home intimately depend upon how they deal with their illness inside the family'*.[3]

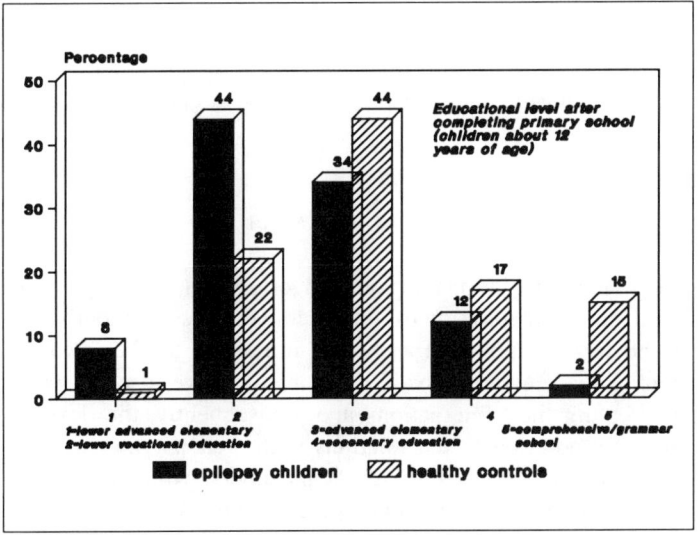

Figure 6a. Educational level of children with epilepsy and healthy controls

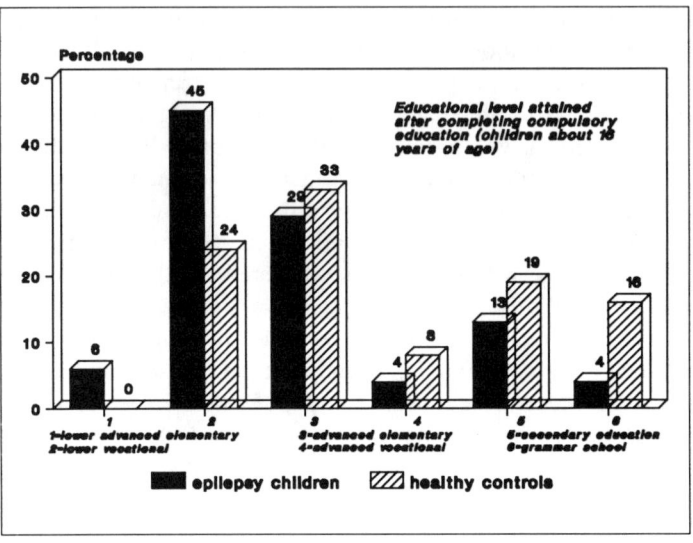

Figure 6b. Educational level of children with epilepsy and healthy controls

FINAL REMARKS

Experiences and interactions inside and outside the family influence the mental and social maturation of children. This is true both for healthy children and for children with a chronic illness. In the latter case, children themselves, as well as their social environment have to cope with an extra problem affecting their interactions and, consequently, the development of the children. In addition to the relations with their parents, children have a multitude of relationships with 'significant others' who all will affect -to a greater or lesser extent- their development. Parents are not the only ones who determine how their children will develop, they are not the single factor, determining growth and maturation.[42,75-78] Although in many studies the adverse impact of epilepsy on the psychosocial development and adjustment of affected children and families is shown, this should not be exaggerated. Three factors must be mentioned. *First*, it should be noticed that most children with 'epilepsy' who don't have additional neurological deficits function normally and so do their families![39,40,79] *Second*, changes and deviations from normal are also found among children with other chronic conditions and their families. Stein and Jessop[4] state that there often is '...*more variability within diagnostic groupings than between them and suggest that diagnosis is not a helpful categorization*'. However, studies about the effect of the severity of the epilepsy are still inconclusive, but it is assumed that many of the problems mentioned, will be aggravated in case of the more severe, chronic or refractory types of epilepsy, mostly with additional neurological deficits.[30,36,37,40,62,80] Third, there is scarce empirical evidence, that child functioning is not only affected when the child itself has epilepsy but also when one of the parents suffers from epilepsy (or another chronic condition): more school problems (failure to pass a grade), concentration problems during periods of seizures of their parent, social isolation, and feelings of fear, shame and anxiety were observed in these children.[81-87]

Many of the successes or failures of the child in situations outside the home, intimately depend upon how their illness is perceived and managed inside the family, and vice versa. This is true both for the social relations of the child and for its educational achievements.[3,42,60] Without the help of others it may be rather difficult for parents to change this process. Finding solutions on their own for these problems may be asking to much from them. In addition, the multi-dimensionality of a chronic illness should be taken into an account. Coping strategies, adequate for one problem or for one family member, are not necessarily adequate for other problems and/or other family members. Therefore, professional as well as informal help, help from teachers, neighbours, family but without any doubt also from the treating physician, is often indispensable. In particular, both patient and parent information and guidance will have beneficial effects on parents and child. And so does social support (in the areas of emotional, appraisal, practical and informative support) from family, friends, or neighbours which may help parents to find a new adaptation to the illness of their child contributing to the family's and child's quality of life.[88-89] This is in line with the literature on social support, pointing at the beneficial as at the effects of social support on the well-being and quality of life of patients and families in many respects.[90] Encouragement and sharing of responsibilities, appraisal and information, instrumental and emotional support, both from significant lay persons and from health professionals, may relieve parents, and mothers in the first place, from the burden to manage with the illness of their child. Particularly in 'periods of transition' this may be important: the period of uncertainty before the moment of diagnosis, first adaptation, and furthermore in moments or periods when 'new choices' have to be made concerning the future of the child.[10,42,91] In this respect the phase of (pre-) adolescence may be considered very important, for parents as well as for the child itself. *'In this period, a person becomes conscious of the limitations imposed by the disease on his/her life habits and career choices'*.[89] Adjustment problems and poor compliance with the treatment regimen are more frequently reported for adolescents than for younger children.

In conclusion, referring to Upton,[88] there is ample evidence that psychological adjustment in children with epilepsy and their parents is not simply a result of medical and biological variables. Psychosocial factors often play an (very) important role in this respect.

REFERENCES

1. Anderson R, Bury M, editors. Living with chronic illness. The experience of patients and their families. London (etc): Unwin Hyman, 1988.
2. Roth JA, Conrad P, editors. Research in the sociology of health care. A research annual, volume 6. The experience and management of chronic illness. Greenwich/London: JAI Press Inc., 1987.
3. Ferrari M, Matthews WS, Barabas G. The family and the child with epilepsy. Fam. Process, 1983; 22: 53-59.
4. Stein RE, Jessop DJ. What diagnosis does tell: the case for a non-categorical approach to chronic illness in childhood. Soc. Sci. & Med., 1989; 29: 769-778.
5. Turk DC. Factors influencing the adaptive processes with chronic illness. In: Sarason A, Spielberger CD, editors. Stress and anxiety, volume 6. Washington DC: Halsted Press, 1979: 291.
6. Litman TJ. The family as a basic unit in health and medical care: a social-behavioral overview. Soc. Sci. & Med., 1974; 8: 495-519.
7. McEwan PJM. The social approach to social health studies. Soc. Sci. & Med., 1974; 8: 487-493.
8. Schwenk TL, Hughes CC. The family as patient in family medicine. Soc. Sci. & Med., 1983; 17: 1-16.
9. Shapiro J. Family reactions and coping strategies in response to the physically ill or handicapped child: a review. Soc. Sci. & Med., 1983; 17: 913-931.
10. Doherty WJ, Campbell TL. Families and health. Family studies text series, no. 10. Newbury Park (etc): Sage Publications, 1988.
11. World Health Organization: Construction of the World Health Organization. Basic documents. Geneva: World Health Organization, 1948.
12. Suurmeijer ThPBM. Kwaliteit van zorg: percepties en prioriteiten. Epilepsie Bull., 1992; 20: 3-10.
13. Suurmeijer ThPBM, Hermann B. Epilepsie en kwaliteit van leven. Epilepsie Bull., 1992; 20: 3-9.
14. Boshes LD, Kienast HW. Community aspects of epilepsy - a modern reappraisal. Epilepsia, 1972; 13: 31-32.
15. Renier WO. De bijzondere plaats van het kind met epilepsie. Epilepsie Bull., 1990; 18: 25-27.
16. Smits H. Een katamnestisch onderzoek van een groep patiënten uit het Instituut voor Epilepsiebestrijding te Heemstede (dissertation). Leiden: Stafleu's Wetenschappelijke Uitgeverij NV, 1970.
17. Levin R, Banks S, Berg B. Psychosocial dimensions of epilepsy: a review of the literature. Epilepsia, 1988; 29: 805-816.
18. Swaak AF. Ongeneeslijke ziekte heeft grote emotionele invloed op ontwikkeling van het kind. MGZ., 1977; 5: 20-21.
19. Gorter KA. Zorgen voor gehandicapte gezinsleden dissertation. The Hague: NIMAWO, 1988.
20. Overweg J, Withdrawal of antiepileptic drugs in seizure-free adult patients. Prediction of outcome (dissertation). Amsterdam: Rodopi, 1985.
21. Rutgers M. De epidemiologie van de epilepsieën. Epilepsie Bull., 1988; 17: 3-14.

22. Shorvon SD. Epidemiology, classification, natural history, and genetics of epilepsy. In: Epilepsy; A Lancet Review. The Lancet, 1990; 3-6.
23. Central Bureau for Statistics / Nederlands Instituut voor Maatschappelijk Werk Onderzoek. Lichamelijke beperkingen bij de Nederlandse bevolking 1986/1988. The Hague: SDU Uitgeverij, 1990.
24. Metsemakers J, Höppener P, Knottnerus A, et al. Gezondheidsproblemen en diagnosen in de huisartspraktijk. Een halfjaarlijkse rapportage van het Registratienet Huisartspraktijken - RNH. Maastricht: Rijksuniversiteit Limburg, 1990.
25. Nijmeegs Universitair Huisartsen Instituut. Morbidity from four general practices. Nijmegen: N.U.H.I., 1985.
26. Central Bureau for Statistics. Statistisch Jaarboek. The Hague: SDU Uitgeverij, 1993.
27. Scambler G, Hopkins A. Accommodating epilepsy in families. In: Anderson R, Bury M, editors. Living with chronic illness. The experience of patients and their families. London (etc): Unwin Hyman, 1988: 156-176.
28. Pless IB, Roghman KJ. Chronic illness and its consequences: observations based on three epidemiologic surveys. J. Pediatr., 1971; 79: 351-359.
29. Blaxter M. The meaning of disability. A sociological study of impairment. London: Heinemann, 1980 (pbk).
30. Goldin GJ, Perry SL, Margolin RJ, et al. The rehabilitation of the young epileptic. Dimensions and dynamics. Toronto/London: Lexington Books, 1971.
31. Austin JK, McBride AB, Davis HW. Parental attitude and adjustment to childhood epilepsy. Nurs. Res., 1984; 33: 92-96.
32. Austin JK. Comparison of child adaptation to epilepsy and asthma. J. Child Adolescence Psychiatr. Ment. Health Nurs., 1989; 2: 139-144.
33. Austin JK. Smith MS, McNelis AM. Childhood epilepsy and asthma: comparison of quality of life. Epilepsia, 1994; 35: 608-615.
34. Austin JK, Risinger MW, Beckett LA. Correlates of behavior problems in children with epilepsy. Epilepsia, 1992; 33: 1115-1122.
35. Hoare P. Adults' attitudes to children with epilepsy: the use of a visual analogue scale questionnaire. J. Psychosom. Res. 1986; 30: 471-479.
36. Hoare P. Children with epilepsy and their families. J. Child Psychol. Psychiatry, 1987; 28; 651-655.
37. Hoare P, Kerley S. Psychosocial adjustment of children with chronic epilepsy and their families. Dev. Med. Child Neurol., 1991; 33: 201-215.
38. Mulder HC, Suurmeijer ThPBM. Families with a child with epilepsy: a sociological contribution. J. Biosoc. Sci., 1977; 9: 13-24.
39. Suurmeijer ThPBM. Kinderen met epilepsie. De invloed van een ziekte op kind en gezin (dissertation). Groningen: Druk Veenstra Visser Offset, 1980.
40. Laybourn A, Hill M. Children with epilepsy and their families: needs and services. Child Care Health Dev., 1994; 20: 1-14.
41. Suurmeijer ThPBM. Epilepsy and the meaning of self-help groups and organizations. In: Canger R, Loeber JN, Castellano F, editors. Epilepsy and society: realities and prospects. International Congress Series 802. Amsterdam/Oxford/New York: Elsevier Science Publishers, 1988: 81-91.
42. Suurmeijer ThPBM. Ouders van kinderen met een chronische aandoening. Gezondheid & Samenleving, 1985; 6: 113-122.
43. Schneider JW, Conrad P. Medical and sociological typologies: the case of epilepsy. Soc. Sci. & Med., 1983; 15A: 211-219.

44. Knafl KA, Gallo AM, Breitmayer BJ, et al. Family response to a child's chronic illness: a description of major defining themes. In: Key aspects of caring for the chronically ill: hospital and home. New York: Springer Pub. Co, 1993: 290-303.
45. Höppener RJEA. Epilepsy and alcohol. The influence of social alcohol intake on seizures and treatment in epilepsy (dissertation). Barneveld: BDU, 1981.
46. Suurmeijer ThPBM, and Van den Nieuwenhuijzen WHG. De invloed van epilepsie op kind, gezin en leefmilieu. In: Knaven FHJ, Meinardi H, Peper C, editors. Integrale epilepsiebehandeling. Amsterdam/ Brussel: Elsevier, 1986: 127-148.
47. Van den Nieuwenhuijzen WHG. Kinderen met epilepsie: wat verwachten ouders nog meer van de arts? Epicare, 1993; 4: 4-13.
48. Brown S, Betts T, Hall B, et al. An epilepsy needs document. Seizure, 1993; 2: 91-103.
49. Safilios-Rothschild C. The sociology and social psychology of disability and rehabilitation. New York: Random House, 1970.
50. Freidson E. Profession of medicine. New York: Dodd, Mead and Company, 1973.
51. Richardson DW, Friedman SB. Psychosocial problems of the adolescent patient with epilepsy. Clin. Pediatr., 1974; 13: 121-126.
52. Gorter KA. Het dagelijks leven van mensen met epilepsie. The Hague: MINAWO, 1981.
53. McAnarney ER. Social maturation. A challenge for handicapped and chronically ill adolescents. J. Adolescent Health Care, 1985; 6: 90-101.
54. Clement MJ, Wallace SJ. A survey of adolescents with epilepsy. Dev. Med. Child Neurol. 1990; 32: 849-857.
55. Pazzaglia P, Frank-Pazzaglia L. Record in grade school of pupils with epilepsy: an epidemiological study. Epilepsia, 1976; 17: 361-366.
56. Tanganelli P, Regesta G. Descriptive and comparative study of epilepsy in a health district of Genoa, Italy, with evaluation of medical and social aspects. Epilepsia, 1994; 35 (suppl 7): 17.
57. Scambler G, Hopkins A. Being epileptic: coming to terms with stigma. Sociol. Health Ill, 1986; 8: 26-43.
58. Scambler G, Hopkins A. Generating a model of epileptic stigma: the role of qualitative analysis. Soc. Sci. & Med., 1990; 30: 1187-1194.
59. Jacoby A. Felt versus enacted stigma: a concept revisited. Evidence from a study of people with epilepsy in remission. Soc. Sci. & Med., 1994; 38: 269-274.
60. Suurmeijer ThPBM. Epilepsy and education. J. Clin. Pract., 1982; Symposium suppl 18 (2nd ed); 261-268.
61. Aldenkamp AP. Learning disabilities in children with epilepsy. In: Aldenkamp AP, Alpherts WCJ, Meinardi H, et al, editors. Education and epilepsy. Amsterdam/Lisse: Swets & Zeitlinger; Berwyn, PA: Swets North America, 1987: 21-37.
62. Bagley C. The social psychology of the child with epilepsy. London: Routledge & Kegan Paul, 1971.
63. Birch HG, Thomas A. Behavioral development in brain-damaged children. Archives of General Psychiatry, 1964; 11: 596-603.
64. Eisenberg L. Behavioral manifestations of cerebral damage in childhood. In: Birch HG, editor. Brain damage in children: the biological and social

aspects. Baltimore: The Williams and Wilkens Co, 1964: 61-76.
65. Richardson SA. The social environment and individual functioning. In: Birch HG, editor. Brain damage in children: the biological and social aspects. Baltimore: The Williams and Wilkens Co, 1964: 100-117.
66. Mitchell WG, Chavez JM, Lee H, et al. Academic underachievement in children with epilepsy. J. Child. Neurol., 1991; 6: 65-72.
67. Bjegović L. School and work efficiency of the epileptic child. In: Meinardi H, Rowan AJ, editors. Advances in epileptology 1977. Psychology, pharmacotherapy and new diagnostic approaches. Amsterdam/Lisse: Swets & Zeitlinger, 1978: 42-43.
68. Green JB, Hartlage LC. Comparative performance of epileptic and non-epileptic children and adolescents. Dis. Nerv. Syst., 1971; 32: 418-421.
69. Hartlage LC, Green JB. The relation of parental attitudes to academic and social achievements in epileptic children. Epilepsia, 1972; 13: 21-25.
70. Long CG, Moore JR. Parental expectations for their epileptic children. J. Child Psychol. Psychiatry, 1979; 20: 299-312.
71. Schwager HJ. Education of children and adolescents suffering from epilepsy. In: Aldenkamp AP, Alpherts WCJ, Meinardi H, et al, editors. Education and epilepsy. Amsterdam/Lisse: Swets & Zeitlinger; Berwyn, PA: Swets North America, 1987: 81-88.
72. Suurmeijer ThPBM, Van Dam A, Blijham M. Socialization of the child with epilepsy and school achievement. In: Meinardi H, Rowan AJ, editors. Advances in epileptology 1977. Psychology, pharmacotherapy and new diagnostic approaches. Amsterdam/Lisse: Swets & Zeitlinger, 1978: 48-53.
73. Suurmeijer ThPBM. Education of children suffering from epilepsy: a follow-up study. In: Wolf P, Dam M, Janz D, et al, editors. Advances in epileptology. XVIth Epilepsy International Symposium. New York: Raven Press, 1987: 615-619.
74. Suurmeijer ThPBM. Treatment, seizure-free periods, and educational achievements: a follow-up study among children with epilepsy and health children. Fam. Pract., 1991; 8: 320-328.
75. Darling RB. Families against societies. A study of reactions to children with birth defects. London/Beverley Hiils: Sage Publications, 1979.
76. Hewett S. The family and the handicapped child. London: Allen and Unwin, 1970.
77. Sallou C. Ouders en kind in hetzelfde schuitje. Psychodynamische ontwikkeling en de rol van de ouderstructuur. Epilepsie Bull., 1990; 18: 16-18.
78. Suurmeijer ThPBM. Ouders en kind in hetzelfde schuitje. Ja, maar wat voor schuitje? Epilepsie Bull., 1990; 18: 28-29.
79. Austin JK, McDermott N. Parental attitudes and coping behaviors in families of children with epilepsy. J. Neurosci. Nurs., 1988; 20: 174-179.
80. Hoare P. The quality of life of children with chronic epilepsy and their families. Seizure, 1994; 2: 269-275.
81. Aldenkamp AP, Bijvoet ME, Heisen ThWM, et al. The influence of epilepsy in parents on psychosocial functioning of their children. In: Manelis J, Bental E, Loeber J, et al, editors. Advances in epileptology, Vol 17. New York: Raven Press, 1989: 450-456.
82. Aldenkamp AP, Suurmeijer ThPBM, Bijvoet ME, et al. Emotional and social reactions of children to epilepsy in a parent. Fam. Pract., 1990; 7: 110-115.
83. Lechtenberg R, Akner L. Psychologic adaptation of children to epilepsy in a parent. Epilepsia, 1984; 25: 40-45.

84. LeClere FB, Kowalewski BM. Disability in the family - the effects on childrens well-being. J. Marriage Fam., 1994: 56: 457-468.
85. Chaplin JE, Yepez R, Shorvon S, et al. A quantitative approach to measuring the social effects of epilepsy. Neuroepidemiology,1990; 9: 151-158.
86. Chaplin JE, Yepez R, Shorvon SD, et al. National general practice study of epilepsy: the social and psychological effects of a recent diagnosis of epilepsy. BMJ, 1992; 304: 1416-1418.
87. Jacoby A. Epilepsy and the quality of everyday life: findings from a study of people with well controlled epilepsy. Soc. Sci. & Med., 1992; 34: 657-666.
88. Upton D. Social support and emotional adjustment in people with chronic epilepsy. J. Epilepsy, 1993; 6: 105-111.
89. Pelletier L, Godin G, Lepage L, et al. Social support received by mothers of chronically ill children. Child Care Health Dev., 1994; 20: 115-131.
90. Krol B, Sanderman R, Suurmeijer ThPBM. Social Support, Rheumatoid Arthritis and Quality of Life: Concepts, Measurement and Research. Patient Educ. and Couns., 1993; 20: 101-120.
91. Manificat S, Guillaud-Bataille JM, Dazord A. La qualité de vie chez l'enfant atteint de maladie chronique. Revue de la litterature et aspects conceptuels. Pediatrie, 1993; 48: 519-527.

16 Counselling and rehabilitation; the clinician point of view

MATTI SILLANPÄÄ

Department of Child Neurology, University of Turku Hospital TYKS, Turku, Finland

Counselling and rehabilitation of a child and its family starts at the time of diagnosis. Treatment, management, counselling, habilitation and rehabilitation are all different processes that may be distinguished but may never be separated. A confident relationship between patient and physician is the cornerstone of a successful rehabilitation. Rehabilitation is defined here as the total of all efforts to aid the child and his/her family to meet and solve the daily life problems, given their possibilities.

INTRODUCTION

The onset of epilepsy often requires a drastic change of social attitudes and behaviour for both the child and his family. Feelings of dread, guilt, anger, shame and anxiety are not an exception and changes in daily routines are needed. As a result, the patient and his family often show symptoms of acute crisis during the first phase after the diagnosis, characterized by disorganized behaviour, defence reactions (such as denial) and psychosomatic reactions. These acute reactions may be followed by acceptance, followed by readaptation to the new circumstances.

Some children with epilepsy do not have other medical or psychosocial problems. For them epilepsy is more an inconvenience than a disability. However, many other children have associated impairments or handicaps. The epilepsy and the associated impairments usually have an organic aetiology in common, but secondary disabilities may also occur.

In our study on 203 unselected patients with childhood epilepsy, mild to severe handicaps were found in 37% of the sample.[1] During follow-up 60% showed mild signs of deterioration (five points or more decline on an intelligence tests). Similar results have been reported in other studies.[2-5]

In all circumstances, it is important that the diagnosis of epilepsy is confirmed before the patient and his family is given further information to avoid unnecessary anxiety. Moreover the child and his family must be allowed to have an active role during the process following the diagnosis.

After the first seizure, the patient and the family should always be informed what to do and where to establish contact in case of further seizures. They should also, from the beginning, become accustomed to use a seizure diary. Counselling of the patient and the family in this phase includes reassurance and advice on e.g. rectal administration and reassurance of the low risk of seizures for life and health. In addition it is often helpful to provide the child and its family with a written explanation about the necessary investigations, diagnostic procedures and treatment. This makes it much easier for them to explain the situation to relatives, friends and other persons that are involved, such as the school teacher.

THE START OF TREATMENT

Drug treatment

After recurrent unprovoked seizures, drug therapy must be considered. It is however important that the physician does not simply 'prescribe' a drug. Patients and parents often have a negative attitude towards therapy and are worried about side-effects. Especially the effects on cognitive function[6-13] which may have an impact on daily life function. The risks for such side-effects have to be discussed, as well as the advantages and disadvantages of drug therapy.

It is the patient and the family who ultimately decide whether or not to take the medication and to comply in other respects. Patients with chronic diseases miss 20-50% of scheduled physician appointments and about half of them do not take their medication as instructed.[14] A number of studies suggest that physicians often overestimate compliance of their patients.[15-18] Methods to check compliance, such as tablet count or blood levels often fail. Poor compliance does not seem to depend on any particular personality traits[19] but is correlated with the type of disease and type of drug therapy. Patients with a chronic epilepsy, polytherapy, multiple dosing, adverse effects and low seizure frequency are typically patients with a high risk for non-compliance.[20,21] Misunderstandings about the incompatibility between alcohol and AED's is reported in adolescents as a reason for not taking the regular doses during week-ends.

To improve compliance, the importance of regular use of medication should be discussed in detail with the parents, and, if possible with the children. This discussion should also focus on the risks of non-compliance. Unintentional errors may be avoided by using medication containers or a medication diary. Moreover, the daily medication schedule may be alingned with daily routines such as the meals.[22]

Follow-up visits

The first follow-up visit should take place 4-6 weeks after the initial visit. The patient should, however, have a possibility to contact the doctor or the hospital any time. Frequent visits are preferable in the initial phase, after the onset of treatment as the patients and parents will have many questions about the seizures, the prognosis and the therapy. If there are no additional problems, the visits may be scheduled with longer intervals, but a minimum of one visit per year must be maintained. In the meantime, the patient must have the opportunity to contact the physician.

Patients and parents often rely on results of laboratory examinations. Regular blood samples are therefore often requested, but it should be stressed to the parents and the child that the physician does not treat blood levels, EEG-abnormalities or results of brain imaging, but the whole clinical situation.

SOCIAL REHABILITATION AND COUNSELLING

Methods of counselling

Epilepsy is characterized by episodes with sudden loss of control.[23] Even when the seizures are controlled, the epilepsy may still disrupt family life.[24,25] Counselling of the child and the family should always be a tailormade process. Counselling must be offered frequently at the onset of treatment, possibly starting with separate meetings with the child, parents and, if needed, with other family members. The problems may then be discussed confidentially and fears and misconceptions can be disclosed. A series of discussions with the patient and family members is then followed in an attempt to reach consensus about the daily life problems that the child and the family have mentioned during the separate meetings. This process improves compliance, prevents fears and anxiety and supports the self-esteem of the child.

Sometimes it is beneficial to have group sessions on problems, common to many patients. The participants of such group meetings may be patients only, patients and their family, or parents only. Voluntary organisations have their role in complementing public health care, as they may offer opportunities to bring people with similar problems together. This may be realized with self-help groups.

Social competence

The development of self-esteem is influenced by the way a child is disciplined. Each child has to have opportunities to experience responsibility, making choices and understanding their consequences. Children with epilepsy may have many negative feelings, e.g. shame, anger, fear for stigmatization, concern about the loss of self-control, and resentment over dependency.[27] Factors causing social incompetence include biological factors (such as poor seizure control), psychosocial factors (separation from one of the parents) and medication (adverse effects).[28] Several studies have shown that chronic epilepsy has an effect on the development of the personality and 'being an epileptic' may become part of the personality.[29,30] After the onset of epilepsy, changes in the role as a family member may occur. The child may adopt a role as a 'sick person', resulting in a more passive role in family interaction and in decision making, while parental control over the child tends to increase.[31] Children with epilepsy are generally more dependent on and overprotected by their parents.[32-34] One factor that is involved in this process, is the change of parental expectations after the onset of epilepsy. Children with epilepsy are perceived by their parents to be less adept in school performance and sports, to be less reliable, and more likely to have emotional problems.[35,36]

Information to the family members

Before the parents can inform others about the epilepsy, they must have a clear understanding of the problem. Next the child should be informed. Naturally, the child's developmental and emotional problems must be taken into account when estimating his/her ability to understand the information.

Siblings may feel neglected when much of the parental attention is focused on the child with epilepsy. Siblings may go through the same feelings such as anger, fear for loss of their parents' love, they may feel guilty about the epilepsy or they may fear that they also will develop epilepsy. Parents should be encouraged and helped to discuss these feelings and problems thoroughly. A group with children with similar problems may be helpful.

Education and work

Children with epilepsy may have learning problems. In a Finnish study[37] children aged 8-19, classified according to the WHO classification[38] were followed during four years. One third (36%) of these children was in special education. The percentage of children with epilepsy referred to special education was higher than children with cerebral palsy or minimal brain dysfunction, but lower compared to children with bronchial asthma, diabetes mellitus or heart disease.

In an other epidemiological study[39] 147 children were followed during 30 years. In this study 26% had no basic education, 6% had uncompleted primary education, 44% had completed primary education, 11% completed lower secondary education and 13% passed the examinations. The factors that correlated with uncompleted education were mental retardation, organic aetiology and a history with status epilepticus. The multifactorial basis for academic underachievement has been confirmed in many studies.[40,41]

A multidisciplinary approach is necessary to cover all areas where special education and rehabilitation is needed. Such assessment prepares the design of an individualized 'tailor-made' educational plan.

Table 1 illustrates a number of disciplines that may be involved in rehabilitation-assessment.[42-46]

Table 1. Multidisciplinary approach to the assessment of a child with epilepsy

Area	Specialist	Skills assessed
Medical	Child neurologist	Brain maturation, developmental and neurological deficits, aetiological investigations, conclusions and plans and recommendations for future rehabilitation measures and follow-up visits
	Child psychiatrist	Psychiatric problems, family mental health, sociopsychiatric aspects
	Audiologist	Hearing
	Ophthalmologist	Vision
Social self-help	Habilitation councelor, social worker	Family environment, functioning, stresses, coping capacity
Ward behaviour	Nursing staff	Observations of activities of daily living, interaction with other children in play and other coexistence, sleep
Cognitive, learning, behaviour	Neuropsychologist Educational specialist	IQ level, learning styles, achievement, thinking, reasoning, understanding, motor, sensory and perceptual planning, detailed evaluation of attention, memory, and sensory processing, especially in relation to antiepileptic drugs
Language, speech	Speech therapist	Receptive and expressive speech, narrative speech, vocabulary, verbal abilities, pronounciation, articulation
Fine motor, perceptual and sensory processing	Occupational therapist	Receiving and using sensory input for accurate movement and organization of behaviour, using small muscles of the hands, face and eye for skilled functions, using small and large muscles in combination to perform activities in daily life
Gross motor	Physical therapist	Use of large muscles to develop quality of movement patterns and to increase strength and endurance in walking, running, skipping, balance, throwing, kicking, coordination

Adolescents with epilepsy have a high risk for developing psychosocial problems. Such problems may be patient-related or environment-related. Patient-related problems may be secondary disabilities, cognitive impairment, severe and/or frequent seizures, complex AED treatment. Environmental factors may pertain to parental fears, parental style, or failure to cope with peer group demands.[47] Vocational guidance for adolescents contains, more or less, the same type of assessment as needed for proper education in younger children. It is important to emphasize that the risks of people in working life are overestimated. Accidents at work and absenteeism are not seen more frequently in people with epilepsy than in other workers. Life style may, however, interfere with the epilepsy in adolescents. Sleep deprivation, often combined with the use of alcohol and non-compliance may provoke seizures. Irregular meals may cause fluctuations in tissue concentrations of antiepileptic drugs. These aspects should be considered in counselling children and their parents.

REFERENCES

1. Sillanpää M. Social functioning and seizure status of young adults with onset of epilepsy in childhood. An epidemiological 20-year follow-up study. Acta Neurol. Scand., 1983; 68 (96S): 1-81.
2. Schlack HG. Zur Prognose der Intelligenz- und Sozialentwicklung anfallskranker Kinder. Monatschr. Kinderheilk., 1974; 122: 676-678.
3. Steinhäuser VA, Wagner KD, Kulz J. Longitudinalstudie zur Entwicklung anfallskranker Vorschulkinder. Kinderärztl. Prax., 1976; 11: 494-500.
4. Rodin EA, Schmaltz S, Twitty G. Intellectual functions of patients with childhood-onset epilepsy. Devel. Med. Child Neurol., 1986; 28: 25-33.
5. Bourgeois BFD, Prensky AL, Palkes HS, et al. Intelligence in epilepsy: a prospective study in children. Ann. Neurol., 1983; 14: 483-444.
6. Nolte R, Wetzel B, Brugmann G, et al. Effects of phenytoin- and primidone-monotherapy on mental performance in children. In: Johannessen SI, Morselli PL, Pippenger CE, et al. Antiepileptic therapy: Advances in drug monitoring. New York: Raven Press 1980: 81-86.
7. Stores G, Hart J. Reading skills of children with generalised or focal epilepsy attending ordinary school. Devel. Med. Child Neurol., 1976; 18: 705-716.
8. Hellstöm B, Barlach-Christoffersen M. Influence of phenobarbital on the psychomotor development and behaviour in preschool children with convulsions. Neuropädiatrie, 1980; 11: 151-160.
9. Guey J. Charles C, Coquery C, et al. Study of psychological effects of ethosuximide (Zarontin) on 25 children suffering from petit mal epilepsy. Epilepsia, 1967; 8: 129-141.
10. Trimble MR, Corbett JA. Behavioural and cognitive disturbances in epileptic children. Irish Med. J., 1980; 73 (10S); 21-28.
11. Schain RJ, Ward JW, Guthrie D. Carbamazepine as an anticonvulsant in children. Neurology, 1977; 27: 476-480.

12. Barnes SE, Bower BD. Sodium valproate in the treatment of intractable childhood epilepsy. Devel. Med. Child Neurol., 1975; 17: 175-181.
13. Browne TR, Dreifuss FE, Dyken PR, et al. Ethosuximide in the treatment of absence (petit mal) seizures. Neurology, 1975; 25: 515-524.
14. Sackett DL. The magnitude of compliance and non-compliance. In: Sackett DL, Haynes RB, editors. Compliance with therapeutic regimens. San Francisco: Jossey-Bass 1976: 189-195.
15. Mushlin AI, Appel FA. Diagnosing patient noncompliance. Arch. Int. Med., 1977; 137: 318-321.
16. Caron HW, Roth HP. Patients' cooperation with a medical regimen. JAMA, 1968; 203: 922-926.
17. Charney E, Bynum R, Eldreedge D, et al. How well do patients take oral penicillin? A collaborative study in private practice. Pediatrics, 1967; 40: 188-192.
18. Rickels K, Priscol E. Assessment of dosage deviation in outpatient drug research. J. Clin. Pharmacol., 1970; 10: 153-160.
19. Haynes RB, Taylor OW, Sackett D, editors. Compliance in health care. Baltimore: John Hopkins University Press, 1979.
20. Epstein LH, Cluss PA. A behavioural medicine perspective on adherence to long-term medical regimens. J. Consult. Clin., 1982; 50: 950-971.
21. Peterson GM, McLean S, Millengen KS. Determinants of patient compliance with anticonvulsant therapy. Epilepsia, 1982; 23: 607-613.
22. Sackett DL. The hypertensive patient: 5. Compliance with therapy. Can. Med. Ass. J., 1979; 121: 259-261.
23. Sahlholdt L. Psychosocial consequences of intractable epilepsy in children. In: Johannessen SI, Sillanpää M, Osterman PO, Gram L, editors. Intractable epilepsy. Petersfield: Wrightson Biomedical Publishing, in press.
24. Berg BO. Prognosis of childhood epilepsy - another look. New Engl. J. Med., 1982; 306: 861-862.
25. Schachter SC. Advances in the assessment of refractory epilepsy. Epilepsia, 1993; 34 (5S): 24-37.
26. Appolone C. Preventive social work intervention with families of children with epilepsy. Soc. Work Health Care, 1978; 4: 139-148.
27. Horowitz MJ. Psychosocial processes induced by illness, injury and loss. In Millon T, Green C, Meather R, editors. Handbook of clinical health psychology. New York: Plenum, 1982: 53-67.
28. Hermann BP, Whitman S, Dell J. Correlates of behaviour problems and social competence in children with epilepsy, aged 6-11. In: Hermann BP, Seidenberg M, editors. Childhood epilepsies: Neuropsychological, psychosocial and intervention aspects. Chichester: John Wiley & Sons, 1989: 143-157.
29. Taylor DC. Psychosocial components of childhood epilepsy. In: Hermann BP, Seidenberg M, editors. Childhood epilepsies: Neuropsychological, psychosocial and invention aspects. Chichester: John Wiley & Sons, 1989: 119-142.
30. Taylor DC. The components of sickness: diseases, illnesses and predicamants. Lancet, 1979; 2: 1008-1010.
31. Ritchie K. Interaction in families of epileptic children. J. Child. Psychol. Psychiatry, 1981; 22: 65-71.
32. Mulder HC, Suurmeijer, TBPM. Families with a child with epilepsy: a sociological contribution. J. Biosoc. Sci., 1977; 9: 13-24.

33. Stores G, Piran N. Dependency of different types of schoolchildren with epilepsy. Psychol. Med., 1978; 8: 441-445.
34. Ferrari M, Matthews WS, Barabas G. The family and the child with epilepsy. Family Process, 1983; 22: 53-59.
35. Ferrari M. Epilepsy and its effects on the family. In: Hermann BP, Seidenberg M, editors. Childhood epilepsies: Neuropsychological, psychosocial and intervention aspects. Chichester: John Wiley & Sons, 1989: 159-172.
36. Safilios-Rothchild C. The sociology and social psychology of disability and rehabilitation. New York: Random House, 1970.
37. Sillanpää M. Social adjustment and functioning of chronically ill and impaired children and adolescents. Acta Paediatr. Scand., 1987; (340S): 1-70.
38. WHO. Classification of impairments, disabilities and handicaps. A manual of classification relating to the consequences of disease. Geneva: World Health Organization, 1980.
39. Sillanpää M. Children with epilepsy as adults: Outcome after 30 years of follow-up. Acta Paediatr. Scand., 1990; (368S): 1-78.
40. Hermann BP, Whitman S. Behavioural and personality correlates of epilepsy: A review, methodological critique and conceptual model. Psychol. Bull., 1984; 95: 451-497.
41. Seidenberg M. Academic achievement and school performance of children with epilepsy. In: Hermann BP, Seidenberg M, editors. Childhood epilepsies: neuropsychological, psychosocial and intervention aspects. Chichester: John Wiley & Sons, 1989: 105-118.
42. Sillanpää M. Epilepsy in children: Prevalence, disability and handicap. Epilepsia, 1992; 33: 444-449.
43. Diamond L, Anderson S, Berg H, et al. An introduction to special education. In: Reisner H, editor. Children with epilepsy. A parent's guide. Washington DC.: Woodbine House, 1988: 171.
44. Bagley CR. The educational performance of children with epilepsy. Brit. J. Educ. Psychol., 1970; 40: 82-83.
45. Rutter M, Graham P, Yule WA. Neuropsychiatric study in childhood. London: SIMP/Heinemann, 1970.
46. Yule W. Educational achievement. In: Kulig BM, Meinardi H, Stores G, editors. Epilepsy and behaviour. Lisse, The Netherlands: Swets & Zeitlinger, 1980: 162-168.
47. Fraser RT, Clemmons DC. Vocational and psychosocial interventions for youths, with seizure disorders. In: Hermann BP, Seidenberg M, editors. Childhood epilepsies: Neuropsychological, psychosocial and interventions aspects. Chichester: John Wiley & Sons, 1989: 201-219.
48. Berent S, Sckallares JC. Clinical monitoring of children with epilepsy: a neurologic and neuropsychological perspective. In: Hermann BP, Seidenberg M, editors. Childhood epilepsies: Neuropsychological, psychosocial and interventions aspects. Chichester: John Wiley & Sons, 1989: 15-31.

17 Assessment of quality of life in children and adolescents with epilepsy

GUS A. BAKER
University Department of Neurosciences, Walton Centre for Neurology and Neurosurgery, Liverpool, UK

ANN JACOBY
Centre for Health Services Research, University of Newcastle upon Tyne, Newcastle/Tyne, UK

In epilepsy 'quality of life' has come to be seen as an important outcome of health care. Definitions of the term 'quality of life' vary, depending on the context in which they are applied, but it is generally agreed that any assessment of quality of life should include measures of physical, social and psychological functioning. There are a number of important methodological issues in assessing quality of life in any age group, including who should make the assessment, what measures should be used to make it, and how to ensure that such measures are psychometrically sound. In measuring quality of life in children and adolescents, some further methodological considerations have to be taken into account. A number of studies have looked at specific aspects of quality of life in children and young people, though few have involved a comprehensive assessment. The findings from research into quality of life have important implications for clinical practice, since they enhance understanding of the impact of the condition and its treatment. There is a need to develop methodologically sound quality of life measures for children and adolescents, which can be used to assess the outcome of care.

INTRODUCTION

What is known about the impact of epilepsy on cognitive and emotional development, social integration and overall 'quality of life' of children and adolescents has been discussed in earlier chapters. In this chapter, a number of important methodological considerations in assessing such 'quality of life' issues will be discussed. Addressing these methodological issues satisfactorily will ensure that meaningful information about quality of life will be collected, which will enhance understanding of the overall impact of epilepsy and thus enable clinicians to target care more effectively. We will begin by discussing the concept and definition of quality of life (QOL), and then consider a number of measurement issues, some applicable to assessing QOL in all age groups and some relating specifically to assessing QOL in young people. The application of QOL measures in studies of young people with epilepsy will be reviewed, and the relevance of their findings to clinical practice will be considered.

What is meant by the term 'quality of life'?

For people with a chronic condition, where cure is not a generally attainable goal and therapy is often prolonged, quality of life has come to be seen as an important outcome of health care. However the concept of quality of life remains vague and there is little agreement about its definition.[1] It has been proposed that quality of life should be considered as an affective response to one's role situations and values;[2] as satisfaction with physical, psychosocial and economic needs;[3] and as the discrepancy between an individual's desired and actual circumstances.[4]

Definitions of quality of life have varied according to their application,[5] but a good general definition is suggested by Fallowfield,[1] who states that quality of life is *"a complex amalgam of satisfactory functioning in terms of physical, social, psychological and vocational well-being"*. Recently, emphasis has been placed on the importance of illness and its contribution to patients' perception of their quality of life overall, and subsequently interest has focused on the concept of health-related quality of life (HRQOL). The term *"health-related quality of life"* has been defined as *"the functional effect of an illness, and its consequent therapy upon the patient, as perceived by the patient"*.[6]

Why is it important to measure quality of life?

Health-related quality of life measures have evolved and expanded rapidly over the last two decades for chronic conditions such as cancer, arthritis, diabetes and cardiovascular disease.[5] Until recently, however, little attention has been paid to developing parallel measures for epilepsy, as demonstrated by a search of the epilepsy literature (figure 1).

It is increasingly accepted that quality of life measures are important in addressing particular clinical questions such as when to start or stop antiepileptic drug treatment;[7] the efficacy of novel antiepileptic drug treatment;[8] and the outcome of surgical intervention.[9] Understanding patients' perceptions of the impact of particular treatments contributes to the overall assessment of their effectiveness, and this applies as much to studies involving children and adolescents as adults.

METHODOLOGICAL ISSUES IN MEASURING QUALITY OF LIFE

There are a number of methodological issues common to QOL assessment in both adults and children, including whether the assessment should be made by the physician or the patient, whether to use generic or disease-specific measures, and the need for measures that are psychometrically sound. Quality of life measures developed for use with adult populations cannot easily be applied to children, without substantial modification.[10]

There are a number of issues relating specifically to the measurement of QOL in children and adolescents, since this must be done against a background of developmental change and non-predictability.[11]

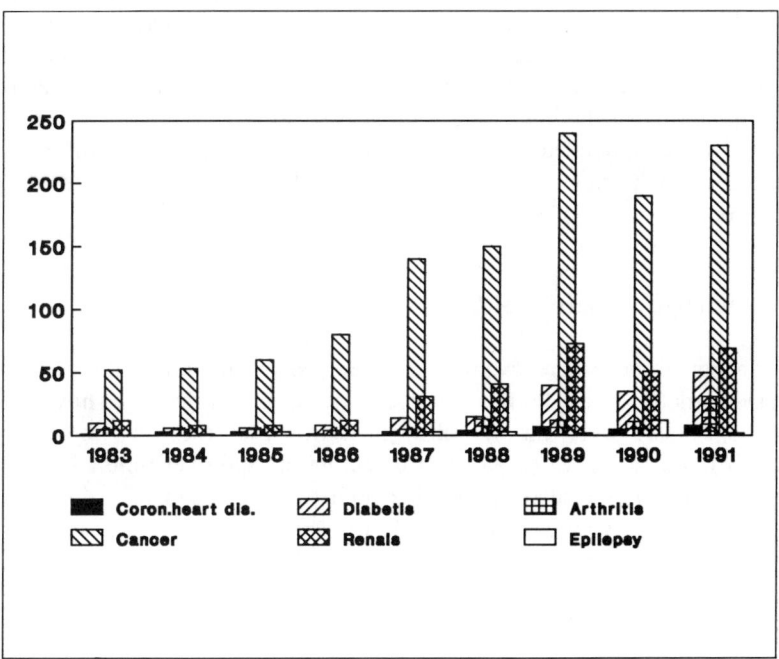

Figure 1. Trends in Quality of Life research for chronic conditions.

Quality of life assessments subjective or objective?

Judgements about 'quality of life' are now recognised as essentially subjective, and a number of studies have shown there is little agreement between physicians and, more importantly, between patients and physicians, even for apparently 'objective' measures.[12] As a consequence, the majority of QOL assessment tools are administered to patients themselves, and have been developed on the basis of patients' perceptions of the impact of their condition and its treatment.[13]

Generic versus disease-specific measures?

One important consideration is whether to employ a generic measure which will cover a broad range of domains and can be applied to a number of different conditions.[14-16] The advantage of this approach is that it allows for comparisons across different diseases. A major limitation, however, is that a generic measure may not be sensitive enough to capture the unique aspects of a disease or condition; nor may it be sensitive to measuring changes that occur as a result of the treatment of the condition.

In contrast, even though its applicability to other conditions may be limited, a disease-specific measure can focus on areas of particular relevance, and it has been argued that disease-specific measures are most appropriate for assessing treatment effects in clinical trials.[17] To overcome the limitations of both generic and disease-specific measures, it is now generally recommended that, where possible, a combination of both be used. One such example is the work of Vickery et al.[9] in assessing the outcome of surgery for resistant epilepsy.

Psychometric properties of QOL measures

We do not propose here to provide a detailed discussion of the principles of psychometric testing[1], but quality of life measures must be shown to be reliable, (stable over time); valid (measure what they are intended to measure); and sensitive (capable of detecting change over time). Any QOL questionnaire should also be sensible both to patient and investigator, and it should be relatively easy and simple to complete.

Standardisation of measures

There are a number of different initiatives ongoing to develop 'quality of life' measures for use in clinical trials of novel antiepileptic drugs.[18-20] However, each of these various initiatives employs different measures, and where different measures are used, it can be difficult, if not impossible, to make meaningful cross-trial comparisons. This raises the question of whether there is, therefore, a need for standardised quality of life measures which would allow comparisons to be made more readily. Nonetheless, we would argue that *no one* measure will ever satisfactorily address all QOL issues in epilepsy - the nature of the questions asked of patients, and the measures used to ask them will always be highly dependent on the research question to be answered and should vary accordingly.

What domains should be measured?

A criticism of previous research has been the inappropriate use of QOL measures, arising from a failure to provide a clear definition of what was meant by the term 'quality of life' and the hypotheses about what changes, if any, were expected.[5] A further criticism we would make is that some investigators have chosen to measure only discreet elements of quality of life yet have claimed to be measuring quality of life overall.

It is generally agreed that QOL questionnaires should contain, at a minimum, measures of physical and social functioning and psychological well-being. Precisely which measures will depend on a number of factors including; the age of the study group, their particular health condition and what questions are being addressed. For example, questions about employment will have less relevance for elderly people than for adults of employable age, and will be irrelevant to children of school age. Likewise, in a clinical trial of a novel anti epileptic drug with a relatively short timespan, there will be little expectation of change in social circumstances as a result of treatment, and therefore collecting information about this will be redundant.

Issues specific to measuring quality of life in children and adolescents

In children, each different developmental stage will bring its own unique problems. Eisner[21] has pointed out that the rapid physical, cognitive and emotional changes that occur through childhood create a number of theoretical difficulties in the measurement of quality of life. For this reason, QOL studies of young people should probably focus on a fairly narrow age-range. An alternative approach, adopted by Stein and Jones Jessop[22] in developing a functional status scale for children, would be to develop a set of core items applicable to all children, together with subsets of age-specific items. To our knowledge, all the studies to date of QOL issues for children and adolescents with epilepsy have focused on school-age children. QOL data about children with epilepsy have, almost without exception, been obtained from parents rather than from children themselves. Recently, however, some attempts have been made to collect information from children and adolescents directly.[23]

APPLICATION OF QOL MEASURES IN STUDIES OF CHILDREN AND ADOLESCENTS WITH EPILEPSY

A number of studies have looked at specific aspects of quality of life in young people with epilepsy, particularly the psychological,[24] neuropsychological,[25] developmental and educational[26,27] aspects. Research has also focused on issues such as self-esteem,[28] locus of control,[29] levels of independence[28] and feelings of stigma.[30] Relatively few studies, in our view, have involved a comprehensive QOL assessment. In this section, we review some examples of studies, both descriptive and experimental, which have done so.

Descriptive studies

Austin and Dunn[23] have compared the quality of life of 136 children with newly diagnosed epilepsy with that of their siblings, using a questionnaire that examined four domains: physical, psychological, social and educational functioning. Parents were asked to assess their child's functioning at the time of their first seizure and six months prior to seizure activity beginning. Children with epilepsy were reported as having significantly more thought and attention problems, and delinquent and aggressive behaviour when compared with their siblings. These problems were significantly reduced one month after the first seizure and coincided with a reduction in parental anxiety and an increase in parental supervision and family activity. In a recent community study of 'quality of life' and 'quality of services' for an unselected population of people with epilepsy,[31] postal questionnaires were sent to the parents of 93 children with epilepsy, aged between five and fifteen. The questionnaire included scales and items addressing four domains: treatment; health status; psychiatric morbidity; and the perceived impact of epilepsy on social functioning (Table 1). Results of this study highlighted a number of important issues, including: high rates of reported side effects associated with AED's; high levels of psychiatric morbidity; and a clear relationship between seizure frequency and parents' perceptions of the impact of epilepsy on the quality of their child's daily life (Table 2).

Table 1. Perceived impact of epilepsy and treatment on children's psychosocial function

	Seizure frequency:	
	Seizure-free in last 12 months (n=26)	One or more seizures in last month (n=36)
Epilepsy affects a lot/some:		
Relationship with parent respondent	15%	25%
Relationship with respondent's spouse	12%	17%
Relationship with siblings	12%	28%
Relationship with other children	12%	31%
Hobbies and interests	12%	31%
Schooling	23%	50%
Overall health	4%	39%
Feelings about self	19%	31%
Plans and ambitions for the future	24%	50%

Table 2. QOL measures used in the UK community study

	Children, 5-15 years	Adolescents, 16-18 years
Physical functioning:		
Seizure severity (36)	*	*
Current AED therapy	*	*
Associated AED side-effects (37)	*	*
Compliance with treatment	*	*
Long-term health problems	*	*
Overall health	*	*
Psychosocial functioning:		
Behavioural symptoms checklist (26)	*	-
Impact of epilepsy scale (38)	* (child version)	* (adult version)
Hospital Anxiety and Depress. scale (39)	-	*
Life fulfilment scale (40)	-	*

* Scale included in the questionnaire

Adolescents aged 16-18 years completed the same questionnaire as was sent to adults in the study, which contained measures of seizure frequency and severity, the side effects of antiepileptic drug treatment, life fulfilment, the impact of epilepsy, anxiety and depression and the perceived stigma of epilepsy.

Experimental studies involving children and adolescents

Measurement of the outcome of treatment in this age group has generally focused on seizure frequency or the side effects of antiepileptic drug treatment[32] and has only rarely included any QOL assessment, a shortcoming attributed to the lack of appropriate and validated scales.[10] While seizure frequency and AED side-effects are clearly important parameters for assessing the efficacy of treatment from a clinical or pharmaceutical perspective, children and parents may consider improvements in cognitive performance, behaviour and mood equally as important, particularly in the severe epileptic disorders, where the goal of seizure freedom is unrealistic.

One recently reported study that included outcome measures other than seizure frequency or cognitive effects is the study of the efficacy of Felbamate in children with the Lennox-Gastaut syndrome.[33] The authors used a quality of life measure that incorporated parents' evaluation of alertness, general well-being, verbal abilities and seizure control. The results from this study were favourable, in that parents whose children were on the active drug rated them as having better quality of life than those on comparative treatment, a finding supported by physicians' ratings. One criticism of this study is that no evidence of the psychometric properties of the measures used has been published.

In a clinical trial to assess the efficacy of Lamotrigine in treatment of the severe epileptic disorders of childhood[34] a questionnaire, intended for completion by parents, was developed from three sources: clinical anecdotal evidence from an open study of 51 children with intractable epilepsy and learning difficulties, in depth interviews with parents of children with epilepsy and learning difficulties to identify key 'quality of life' issues, and discussion with expert physicians in the field of paediatric neurology. The psychometric properties of the questionnaire were assessed in a cohort of the parents of 50 children, and it was found to have good content and construct validity and high reliability. The contents of the questionnaire are presented in Table 3.

Table 3. QOL measures included in trials of novel antiepileptics for treatment of Lennox-Gastaut and other severe epileptic disorders of childhood

Felbamate Study	Lamotrigine Study
Alertness	Seizure severity
Verbal responsiveness	Seizure-related injuries
General well-being	AED side-effects
Seizure control	Behavioural disturbance
	Mood disturbance
	Overall health
	Overall QOL
	Parental anxiety

RELEVANCE OF QOL RESEARCH TO CLINICAL PRACTICE

Quality of life measures can be used to enhance our understanding of the impact of epilepsy and its treatment and may enable us to chart the development of psychopathology in children and adolescents who develop this condition. They are valuable as outcome measures in studies of pharmacological and surgical interventions, particularly where complete seizure control cannot be achieved. Another important aspect of assessing quality of life in such studies is that it allows parents and children to be actively involved in the assessment of particular treatment programmes, creating an opportunity for care which is genuinely 'shared' between physician, parent and child.

Dulac & Bulteau[10] state that while much is known about the epidemiology of childhood epilepsy, the "QOL of the whole population of epileptic children, in a given area and in a given age range, remains unexplored" and this lack of information is likely to contribute to the inadequate provision of appropriate specialised services for this population. Descriptive studies of the quality of life in children and adolescents with epilepsy may be valuable in identifying specific problem areas, and such information could be used by public health services to determine priorities about care and to suggest potential improvements.

Findings from QOL studies can also be used to improve knowledge of the impact of epilepsy both at home and at school. At home, not only the child him or herself, but also parents and siblings need to understand the meaning of epilepsy and its implications for QOL. At school, teachers' knowledge about epilepsy needs to be increased, so that the restrictions often imposed by a diagnosis of epilepsy are minimised.

The treatment of children and adolescents with epilepsy may involve a number of professionals including the general physician, specialist nurses and the multidisciplinary hospital team including neurologists, neuropsychologists, speech therapists, educational psychologists and social workers. Quality of life measures can be useful to identify children's and parents' concerns about epilepsy - and these concerns can then become the target for interventions from health care specialists helping both children and their families to better adjust to their condition.

CONCLUSIONS

Sabbeth and Stein[35] have argued that it is no longer sufficient *"to provide biomedical care and technically sophisticated interventions, without appropriate attention to the psychological and social impact both of the child's condition and treatment"*. While there are a number of important methodological difficulties in assessing quality of life in children and adolescents, just as in adults, we believe that the value of doing so is clear. Developing methodologically sound measures, which take account of the particular issues of childhood and adolescence, is not a simple task, but their inclusion as an outcome of care will provide an important extra dimension in its assessment.

Acknowledgements
We would like to thank Professor David Chadwick, for his support for our work, and all the individuals, of all ages, who have completed QOL questionnaires for us. Our thanks also to Wellcome Trust and the Wellcome Foundation, for funding the two studies we have reported here.

REFERENCES

1. Fallowfield L. Quality of life: The missing measurement in health care, London: Souvenir Press, 1990.
2. Andrews FM, Withey SB. Social Indicators of Well-being. New York: Plenum Press, 1976.
3. Bubolz M, Eicher JB, Evers SJ, et al. A human ecological approach to quality of life: conceptual framework and results of a preliminary study.

Social Indicators Research, 1980; 7: 103-106.
4. Calman KC. Quality of life in cancer patients - a hypothesis. Journal of Medical Ethics, 1984; 10: 124-127.
5. Spilker B. Quality of life assessments in clinical trials. New York Raven Press, 1990.
6. Schipper H, Clinch J, Powell V. Definitions and conceptual issues. In: In: Spilker B, editor. Quality of life assessments in clinical trials. New York: Raven Press, 1990.
7. Jacoby A, Johnson AL, Chadwick DW. Psychosocial outcomes of antiepileptic drug discontinuation. Epilepsia; 1992; 33: 1123-1131.
8. Smith DF, Baker GA, Davis G, et al. Outcomes of add-on treatment with lamotrigine in partial epilepsy. Epilepsia, 1993; 34: 312-322.
9. Vickrey BG, Hays RD, Graber J, et al. A health-related quality of life instrument for patients evaluated for epilepsy surgery. Medical Care, 1992; 30: 299-319.
10. Dulac O, Bulteau C. Quality of life in children with epilepsy. In: Quality of life and quality of care in epilepsy: Update 1993. Chadwick DW, Baker GA, Jacoby A, editors. London RSM 1993.
11. Eisner C, et al. Severity of asthma and parental discipline practices. Patient Education and Counselling, 1991; 17: 227-233.
12. Slevin MI, Plant H, Lynch D, et al. Who should measure quality of life, doctor or patient? British Journal of Cancer, 1988; 57: 109-112.
13. Bowling A. Measuring health: a review of quality of life measurement scales. Milton Keynes: OUP 1991.
14. Bergner M, Bobbitt RA, Carter WB, et al. The Sickness Impact Profile: development and final revision of a health status measure. Medical Care, 1981; 19: 787-805.
15. Hunt SM, McKenna SP, McEwan J, et al. A quantitative approach to perceived health status: a validation study. Journal of Epidemiology and Public Health, 1980; 34: 281-286.
16. Ware JE, Sherbourne CD. The MOS 36-item short-form health survey (SF-36) I. Conceptual framework and item selection. Medical Care, 1992; 30: 473-481.
17. Guyatt GH, Bombardier C, Tugwell PX. Measuring disease-specific quality of life in clinical trials. Canadian Medical Association Journal, 1986; 134: 889-895.
18. Baker GA, Smith DF, Dewey M, et al. The initial development of a health-related quality of life model as an outcome measure in epilepsy. Epilepsy Research, 1993; 16: 65-81.
19. Devinsky O. Clinical uses of the quality of life in epilepsy inventory. Epilepsia, 1993; 34 Suppl. 4: 39-44.
20. Wagner AK, Keller S, Kosinski M, et al. Advances in methods for assessing the impact of epilepsy and antiepileptic drug therapy on patients health related quality of life. Quality of life Research. In press.
21. Eisner C. Growing up with a chronic disease: the impact on children and their families. London: Jessica Langley, 1993.
22. Stein REK, Jones Jessop D. Functional Status II (R): a measure of child health status. Medical Care, 1990; 28: 1041-1055.
23. Austin JK, Dunn DW. Children with newly diagnosed epilepsy: impact on quality of life. In: Quality of life and quality of care in epilepsy . Update 1993. Chadwick DW, Baker GA, Jacoby A, editors. London RSM 1993.

24. Matthews WS, Barabas G, Ferrari M. Emotional concomitants of childhood epilepsy. Epilepsia, 1982; 23: 671-681.
25. Cull CA. Cognitive function and behaviour in children. In: Trimble MR, Reynolds EH, editors. Epilepsy behaviour and cognitive function. Chichester: John Wiley, 1988.
26. Rutter M, et al. A neuropsychiatric study in childhood. In: Clinics in developmental medicine Nos 35/36. London: Spastics International and Heinmann Medical, 1970.
27. Seidenberg M. Academic achievement and school performance of children with epilepsy. In: Hermann BP, Seidenberg M, editors. Childhood epilepsies: neuropsychological psychosocial and intervention aspects. Chichester: John Wiley, 1989.
28. Hoare P, Kerley S. Psychosocial adjustment of children with chronic epilepsy and their families. Developmental Medicine and Child Neurology, 1991; 33: 201-215.
29. Matthews WS, Barabas G. Perceptions of control among children with epilepsy. In: Whitman S, Hermann BP, editors. Psychopathology in Epilepsy. Oxford: OUP, 1986.
30. West P. The social meaning of epilepsy: stigma as a potential explanation for psychopathology in children. In: Whitman S, Hermann BP, editors. Psychopathology in Epilepsy. Oxford: OUP, 1986.
31. Jacoby A. Quality of life and care in epilepsy. In: Chadwick DW, Baker GA, Jacoby A, editors. Quality of life and quality of care in epilepsy. Update 1993. London: RSM, 1993.
32. Livingstone JH, et al. Vigabatrin in the treatment of epilepsy in children. British Journal of Clinical Pharmacology, 1989; 27 Suppl. 1: 109-112.
33. Felbemate study group. Efficacy of Felbemate in childhood epileptic encepthalopathology (Lennox-Gastaut syndrome). New England Journal of Medicine, 1993; 328: 29-33.
34. Baker GA, Jacoby A, Dewey M, et al. Development of an instrument to assess quality of life in children with severe epilepsy and learning disability. (In preparation).
35. Sabbeth BF, Stein REK. Mental health referral: a weak link in the comprehensive care of children with a chronic physical illness. Developmental and Behavioural Paediatrics, 1990; 11: 73-78.
36. Baker GA, Smith DF, Dewey M, et al. The development of a seizure severity scale as an outcome measure in epilepsy. Epilepsy Research, 1991; 8: 245-251.
37. Baker GA, Jacoby A, Francis P, et al. The development of a patient-based adverse drug events profile as an outcome measure in epilepsy. (In preparation).
38. Jacoby A, Baker GA, Smith DF, et al. Measuring the impact of epilepsy: the development of a novel scale. Epilepsy Research, 1993; 16: 83-88.
39. Zigmond AS, Snaith RP. The Hospital Anxiety and Depression Scale. Acta Psychiatrica, 1983; 67: 361-370.
40. Baker GA, Jacoby A, Smith DF, et al. Development of a novel scale to assess life fulfilment as part of the further refinement of a quality of life model for epilepsy. Epilepsia, 1994; 35: 591-596.

18 Assessment of quality of epilepsy care from child to adult; the clinimetric point of view

HARRY MEINARDI

Department of Neurology, University of Nijmegen, The Netherlands

Clinimetrics pertains to the use of quantitative indices for health status. This approach may be of great value for epilepsy as it allows us to use comprehensive measures to follow both clinical symptoms, treatment effects and psychosocial outcomes. This approach holds a promise for trials on new antiepileptic drugs as it generates indices that guarantee comparability between centres.

INTRODUCTION

All therapeutic interventions implicitly require some form of evaluation of results. The need to devote a chapter to this self-evident aspect of epilepsy care is due to the differences in thoroughness of the evaluation, its lack of a unique interpretation and the differences in scope. Without further definitions, statements that the patient's condition has "improved" or "got worse" are meaningless. It is also necessary to be clear about the aspects of the disease that the treatment is supposed to deal with. Furthermore, while the treatment may free a patient from the trouble caused by the disease, the ensuing side-effects of the treatment may have caused the patient to jump from the fire into the frying pan. In a condition like epilepsy, counting seizures would seem an objective and simple procedure to keep track of treatment outcome. However, as will be elaborated below, this is only partly true. The impact of epilepsy is not confined to brief paroxysmal interruptions of brain-functioning due to excessive neuronal discharges. The sheer fact that the person with epilepsy can no longer trust the continuous integrity of the functioning of his brain, and in particular the breakdown without warning, has an important psychological impact. Also, unavoidably, the social 'environment' will respond to the fact that someone has epilepsy and, unfortunately, this may also have a negative influence on the well-being of the patient. Specific treatments may be directed at counteracting these pathogenic factors. However, controlling seizures by itself may have a beneficial effect. Whether or not this is the case has to be evaluated and recorded. The technique to develop comprehensive quantitative indices of health status has been called *clinimetry* by Feinstein.[1] Also, in the field of epileptology, clinimetric systems are presently developed.

SEIZURE FREQUENCY AS EFFECT PARAMETER

Studies on the efficacy of potential antiepileptic drugs like to report results in terms of the number of individuals who experienced a 100%, >75%, or >50% reduction of seizure frequency during the treatment phase as compared with a base line period. These studies select patients who have frequent seizures. In clinical practice for the patient, however, the actual number of seizures that no longer occur may be of greater importance than the percentage reduction. For example a 50% seizure reduction may mean for one patient a pleasant decrease from 50 to 25 seizures per year while another patient may have two seizures instead of four per year and therefore still does not fulfil the requirements for a drivers license. Patients often suffer from different seizure types. A patient with generalized epilepsy may have absence seizures and tonic-clonic seizures. A patient with partial epilepsy may have simple and/or complex partial seizures and/or generalized tonic-clonic seizures. A reduction in frequency of one type may be accompanied by an increase in frequency of the other.

CLINIMETRIC APPROACH

When the Veteran Affairs group (VA group) decided to develop a clinimetric system to score treatment efficacy, their index of seizures (IS) accounted for the difference in prevalence of these seizures.[2] A clinimetric system is a coherent system of validated indices which quantitatively describe clinical symptoms and signs, and which can be summarized by a composite index epitomizing the overall situation of the patient. The VA-group also introduced a measure of seizure severity by modifying the IS by various factors thus arriving at the index of seizure activity (SA). Furthermore, the condition of the patient is undeniably determined by the presence or absence of side-effects of the treatment. For this purpose separate indices were developed for the dose-dependent neurotoxic symptoms (NTX), and for the systemic, usually idiosyncratic, symptoms (STX). The indices were summarized by the Composite Index of Impairment (CII), where CII= SA+NTX+STX.

In the VA-group consensus was reached that a CII>50 meant unsatisfactory result of treatment and the need to look for a new treatment strategy. In their case, dealing with a blinded comparison of four antiepileptic drugs, the CII>50 was a signal to switch over to the next drug of the series.

SEIZURE SEVERITY

Notwithstanding that the VA indices encompassed a seizure activity index, one can argue that the impact of the seizure on the patients is not assessed by this index as the index accounts for factors modifying the seizure types but does not reckon with the actual outcome. Actual outcome is likely to be individualized due to non-seizure related constitutional and environmental factors. The same generalized tonic-clonic seizure will have different physical impact depending on whether the patient is a young healthy person or an older one suffering from osteoporosis. Similarly a complex partial seizure with automatisms will have a different psychological impact on a youngster who gets a seizure in school, an executive who is hit during a business meeting or a multiply handicapped in a home for mentally retarded. Attempts to address this problem are roughly separable in patient-based seizure severity scales and observer-based seizure severity scales. Thorbecke responded to the need of job-counselling by developing a scale which would assist with the judgment whether a person with epilepsy interested in specific functions of the electronics industry would be well-advised to pursue such a career. Also the Chalfont Seizure Severity scale is an observer-based scale, but includes features of a patient-based scale. The originators intended to develop a tool which can be used in the evaluation of medical and surgical treatment of epilepsy both under strict experimental conditions and in a routine clinical setting. A patient-based seizure severity scale was included in a questionnaire assessing the change of quality of life of patients exposed to an experimental antiepileptic drug.[3]

NEUROTOXICITY AND SYSTEMIC TOXICITY

Pharmacotherapy is still the major remedy for epilepsy. In The Netherlands, with an estimated incidence of 6000 new cases of epilepsy per year annually, only about 30 epilepsy-specific neurosurgical interventions take place. In the majority of the patients, pharmacotherapy is used. Such therapy is known to have adverse as well as beneficial effects. When quantitative data have been used to describe adverse effects, those are intrinsically quantified, because they are based upon laboratory findings. Also when assessing cognitive functions standardized quantitative tests are generally used. When semi-quantification is employed, categories are usually poorly characterised, like mild, moderate, or severe, with no explanation of what exactly is meant by these terms.[4] Often, in particular in the case of polytherapy, no reference is made to the strength of the antiepileptic medication which caused the side-effects. So one can read about trials of new antiepileptic drugs which in phase III are usually tested as add-on medication in therapy-resistant patients, and these trials report that certain percentages of the patients receiving increasing amounts of the new drug presented with side-effects. However, little or no information is provided about the strength of the antiepileptic medication to which the new drug was added. In the same vein often reference is made of an association between polypharmacy and increased rates of behavioural problems in children.[5,6]

MEASURE OF MEDICATION STRENGTH

Lammers et al. [7] have suggested that the strength of medication can be computed by adding the dimensionless Prescribed Daily Dose/Defined Daily Dose (PDD/DDD) ratios of the components. The Defined Daily Dose is developed based on the assumed average effective dose per day for the drug used in its main indication in adults. The Defined Daily Dose is expressed in amount of the active substance. For each drug a Defined Daily Dose is assigned. A written dose-document is prepared by WHO-Oslo based on international textbooks, journals and documentation approved by drug control authorities.[8] In Table 1 the defined doses of commonly used antiepileptic drugs are presented.

Table 1. Defined daily dose per antiepileptic drug

Antiepileptic Drug	Defined Daily Dose (mg/day)
CBZ	1000
CLB	20
CZP	8
ESM	1250
OCB	1000
PHB	100
PHT	300
VPA	1500

CBZ = carbamazepine; CLB= clobazam; CZP= clonazepam;
ESM = ethosuximide; OCB= oxcarbazepine; PHB= phenobarbitone;
PHT = phenytoin; VPA= valproate.

In an observational study Lammers et al. [9] showed that the prevalence and severity of toxicity did not differ significantly if equal PDD/DDD ratios of monotherapy and polytherapy were compared (Table 2 and 3).

QUALITY OF LIFE

Almost per definition a disease will influence aspects of quality of life. In this volume Baker and Jacoby extensively discuss the purpose, the design, and the application of scales which measure quality of life of persons with epilepsy. They point out that there is not an one and only scale, but for each question an investigator wishes to examine, an appropriate set should be revised.

Table 2. Prevalence and severity of neurological adverse effects per dose for patients on monotherapy and polytherapy

PDD/DDD ratio	Monotherapy ncs				Polytherapy ncs			
	N	%	ms	25-75	N	%	ms	25-75
0.01-0.33	10	50	5	(5- 5)	-	-		
0.34-0.66	62	60	10	(5-15)	2	50	35	(35-35)
0.67-1.00	64	72	10	(8-20)	16	75	9	(5-15)
1.01-1.33	18	78	10	(5-20)	17	82	16	(5-28)
1.34-1.66	4	75	15	(10-30)	45	63	15	(10-25)
1.67-2.00	3	67	17	(5-30)	48	81	15	(13-25)

ncs = neurotoxicity cumulative score
N = number of patients per stratum
% = percentage of patients per stratum with adverse effects
ms = median score
25-75 = 25th and 75th percentile

Table 3. Prevalence and severity of systemic adverse effects per dose for patients on monotherapy and polytherapy

PDD/DDD ratio	Monotherapy scs				Polytherapy scs			
	N	%	ms	25-75	N	%	ms	25-75
0.01-0.33	10	10	3	(3- 3)	-	-	-	
0.34-0.66	62	23	10	(5-15)	2	0	-	
0.67-1.00	64	22	6	(3-15)	16	31	10	(10-20)
1.01-1.33	18	33	5	(5-15)	17	24	25	(16-27)
1.34-1.66	4	25	50	(50-50)	45	23	10	(8-15)
1.67-2.00	3	33	10	(10-10)	48	29	12	(5-25)

scs = systemic toxicity cumulative score
N = number of patients per stratum
% = percentage of patients per stratum with adverse effects
ms = median score
25-75 = 25th and 75th percentile

Care should be taken to measure separately the patients self-perceived quality of life, and the assessment of the quality of life which an observer would have expected to be attainable for this person if epilepsy, with all its implications, had not intervened. For example, the advent of antiepileptic drug monitoring showed that several seizure-free patients had excessive serum levels of phenobarbitone. These patients were quite satisfied with their quality of life. However, when the phenobarbitone dosage was carefully reduced, bringing the toxic serum level back to an average therapeutic level, these patients suddenly saw life with different eyes. However, as also emphasized by Baker and Jacoby, whether the scales are patient-based or observer-based they should be proven to be reliable, valid, sensitive and sensible.

REFERENCES

1. Feinstein AR. Clinimetrics. New Haven and London: Yale University Press, 1987.
2. Cramer JA, Smith DB, Mattson RH, et al. A method of quantification for the evaluation of antiepileptic drug therapy. Neurology, 1983; 33(S1): 26-37.
3. Baker GA, Jacoby A, Dewey M, et al. Development of an instrument to assess quality of life in children with severe epilepsy and learning disability. Submitted to Epilepsy Research.
4. Lammers MW, Meinardi H. On the reporting of adverse drug events. In: Meinardi H, Cramer JA, Baker GA, et al. editors. Quantitative Assessment in Epilepsy Care. New York: Plenum Press, 1993: 117-122.
5. Hermann BP, Whitman S, Dell J. Correlates of behaviour problems and social competence in children with epilepsy aged 6-11. In: Hermann BP, Seidenberg M, editors. Childhood epilepsies: Neuropsychological, psychosocial and intervention aspects. Chichester: John Wiley and Sons, 1989: 143-158.
6. Cull CA, Trimble MR, Wilson J. Changes in antiepileptic drug regimen and behaviour in children with epilepsy. J. Epilepsy, 1992; 5: 1-9.
7. Lammers MW, Hekster YA, Keyser A, et al. Epilepsy treatment in The Netherlands. Comparison of matched groups of two medical centres. Acta Neurol. Scand., 1994; 89: 415-420.
8. WHO Collaborating Centre for Drugs Statistics Methodology and Nordic Council on Medicines. Guidelines for DDD. Oslo, 1991.
9. Lammers MW, Hekster YA, Keyser A, et al. Monotherapy or Polytherapy for Epilepsy Revisited. A Quantitative Assessment. Submitted to Epilepsia.

Index

A

Abcesses, 187
Abdominal pain, 2
Absence seizures, 4, 7, 10, 12, 25. See also
    Childhood absence epilepsy (CAE);
    Juvenile absence epilepsy (JAE)
  EEG analysis, 83–84
  valproate for, 24, 27
Absorption rate, 102, 103
  carbamazepine, 104
  clobazam, 107
  clonazepam, 107–108
  ethosuximide, 108
  gabapentin, 116
  lamotrigine, 117
  phenobarbital, 109
  phenytoin, 110
  primidone, 112
  valproate, 113
  vigabatrin, 119
Acetazolamide, 145, 154
Acquired epileptic aphasia (LKS), 7, 13, 54–55, 86
Active epilepsy, definition, 19
Adenosine receptors, 105
Adenosine uptake, 107
Adjunctive therapy, 154–155
Adolescence
  epilepsies and syndromes, 14, 26–28, 69–72, 76–81
  psychosocial aspects, 244–247, 254–265, 276
Adrenocorticotropic hormone (ACTH), 11, 145
  dosage, 49
  LGS, 54
  LKS, 55
  West syndrome, 47
AED (antiepileptic drugs). See Treatment
Age factors, 5, 102, 136
Aicardi syndrome (EME), 7, 12, 22, 43, 44–46
Airway obstruction, 132, 133
Alcohol, as precipitating factor, 79, 80
Alpha-methyl-paratyrosine, 48
Amaurosis, 8
Amnesia, 230
Anaemia, 144
Anaesthesia, 192–193
Anatomical substrates, 5
Angelman syndrome, 23
Anoxic seizures, 132, 133
Anterior temporal lobectomy, 193
Antiepileptic drugs. See Treatment
Aphasia (Landau-Kleffner syndrome), 7, 13, 54–55, 86
Aplastic anaemia, 144
Apnea, 2, 132
Arteriovenous malformations, 187
Aspartate, 114, 117
Assessment, 231–233, 275. See also Monitoring
Ataxia, 153
Atonic seizures, 5
Audiologists, 275
Autism, 191
Automatisms, 8
Autopsies, 50

B

Baltic myoclonus, 55
BECT. See Benign childhood epilepsy with centro-temporal spike (BECT)
Behaviour, 51, 84, 162, 239–247
  and seizure type, 240
Benign childhood epilepsy with centro-temporal spike (BECT), 6, 8, 59, 60–64, 86
  treatment, 143
Benign childhood epilepsy with occipital paroxysms (BEOP), 6, 8–9, 59, 64–66
Benign epilepsy associated with multiple spike foci, 60
Benign frontal epilepsy, 60
Benign myoclonic epilepsy in infancy. See Myoclonic epilepsy in infancy (MEI)
Benign neonatal convulsions
  familial, 7, 9, 22
  idiopathic, 22
Benign partial epilepsies, 59–81
Benign partial epilepsy with affective symptoms, 59, 66–68
Benign partial epilepsy with extreme somato-sensory evoked potentials (ESEP), 60, 87
Benign partial seizures of adolescence, 27–28, 60, 69–72
Benzodiazepines, 39, 52, 53, 107
  LKS, 55
  side effects, 53

297

BEOP. See Benign childhood epilepsy with occipital paroxysms (BEOP)
Bioavailability. See Absorption
Blitz-Nick-Salaam Krämpfe. See West syndrome
Blood tests, 152
Bradycardia, 132
Brain development, 36
Brainstem release phenomena, 34
Breath-holding, 2, 133

C

CAE. See Childhood absence epilepsy (CAE)
Calcium channels, 107, 108–109, 110, 111, 114
Calcium influx, 105, 109, 110
Carbamazepine, 52, 104–106
    BECT, 64
    dosages, 105, 106, 147, 149, 294
    GMA, 81
    JME, 79
    LKS, 55
    MEI, 53
    pharmacokinetics, 103, 104
    as second generation drug, 144
    side effects, 53, 144, 148, 150, 152, 162
    cognitive, 162, 167, 169–177
Cardiac syncope, 2
Cardiac output, reduced, 2
Cardiomyopathies, 133
Care, quality of, 291–296
CCS. See Corpus callosum section (CCS)
Cerebral homosiderosis, 196
Cerebral spongiosis, 44
Ceroid lipofuscinosis, 14
Chalfont Seizure Severity scale, 293
Childhood absence epilepsy (CAE), 7, 10, 24, 72–76
    EEG analysis, 83–84
    epidemiology, 24–25, 73
Chloride channel opening, 107, 109
Chronic invasive electrical stimulation, 156, 197
Circadian rhythm, 79, 89
Classification, 1–14, 21
    problems in, 17–18
Clearance rates, 102
Clinical aspects
    of antiepilepsy treatment, 141–157
    BECT, 60–61
    benign partial epilepsy with affective symptoms, 67
    benign partial seizures of adolescence, 69
    BEOP, 65

ESEP, 68
GMA, 80
JAE, 76
JME, 77–78
reading epilepsy, 70–71
for surgery, 188–189
Clinimetrics, 291–296
Clobazam, 107, 145, 154
    dosages, 154, 294
    pharmacokinetics, 103, 107
Clonazepam (CZP), 107–108, 145
    dosages, 37, 108, 154, 294
    pharmacokinetics, 103, 107–108
    for reading epilepsy, 72
    side effects, 162
Clonic convulsions, 5, 33
Cognitive development, 153, 225–234
    assessment of, 283
    EEG analysis, 228–229
    effects of seizures, 227–229
    learning disabilities, 153, 226–227, 232–233, 274–275
Cognitive side effects
    of antiepilepsy drugs, 144, 148, 161–178
    of seizures, 214–215
    of surgery, 213
Communication, 142–143
Concentration, 167. See also Cognitive side effects
Concentration-dependent protein binding. See Protein binding
Concentration-time profile, 101
Consumer attitudes, 142
Convulsive seizures, 4, 6–7
Coping strategies, 253, 254, 265
Corpus callosum section (CCS), 184, 197–198, 200
    neuropsychology of, 218
Cortical excisions, 184–185, 186, 194–196
Cortical migration disorders, 187
Corticosteroids, dosages, 49
Cost
    of medications, 144
    of surgery, 200–201
Councelors, 275
Counselling and rehabilitation, 271–276
CT-scan, 86, 183, 190
Cumulative incidence, 20
Cytochrome p450, 119
CZP. See Clonazepam

D

Data, pooling, 3
Decision making, 167

Defined Daily Dose, 294
Definitions, 18–19
Dendritic arborization, 51
Developmental disorders, 187
Developmental stage of onset, 245–246
Dexamethason, dosages, 49
Diagnosis, 132–143
  BECT, 63–64
  first medical contact, 142–143
Diazepam (DZP), 145
  dosages, 37, 38, 39, 46
  rectal administration, 141, 155–156
Diet, 54, 156
5,5-Diphenylhydantoin. See Phenytoin
Discrimination, 234, 244–245
Disease-specific measure, 281–282
Doose syndrome. See Epilepsy with myoclonic-astatic seizures (EMAS)
Dosages, 101
  carbamazepine, 105, 106, 147, 149, 294
  clobazam, 107, 294
  clonazepam, 108, 154, 294
  compared with adult dosages, 102
  diazepam, 37, 38, 39, 46
  ethosuximide, 109, 147, 294
  formulations, 151
  gabapentin, 147
  lamotrigine, 147
  for neonates, 37
  oxcarbazepine, 147, 149, 294
  phenobarbital, 37, 38, 110, 147, 294
  phenytoin, 37, 38, 110–111, 112, 147, 294
  primidone, 37, 113, 147
  timing, 151
  valproate, 37, 47, 49, 147, 149, 294
  vigabatrin, 49, 147
Down syndrome, 47
Dravet epilepsy. See Myoclonic epilepsy of infancy (MEI)
Drop attacks, 52, 186
Drug resistant cases, 38–39, 185, 186
Drug therapy. See Treatment
DSA (digital subtraction technique), 183, 189, 190
Dysphoria, 52
DZP. See Diazepam

E

Early infantile epileptic encephalopathy (EIEE), 22, 43, 44–46
Early myoclonic encephalopathy (EME), 7, 12, 22, 43, 44–46
  comparison with EIEE, 45

Early partial epilepsy, 46
Eclampsia, 1, 3
ECSWS. See Epilepsy with continuous spike-waves during slow wave sleep (ESES)
Educational aspects, 56, 261–264, 274
  specialists, 275
EEG (encephalography)
  abnormalities and risk of recurrence, 136–137
  BECT, 62–63, 86, 94
  benign partial epilepsy with affective symptoms, 67
  benign partial seizures of adolescence, 70
  BEOP, 65–66
  CAE, 75
  in diagnosis, 133–134
  EIEE, 45
  EME, 34–35, 44
  epilepsia partialis continua, 87–88
  ESEP, 87
  ESES, 86–87
  GMA, 80, 90
  history of, 3
  idiopathic epileptic syndromes, 72, 83–84
  JME, 78, 89–90
  LGS, 51, 85, 93
  LKS, 54–55, 86
  MEI, 50
  of myoclonic epilepsy of childhood, 52, 92
  of neonatal seizures, 34–35
  as presurgical analysis, 190
  reading epilepsy, 71
  relationship between seizures and cognition, 228–229
  West syndrome, 46–47, 84, 93
EEG-polygraphic-video-monitoring systems, 34, 132, 183, 189, 213
EIEE. See Early infantile epileptic encephalopathy (EIEE)
Electrical stimulation, 156, 197
Electrocorticography, 192–193
Elimination half-life, 102, 103, 149, 168
  carbamazepine, 104
  clobazam, 107
  clonazepam, 108
  ethosuximide, 108
  felbamate, 116
  gabapentin, 116
  lamotrigine, 117
  phenobarbital, 109, 110
  phenytoin, 110
  primidone, 112
  valproate, 113, 114
  vigabatrin, 119

Elimination routes, 103
EMAS. See Epilepsy with myoclonic-astatic seizures (EMAS)
EME. See Neonatal myoclonic encephalopathy (EME)
Emotional development, 239–247
Encephalitis, 187
Environmental factors, 24
Enzyme inducer, 102
Epidemiology, 17–29, 186
　CAE, 73
　GMA, 79–80
　guidelines for research, 18–28
　JAE, 76
　reading epilepsy, 70
Epilepsia partialis continua, 28, 43, 55, 87
　EEG analysis, 87–88
　surgery, 200
Epilepsy
　classification, 5–14
　definition, 18–19
　epidemiology, 17–29
　in remission off treatment, 19, 135
　in remission on treatment, 19
　without generalized or focal features, 7
Epilepsy with continuous spike-waves during slow wave sleep (ESES), 7, 13, 26, 43, 54–55
　EEG analysis, 86–87, 95
Epilepsy with grand mal on awakening (GMA), 10–11, 27, 79–81
　EEG analysis, 84, 90
Epilepsy with myoclonic-astatic seizures (EMAS), 7, 12, 24, 51, 52
Epileptogenesis, 1
Errors in metabolism, 14, 23–24
ESEP (benign partial epilepsy with extreme somato-sensory evoked potentials), 60, 87
ESES. See Epilepsy with continuous spike-waves during slow wave sleep (ESES)
Esophageal spasm, 132, 133
Ethosuximide, 108–109
　CAE, 24
　dosages, 109, 147, 294
　as first generation drug, 144
　JAE, 76
　for myoclonic absences, 25
　pharmacokinetics, 103, 108
　side effects, 148, 162
Etiology, 5, 135, 187
Experimental design, 163–168
Extratemporal excisions, 194, 199

F

Faints, 132, 133
Family, 242–244, 246–247, 252–259, 274
Family history, 5, 134, 136. See also Genetics
　benign partial seizures of adolescence, 69
　EME, 12
　febrile seizures, 9
　IPEC, 59
Fatalities, 45
Febrile seizures, 7, 9
　definition, 19
　epidemiology, 17, 21, 22–23
　reflex anoxic, 133
Felbamate, 115–116
　side effects, 144
Focal cortical dysplasia, 187
Focal seizures, 4, 6, 27
Formulations, 151
Frequency, 19–20
Frontal lobes, 6, 8, 88
　excisions, 194

G

GABA/da/D receptors, 107, 109, 112, 114, 116
GABA-mediated synaptic inhibition, 111
GABA neurotransmitters, 119
Gabapentin, 115, 116–117
　dosages, 147
　as second line agent, 150
　side effects, 148
　as third generation drug, 144
GABA transaminase, 119, 121
Gastro-esophageal reflux, 132, 133
Gastrointestinal side effects, 148
Generalized seizures, 4, 6–7
Genetics, 3, 187
　BECT, 63
　benign partial epilepsy with affective symptoms, 67
　BEOP, 66
　CAE, 73
　JME, 78–79
　reading epilepsy, 70
Glutamate decarboxylase, 121
Glutamate receptors, 110, 116
d-Glyceric acidemia, 14
Glycine encephalopathy, 4, 46
GMA. See Epilepsy with grand mal on awakening (GMA)
Grand mal (GTCS) seizures, 7, 8, 10–11
　on awakening, 10–11, 27, 79–81
　CAE, 76

JAE, 10
JME, 10, 78
LGS, 11
    with reading epilepsy, 71
GTCS. See Grand mal (GTCS) seizures
Guide Bleu, 28
Gyri, 196–197

H

Habituation, 168
Half-life. See Elimination half-life
Hallucinations, 8, 71
Headaches, 2, 8, 133
Healthcare, quality of, 291–296
Health-related quality of life. See Quality of life
Hemiconvulsions hemiplegia epilepsy (HHE) syndrome, 187
Hemidecorticaton, 183
Hemimegalencephaly, 187
Hemispherectomy, 183, 184, 195–196, 199
    neuropsychology of, 217–218
Hepatotoxicity, 144, 148, 152
Histopathology, 202
Historical background
    EEG analysis, 3
    GMA, 79
    JME, 77
    paediatric epilepy surgery, 183–184, 211–212
    petit mal, 72–73
Hydrocephalic attacks, 2
Hyperactivity, 52
Hyperammonaemia, 148
Hyperglycenemia, 14
Hyperventilation, 2, 27, 133
Hypnogogic hallucinations, 2
Hypoalbuminaemia, 110
Hypocalcemia, 1, 36
Hypoglycaemia, 1
Hypomagnesemia, 36
Hyponatremia, 148
Hypoxic-ischaemic encephalopathy, 44
Hypsarrhythmia, 11, 12
    West syndrome, 46, 84
Hysteria, 2

I

IAP (intracarotid amytal procedure), 189, 190–191
Ictogenesis, 1
Idiopathic epileptic syndromes, 5, 6, 7, 47, 72–81
    EEG analysis, 83–84
Idiopathic partial epilepsies in children (IPEC), 25–26, 59
Illusions, 8
Immunoglobulins, 48, 49, 54
Impulsive petit mal. See Juvenile myoclonic epilepsy
Inborn errors of metabolism, 14, 23–24
Incidence, 20–21, 29
    benign partial seizures of adolescence, 69
Incidence density, 20
Index of seizures, 292
Infantile spasms. See West syndrome
Interactions
    carbamazepine, 106
    clobazam, 107
    clonazepam, 108
    ethosuximide, 109
    felbamate, 116
    oxcarbazepine, 119
    phenobarbital, 110
    phenytoin, 106, 112
    primidone, 113
    valproate, 114
International League Against Epilepsy, 5, 17, 18, 101
Intracarotid amytal procedure, 189, 190–191
Intracranial hypertension, 196
Intravenous administration, 155
Intravenous immunoglobulins (IV-Ig), 48, 54
    dosages, 49
IPEC (idiopathic partial epilepsies in children), 25–26, 59
IQ testing, 214, 216, 218, 229
Ischaemic attacks, 133

J

JAE. See Juvenile absence epilepsy (JAE)
JME. See Juvenile myoclonic epilepsy (JME)
Juvenile absence epilepsy (JAE), 7, 10, 27
    epidemiology, 76
    treatment, 143, 144
Juvenile myoclonic epilepsy (JME), 7, 10, 27, 77–79
    EEG analysis, 84, 89–90
    first treatment, 144

K

Ketogenic diet, 156
Kidneys, 108, 109, 111
Kojewnikow's syndrome. See Epilepsia partialis continua

L

Laboratory tests, 152–153
Lactate-pyruvate metabolism, 14
Lafora body epilepsy, 55
Lamotrigine, 53, 115, 117–118, 150
   dosages, 147
   side effects, 148, 167
   as third generation drug, 144
Landau-Kleffner syndrome (LKS), 7, 13, 26, 54–55, 86
   surgical treatment, 187
Language. See Speech
Learning disabilities, 153, 226–227, 232–233, 274–275
Lennox-Gastaut syndrome (LGS), 7, 11, 25, 43, 50–54
   distinction between MEI, 51–52, 53
   surgery, 199
   treatment, 54, 145, 187
LGS. See Lennox-Gastaut syndrome (LGS)
Licensing, of drugs, 144
Lidocaine, dosages, 37
Lifetime prevalence, 19–20
Linguistic processing, 55. See also Speech
Lipophilic substances, 104
Lissencephaly-pachygyria, 13
Liver, 108, 109, 110, 112, 113
   hepatotoxicity, 144, 148, 152
LKS. See Landau-Kleffner syndrome (LKS)
Local seizures, 4, 6, 27
Lorazepam, 37

M

Magnetic resonance imaging, 183, 189, 197–198, 230
   BECT, 86
Magnetoencephalography, 190
Malignant epilepsies, 43–56
Malingering, 2
Measurement indices, 19–20
Medical history. See Historical background
MEG (magnetoencephalography), 190
MEI. See Myoclonic epilepsy in infancy (MEI)
Memory, 167, 169, 190, 216–217
Meningitis, 187
Mental retardation, 45, 47, 50, 136, 191
Mesial temporal sclerosis, 187
Metabolism, errors in, 14, 23–24
Metabolites
   from carbamazepine, 104, 106, 120, 150
   from clobazam, 107
   from oxcarbazepine, 118–119, 120
   from phenobarbital, 109
   from primidone, 112
Methysergide, 48
Microdysgeneses, 187
Microsurgical undercutting, 186
Migraine headaches, 2, 8, 133
Misdiagnosis, 132–134
Mitochondrial encephalomyopathy, 55
Modeling, 5
Mode of action
   carbamazepine, 105
   clobazam, 107
   clonazepam, 108
   ethosuximide, 108–109
   lamotrigine, 118
   oxcarbazepine, 119
   phenobarbital, 109–110
   phenytoin, 111
   primidone, 112
   valproate, 113–114
Monitoring, 34, 132, 151–153
   EEG analysis, 231
   laboratory, 152–153
Monotherapy, 141, 149, 169, 172
Moro-reflex, 46
Motor cortex, 194
MRI (magnetic resonance imaging), 183, 189, 197–198, 230
   BECT, 86
Multiple lobes, 6
Multiple subpial transsection, 196–197, 199
Myoclonic absences, 7, 12, 25. See also Childhood absence epilepsies (CAE); Juvenile absence epilepsies (JAE)
Myoclonic epilepsy in infancy (MEI), 7, 9, 23, 43, 49–50
   distinction between LGS, 51–52, 53
   EEG analysis, 84
Myoclonic epilepsy in non-progressive encephalopathies, 23
Myoclonic epilepsy of childhood, 52
Myoclonic seizures, 5, 7, 9, 12
   EEG analysis, 83–84
Myoclonus syndromes, 2, 14. See also Progressive myoclonus epilepsies

N

Narcolepsy, 2
National Child Development Study, 244
Natural history, 5
Neonatal myoclonic encephalopathy (EME), 7, 12, 22, 43, 44–46
Neonatal seizures, 7, 12–13, 33–39

definition, 19
EEG analysis, 83
investigations, 34–35
Neurofibromatosis, Schwannoma, 187
Neurologists, 275. See also Physicians
Neuronal ceroid lipofuscinosis, 55
Neurophysiology, 83–95
Neuropsychological aspects. See also Memory; Speech
of surgery, 211–220, 217–220
tests, 190–191, 217
Neuropsychologists, 275
Neuroradiology and imaging, 51, 86, 189, 230
availability of, 213
Neurotoxicity, 36, 118, 295
Neutropenia, 148, 152
Newborn seizures. See Neonatal seizures
Nitrazepam, 145
dosages, 49
for West syndrome, 47
Nonconvulsive seizures, 4, 6–7
Nonepileptic seizures, 1–2, 132
definition, 19
Non-idiopathic localization related epilepsies, 28
Nonketotic hyperglycinaemia, 4, 46
Normal-volunteer studies, 166–167
NREM sleep, 83, 84, 85, 88, 90
Nursing, 275
Nystagmus, 153

O

Occipital lobes, 6, 9, 64, 88, 195
Occupational therapists, 275
Oedema, 148
Ohtahara syndrome, 22, 43, 44–46
Ophthalmologists, 275
Overprotection, 256–259
Oxcarbazepine, 115, 118–119
dosages, 147, 149, 294
as second generation drug, 144
side effects, 148, 167

P

Paediatric epilepsy surgery (PES). See Surgery
Palliative surgery, 183, 185–186, 197–198
Parahydroxylation, 109
Paraldehyde
dosages, 37
rectal administration, 141, 156
Parietal lobes, 6, 88, 195

Paroxysmal choreoathetosis, 2
Partial seizures, 4, 6, 27
Pathophysiology
BECT, 63
BEOP, 66
PB. See Phenobarbital
Pediatricians. See Physicians
Periodic spasms, 23. See also West syndrome
Period prevalence, 19
PET. See Positron emission tomography (PET)
Petit mal
historical background, 72–73
impulsive. See Juvenile myoclonic epilepsy
Phacomatoses, 13
Pharamcokinetics
valproate, 101, 113
vigabatrin, 115, 119
Pharmacokinetics, 101–102
carbamazepine, 101
clobazam, 107
clonazepam, 107–108
ethosuximide, 108
felbamate, 115
gabapentin, 115
lamotrigine, 115, 117
non-linear, 101
oxcarbazepine, 115
phenobarbital, 103, 109, 110–111
phenytoin, 101, 103, 110–111
primidone, 112
Phenobarbital (PB)
dosages, 37, 38, 110, 147, 294
as first generation drug, 144
pharmacokinetics, 103, 109, 110–111
as precipitating factor, 52
side effects, 53, 148, 150–151
cognitive, 162, 167, 169–176
Phenobarbitone. See Phenobarbital
Phenylethylmalonamide (PEMA), 112
Phenylketonuria, 14
Phenyltriazine, 117
Phenytoin (PHT), 110–112, 150, 152
dosages, 37, 38, 147, 294
as first generation drug, 144
pharmacokinetics, 103, 110–111
side effects, 53, 148
cognitive, 162, 167–176
Phosphenes, 8
Photosensitive epilepsy, 10, 26–27
PHT. See Phenytoin
Physical therapists, 275
Physicians, 275

communication with patients, 142
    first medical contact, 142–143
Plasma level monitoring, 152
PMD. See Primidone
Point prevalence, 19
Poliodystrophy, 44
Polymicrogyria, 187
Polyspike-wave (PSW), 84
    JME, 89–90
    LGS, 85
Polytherapy, 153–155, 164
Porencephalic cyst, 187
Positron emission tomography (PET), 48,
    51, 183, 189, 230
Postoperative follow up, 202
Posttest studies, 164–165
Potassium currents, 110
Precipitating factors, 11, 27, 52, 77, 79, 80,
    90
Prednison, dosages, 49
Prescribed Daily Dose, 294
Prescribing treatment, 149–151
Prevalence, 19–21, 77
Primidone (PMD), 112–113
    dosages, 37, 113, 147
    as first generation drug, 144
    pharmacokinetics, 103, 112
    side effects, 53, 148, 151
Prognosis
    BECT, 64
    benign partial epilepsy with affective
        symptoms, 67
    benign partial seizures of adolescence, 70
    BEOP, 66
    CAE, 75–76
    GMA, 81
    JAE, 76–77
    JME, 79
    reading epilepsy, 71–72
Progressive cerebral atrophy, 44
Progressive myoclonus epilepsies, 28, 43,
    55–56
Propionic acidopathy, 44
Protein binding, 101, 103
    clonazepam, 108
    lamotrigine, 117
    phenobarbital, 109
    phenytoin, 110
    valproate, 113
Pseudoseizures, 2
Psychiatrists, 275
Psychological aspects, 133, 157
    evaluation, 191
    neuropsychological tests, 190–191
    psychometric testing, 279–287

Psychosocial aspects, 56, 185–186, 214–216,
    226, 239–247
    counselling and rehabilitation, 271–276
    evaluation, 191
    social integration, 251–265
Pyknolepsy. See Absence seizures;
    Childhood absence epilepsy
Pyridoxine
    dosages, 37, 49
    West syndrome, 48
Pyridoxine deficiency and dependency, 14, 36

Q

QOL. See Quality of life
Q.T. syndrome, 133
Quality of life, 185, 231, 251–265
    assessment methodologies, 279–287
    definition, 280
Questionnaires, 283

R

Rage attacks, 2
Ramsay-Hunt syndrome, 14
Rasmussen's chronic encephalitis, 187, 217
Reading epilepsy, 6, 26, 60, 70–72
Recreational drugs, as precipitating factor,
    79
Rectal administration, 141, 155–156, 271
Recurrence, of seizures, 135–137, 212
Reflex breath holding, 2
Reflex syncope, 2
Rehabilitation, 271–276
Remission, 19, 135–137
Research
    cognitive side effects, 161–178
    guidelines for, 18–28
    neonatal seizures, 34–35, 36
    quality of life descriptive studies,
        284–285
    quality of life experimental studies,
        285–286
    trends in, on quality of life, 281
    types of studies, 163–174
Rickets, 148
Risk factors, 136–137, 228
Rolandic epilepsy. See Benign childhood
    epilepsy with centro-temporal spike
    (BECT)

S

Salivary level monitoring, 152
Sandhoff disease, 14

School, 244, 247
  educational level, 261–264
  progress in, 141
Sedation, 144, 148
Seizure activity index, 292–293
Seizure confound, 164
Seizure-free interval, 137
Seizure-related events, 132
Seizures
  anoxic, 132, 133
  BECT, 61–62
  CAE, 73–75
  classification, 2–4, 18
  due to toxins, 7
  effects on academic performance, 215
  effects on cognitive functioning, 214–215, 227–229, 240
  evaluating after surgery, 198–200
  febrile reflex anoxic, 133
  generalized, 4, 6–7
  GMA, 80
  impact of type on behaviour, 240
  index of seizures (IS), 292
  indices for, 291–293
  isolated, 7
  JAE, 76
  LGS, 51, 85
  neonatal, 33–39
  nonepiliptic, 1–2
  partial, 4, 6, 27
  rare, 134–135
  recurrence of, 135–137, 212
  single unprovoked, 131, 134
  subtle, 33
  unclassified, 4, 60
Selfhelp groups, 56
Sensory cortex, 195
Serum concentration, 101, 102, 146
  ethosuximide, 108
  lamotrigine, 150
  oxcarbazepine, 119
  phenytoin, 110, 150
Severe myoclonic epilepsy. See Myoclonic epilepsy of infancy (MEI)
Sex distribution, 10, 11, 12
Shelf life, 102
Shuddering, 2
Sialidosis, 55
Side effects, 47, 52, 131, 295. See also Individual drug names
  behavioural, 51, 84, 148, 162, 239–247
  cognitive, 144, 148, 161–178, 215, 230–231
  consumer attitudes, 142
  cosmetic, 143, 144, 148
  hair loss, 143
  hepatotoxicity, 144, 148, 152
  hyperammonaemia, 148
  hyponatremia, 148
  mood, 144
  neurotoxicity, 36, 118, 295
  neutropenia, 148
  oedema, 148
  rickets, 148
  sedative, 144, 148
  skin reactions, 148, 153
  surgical complications, 201
  teratogenicity, 143, 144, 148
  weight gain, 143
Single photon emission computerized tomography (SECT), 183, 189, 230
Sinus syndrome, 133
Skin reactions, 148, 153
Sleep deprivation
  due to nocturnal seizures, 228
  as precipitating factor, 11, 27, 77, 79, 80
Sleep paralysis, 2
Sleep syndromes, 2, 133
Sleep terrors, 2
Social aspects, 234, 244–245, 255
  integration into, 247, 251–265
  peer interaction, 259–261
Social workers, 275
Socioeconomic status, 233–234
Sodium channels, 111, 114, 117
Sodium valproate. See Valproate
Somnambulism, 2
SPECT (single photon emission computerized tomography), 183, 189, 230
Speech, 190–191, 193, 217, 218
  excisions in language area, 195, 217–218
  linguistic processing, 55
  therapists, 275
Spike-wave bursts (SW), 84
  epilepsia partialis continua, 88
  ESES, 87
  LGS, 85
Spike waves. See Polyspike-wave (PSW); Spike-wave bursts (SW)
Stage of onset, 245–246
Status epilepticus, 19, 85
Stereotaxy functional surgery, 197
Steroids, 11, 145
Stigma of epilepsy, 234, 244–245
Stroke, 187
Studies. See Research
Sturge-Weber syndrome, 187
Sudden infant death syndrome, 132
Support groups, 56, 152, 265

Surface electromyography, 25
Surgery, 11, 183–203
    and behaviour, 241
    candidates for, 184–186
    complications, 201
    contraindications for, 191
    controversies, 212–214
    etiologies and syndromes remediable with, 187
    evaluation after, 198–200
    neuropsychological assessment, 217–220
    non-invasive, 197
    postoperative follow up, 202
    preparations for, 192
    rationale, 214–216
    re-operations, 198, 200
Symptomatic epilepsy, 6
Syncopes, 133
Synergistic therapy, 155
Synthesis, 104, 121

T

Tay-Sachs disease, 14
Temporal cortex, 8, 9
Temporal lobes, 6, 88, 192, 214–215, 229
    excision, 193, 198–199, 216–217
Teratogenicity, 143, 144
Terminology, 5–14
Therapies. See Treatment
Thermatic Apperception Test, 245
Thwarted child syndrome, 2
Thyrotropin-releasing hormone (TRH), 48, 49, 54
Tiagabine, 144
Tics, 133
Timing, of dosages, 151
Tonic-clonic seizures, 5, 25, 27
Tonic convulsions
    definition, 33
    LGS, 51
Topiramate, 144
Toxicity, 7, 102
    hepatotoxicity, 144, 148, 152
    neurotoxicity, 36, 118, 295
Transient focal abnormalities, 84
Trauma, 187
Treatment, 101–122, 246. See also Interactions; Pharmacokinetics
    'add-on,' 144
    adjunctive, 154–155
    alternative, 156–157
    basic principles of, 131–134
    BECT, 64
    and behaviour, 241
    benign partial epilepsy with affective symptoms, 67
    BEOP, 66
    CAE, 75
    clinical aspects, 141–157
    clinimetrics, 291–296
    counselling about, 272
    ESEP, 69
    family involvement in, 143
    first and second line, 145–146, 149–150
    formulations, 151
    GMA, 81
    habituation, 168
    intermittent, 145
    intravenous, 155
    JAE, 76–77
    JME, 79
    medication strength, 294
    neonatal seizures, 35–39
    new drugs, 115–122
    polytherapy, 153–155
    prescribing, 149–151
    psychological, 157
    reading epilepsy, 71–72
    rectal, 141, 155–156, 271
    surgery, 183–203
    synergistic, 155
    West syndrome, 49
    when to start, 134–135
    when to stop, 135–137
    withdrawal from antiepileptic drugs, 131, 137
Tremors, 148, 153
Tuberous sclerosis, 11, 187
Tumours, 187

U

Unclassified seizures, 4, 60
Unemployment, 233
Upper airway obstruction, 132, 133

V

Vagal nerve stimulation, 156, 197
Valproate, 64, 113–115
    CAE, 24
    dosages, 37, 47, 49, 147, 149, 294
    GMA, 81
    JAE, 27, 76
    JME, 79
    for myoclonic absences, 25
    for neonates, 38
    pharmacokinetics, 102, 103, 113
    for reading epilepsy, 72

as second generation drug, 144
side effects, 47, 143–144, 148, 152, 153
   cognitive, 162, 167, 169, 172, 175–177
Valproic acid. See Valproate
Vasovagal syncope, 2
Vertigo, 2
Veteran Affairs group, 292
Video EEG. See EEG-polygraphic-video-monitoring systems
Vigabatrin, 115, 119–122
   dosages, 49, 147
   as precipitating factor, 52
   as second line agent, 150
   side effects, 148, 167
   as third generation drug, 144
   West syndrome, 48
γ-Vinyl GABA. See Vigabatrin
Vocational guidance, 276
Vomiting, 2

Von Muenchausen Syndrome, 2
VPA. See Valproate

W

Warfarin metabolism, 119
Wechsler Memory Scale, 216
West syndrome, 7, 11, 12, 23, 43, 46–49
   characteristics of types, 48
   incidence, 46
   surgery, 199
   treatment, 47, 145
Withdrawal
   from adjunctive treatment, 154–155
   from antiepileptic drugs, 135, 137, 149, 212

X

X-rays, 190